T0192008

Communications in Computer and Information Science 1417

More information about this series at http://www.springer.com/series/7899

Mahua Bhattacharya · Latika Kharb ·
Deepak Chahal (Eds.)

Information, Communication and Computing Technology

6th International Conference, ICICCT 2021
New Delhi, India, May 8, 2021
Revised Selected Papers

Editors
Mahua Bhattacharya
ABV Indian Institute of Information
Technology and Management
Gwalior, India

Deepak Chahal
Jagan Institute of Management Studies
Delhi, India

Latika Kharb
Jagan Institute of Management Studies
Delhi, India

ISSN 1865-0929 ISSN 1865-0937 (electronic)
Communications in Computer and Information Science
ISBN 978-3-030-88377-5 ISBN 978-3-030-88378-2 (eBook)
https://doi.org/10.1007/978-3-030-88378-2

This Springer imprint is published by the registered company Springer Nature Switzerland AG
The registered company address is: Gewerbestrasse 11, 6330 Cham, Switzerland

Preface

The International Conference on Information, Communication and Computing Technology (ICICCT 2021) was held on May 8, 2021, in New Delhi, India. ICICCT 2021 was organized by the Department of Information Technology, Jagan Institute of Management Studies (JIMS) Rohini, New Delhi, India. The conference received 83 submissions and after rigorous reviews 20 papers were selected for this volume. The acceptance rate was around 16.6%. The contributions came from diverse areas of information technology categorized into two tracks, namely (1) Communication and Network Systems and (2) Computational Intelligence Techniques

The aim of ICICCT 2021 was to provide a global platform for researchers, scientists, and practitioners from both academia and industry to present their research and development activities in all the aspects of communication and network systems and computational intelligence techniques.

We thank all the members of the Organizing Committee and the Program Committee for their hard work. We are very grateful to Mahua Bhattacharya, ABV Indian Institute of Information Technology and Management, India, as general chair, and Sami Muhaidat, Khalifa University, Abu Dhabi, UAE, as program chair. We are also grateful to our keynote speakers, Marcin Paprzycki, Systems Research Institute, Polish Academy of Sciences, Poland, and Maria Ganzha, Warsaw University of Technology, Poland. We would like to thank Ferdous Ahmed Barbhuiya, Indian Institute of Technology (IIT) Guwahati, India, as session chair for Track 1, and Narendra Londhe, National Institute of Technology, Raipur, India, as session chair for Track 2.

We also thank all the Technical Program Committee members and referees for their constructive and enlightening reviews on the manuscripts, Springer for publishing the proceedings in the Communications in Computer and Information Science (CCIS) series, and all the authors and participants for their great contributions that made this conference possible.

August 2021

Latika Kharb
Deepak Chahal

Preface

Organization

General Chair

Mahua Bhattacharya	ABV Indian Institute of Information Technology and Management, India

Program Chair

Sami Muhaidat	Khalifa University, Abu Dhabi, UAE

Keynote Speakers

Marcin Paprzycki	Systems Research Institute, Polish Academy of Sciences, Poland
Maria Ganzha	Warsaw University of Technology, Poland

Conference Secretariat

Praveen Arora	Jagan Institute of Management Studies, India

Session Chair for Track 1

Ferdous Ahmed Barbhuiya	Indian Institute of Technology (IIT) Guwahati, India

Session Chair for Track 2

Narendra Londhe	National Institute of Technology, Raipur, India

Program Committee Chairs

Latika Kharb	Jagan Institute of Management Studies, India
Deepak Chahal	Jagan Institute of Management Studies, India

Technical Program Committee

Rastislav Roka	Slovak University of Technology, Slovakia
Siddhivinayak Kulkarni	MIT World Peace University, India
P. Chenna Reddy	Jawaharlal Nehru Technological University Anantapur, India
Razali Yaakob	Universiti Putra Malaysia, Malaysia
Noor Afiza Mohd Ariffin	Universiti Putra Malaysia, Malaysia
Malti Bansal	Delhi Technology University, India

R. Chithra	K.S. Rangasamy College of Technology, India
Homero Toral Cruz	University of Quintana Roo, Mexico
J. Viji Gripsy	PSGR Krishnammal College for Women, India
Boudhir Anouar Abdelhakim	Abdelmalek Essaâdi University, UAE
Muhammed Ali Aydin	Istanbul Cerrahpaşa University, Turkey
Suhair Alshehri	King Abdulaziz University, Saudi Arabia
Dalibor Dobrilovic	University of Novi Sad, Serbia
A. V. Petrashenko	National Technical University of Ukraine, Ukraine
Ali Hussain	Sri Sai Madhari Institute of Science and Technology, India
A. Nagaraju	Central University Rajasthan, India
Cheng-Chi Lee	Fu Jen Catholic University, Taiwan
Apostolos Gkamas	University Ecclesiastical Academy of Vella of Ioannina, Greece
M. A. H. Akhand	Khulna University of Engineering & Technology, Bangladesh
Saad Talib Hasson	University of Babylon, Iraq
Valeri Mladenov	Technical University of Sofia, Bulgaria
Kate Revoredo	Vienna University of Economics and Business, Austria
Dimitris Kanellopoulos	University of Patras, Greece
Samir Kumar Bandyopadhyay	University of Calcutta, India
Baljit Singh Khehra	BBSBEC, India
Nitish Pathak	BVICAM, India
Md Gapar Md Johar	Management Science University, Malaysia
Kathemreddy Ramesh Reddy	Vikrama Simhapuri University, India
Shubhnandan Singh Jamwa	University of Jammu, India
Surjeet Dalal	SRM University Delhi-NCR, India
S. Vasundra	Jawaharlal Nehru Technological University, Anantapur, India
Manoj Patil	North Maharashtra University, India
Rahul Johari	GGSIPU, India
Adeyemi Ikuesan	University of Pretoria, South Africa
Pinaki Chakraborty	Netaji Subhas University of Technology, India
Subrata Nandi	National Institute of Technology, Durgapur, India
Vinod Keshaorao Pachghare	College of Engineering, Pune, India
A. V. Senthil Kumar	Hindusthan College of Arts and Science, India
Khalid Raza	Jamia Milia Islamia, India
G. Vijaya Lakshmi	Vikrama Simhapuri University, India
Parameshachari B. D.	TGSSS Institute of Engineering and Technology for Women, India
E. Grace Mary Kanaga	Karunya University, India
Subalalitha C. N.	SRM University Kanchipuram, India
Niketa Gandhi	Machine Intelligence Research Labs, USA

T. Sobha Rani	University of Hyderabad, India
Zunnun Narmawala	Nirma University, India
Aniruddha Chandra	National Institute of Technology, Durgapur, India
Ashwani Kush	Kurukshetra University, India
Manoj Sahni	Pandit Deendayal Petroleum University, India
Promila Bahadur	Maharishi University of Management, USA
Gajendra Sharma	Kathmandu University, Nepal
Rabindra Bista	Kathmandu University, Nepal
Renuka Mohanraj	Maharishi University of Management, USA
Eduard Babulak	Institute of Technology and Business, Czech Republic
Zoran Bojkovic	University of Belgrade, Serbia
Pradeep Tomar	Gautam Buddha University, India
Arvind Selwal	Central University of Jammu, India
Atif Farid Mohammad	University of North Carolina at Charlotte, USA
Maushumi Barooah	Assam Engineering College, India
Prem Prakash Jayaraman	Swinburne University of Technology, Australia
Kalman Palaggi	University of Szeged, Hungary
J. Vijayakumar	Bharathiar University, India
Jacek Izydorczyk	Silesian University of Technology, Poland
Pamela L. Thompson	University of North Carolina, Charlotte, USA
Arka Prokash Mazumdar	Malaviya National Institute of Technology Jaipur, India
R. Gomathi	Bannari Amman Institute of Technology, India
Zunnun Narmawala	Nirma University, India
Diptendu Sinha Roy	National Institute Technology, Meghalaya, India
Nitin Kumar	National Institute of Technology, Uttarakhand, India
B. Surendiran	National Institute of Technology, Puducherry, India
Parismita Sarma	Guwahati University, India
Manas Ranjan Kabat	VSS University of Technology, India
Anuj Gupta	Chandigarh Engineering College, India
Md. Alimul Haque	Veer Kunwar Singh University, India
Abdullah M. Al BinAli	Taibah University, Saudi Arabia
Subhojit Ghosh	National Institute of Technology, Raipur, India
Rohini Sharma	Panjab University Chandigarh, India
Alessio Bottrigh	University of Eastern Piedmont, Italy
Sunita Sarkar	Assam University, India
Sonal Chawla	Panjab University, India
Anurag Jain	Guru Gobind Singh Indraprastha University, India
Matt Kretchmar	Denison University, USA
Sharad Saxena	Thapar Institute of Engineering and Technology, India
Dushyant Kumar Singh	Motilal Nehru National Institute of Technology Allahabad, India
R. I. Minu	SRM Institute of Science and Technology, India
M. Murali	SRM Institute of Science and Technology, India
Rajesh Mehta	Thapar Institute of Engineering and Technology, India
Vibhav Prakash Singh	Motilal Nehru National Institute of Technology Allahabad, India

Contents

Communication and Network Systems

Performance Enhancement in Big Data by Guided Map Reduce

Himadri Sekhar Ray(✉), Anurag Chakraborty, and Radib Kar

Jadavpur University, Kolkata, India

Abstract. In recent years, due to the emergence of the Internet and communication technology data is generated at an increasing rate. Everything around us is connected to the internet and seems to generate data. Health care is one of the biggest booming data producing sector. To efficiently fetch the required information form the large volume of data, it is required to store the data in a distributed fashion so that the retrieval time should be very less and for that, a massive storage cluster of tens of thousands of connected machines with built-in analytics tools are required to systematically store and fast retrieve of the data. Hadoop, a well-known analytical application, divides the data into small blocks and with the help of Hadoop Distributed File System (HDFS), it distributes among the cluster node for storage and future processing. If there is a need to run a query on a specific item, Hadoop runs its query in all of its thousands of nodes for fetching a particular dataset, which increases the overall execution time. To overcome this issue, we have made an application Node Guided Map-Reduce (NGMR) based on Node.js in a distributed cluster with movable nodes, where this issue has been resolved. To build and run this application and make research on it, a huge amount of capital is required to build a data storage service. Due to this limitation, we have made a small scale model of the replica of a large storage cluster using low-end computational power and limited resource ARM-based single-board computers (Raspberry Pi). The work was done over the past few years is the implementation of the proposed scheme to enhance the performance in big data handling by Guided Map-Reduce with the help of Node.js and to present and create efficient cost-effective data storage and retrieval system for the health-care industry.

Keywords: Big Data · Open source hardware · Raspberry pi · BATMAN-adv · Distributed processing · Node.js · Map-Reduce · Healthcare · Movable nodes · Cluster · Node guided Map-Reduce (NGMR)

1 Introduction

With the recent advancement of distributed processing, healthcare sector is one of the large data-producing sectors which produce a huge number of data every day. One of the major challenges is to make a large scale distributed storage system to efficiently store, process and retrieve the necessary health records. Since its early beginning, the Hadoop Map-Reduce has become a popular tool for storing, managing and processing the massive amount of data. Hadoop Map-Reduce has come up with a highly effective

M. Bhattacharya et al. (Eds.): ICICCT 2021, CCIS 1417, pp. 3–17, 2021.
https://doi.org/10.1007/978-3-030-88378-2_1

framework for storing and analysing Big data. But this type of framework distributes data to all of its nodes and later at the processing time, the user provided Map-reduce queries also execute on all nodes. Thus, sometimes a given queries may be processed on some of the nodes where no data related to the user given query is stored. This phenomenon increases the overall processing time. Therefore, in [1], we have proposed a distributed storage platform and Map-Reduce model that overcome the problem faced in the existing systems. Based on the proposed model, we have developed a framework Node guided Map Reduce (NGMR) using google Map-Reduce paradigm.

The paper describes an efficient big-data handling application NGMR. The framework NGMR guides the Map-Reduce process to run on a specific set of cluster nodes among all nodes in the cluster. For developing and experimenting with the health data, single board computers with open-source hardware are used as distributed nodes. All nodes are connected in infrastructure-less ADHOC mobile network established by BATMAN-ADV [6]. NGMR is developed with Node.js, a free open-source server environment that runs on JavaScript on the server and can run on various platforms like Windows, Linux, Mac OS etc.

The rest of the paper is organized as follows. Section 2 discusses the related works. Section 3 gives a brief idea about the file system of the Node Guided Map Reduce (NGMR) application. Section 4 deals with the setup procedure of Node Guided Map Reduce application cluster. In Sect. 5 the test scenarios and system setup are described for comparing NGMR file system with Hadoop file system. In Sect. 6 data uploading procedure and evaluation result are present to compare the performance with existing framework along with the execution process of user defined queries are elaborated. Section 2 gives a brief idea of the related present day works and finally Sect. 7 concludes and provides the possible future directions (Table 1).

Table 1. List of Abbreviations

Abbreviation	Stands For
NGMR	Node Guided Map-Reduce
MR	Map-Reduce
HDFS	Hadoop Distributed File System
JSON	JavaScript Object Notation

2 Related Work

With the recent advancement of the Big Data solutions, scientists have focused on the application of the Big Data concept following the Map-Reduce paradigm. The Big Data storage model based on Map-Reduce text categorization has been used for forecasting data in the next period of time in a particular region [9]. The two approaches are compared, i.e. Map-Reduce and GPU-Reduce to calculate the performance measurement for searching index file in database query processing. Various studies are made to check the performance of NoSQL database system using the Map-Reduce programming model with single and multiple nodes. It is shown that NoSQL with MAP-Reduce programming model provides better performance gain compared to relational database systems [10].

The Hadoop framework [8] handles distributed file system with the help of HDFS (Hadoop file system). HDFS runs on master/client architecture. An HDFS cluster consists of a single NameNode or MasterNode and number of DataNodes. DataNode stores the data files in the cluster, however, Master Node does not store any data files. The Master Node manages the file system namespace and regulates access to files which are stored in the DataNodes. Besides, there is a secondary NameNode, usually, one per cluster, which manages storage attached to the nodes and helps the Master Node. If for some reason, the Master Node is down, Secondary NameNode can take the responsibility of handling the task of the Master Node. HDFS exposes a file system namespace and allows user data to be stored in files. Internally, a file is split into one or more blocks and these blocks are stored in a set of DataNodes. By default, HDFS maintains the data block size as 128 MB. The NameNode/MasterNode executes file system namespace operations, like opening, closing, and renaming files and directories. It also determines the mapping of blocks onto DataNodes. The DataNodes are responsible for serving read and write requests, block creation, deletion, and replication upon instruction from the NameNode. A graphical representation of HDFS architecture is shown in Fig. 1.

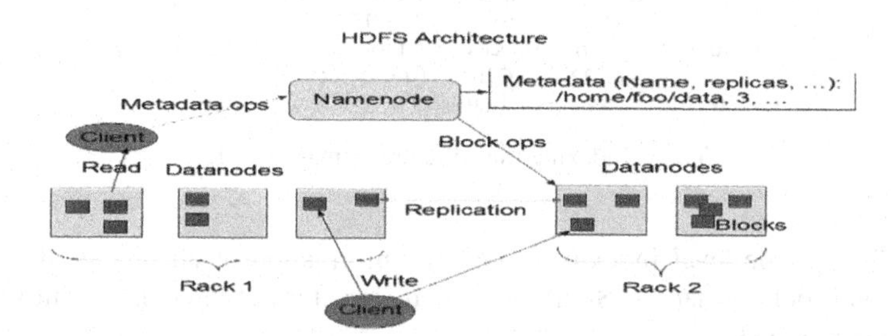

Fig. 1. HDFS Architecture

At the time of data processing in HDFS, the Hadoop framework sends the user given analytics code to all of its nodes to run the user-given queries. Sometimes it may happen that the given query runs on such a node where no data is returned as output of the query execution. Therefore, processing on such nodes increases the overall data retrieval time.

Our work is different from the above-mentioned research works, because we consider the specific issues related to the storage and retrieval of the Big Data and carry the experimental study regarding the performance of NGMR and the well-known big data solution on very low commodity hardware.

3 NGMR File System

NGMR (Node guided Map-Reduce) file system is developed as an efficient distributed file system based on the framework proposed in [1]. The framework follows master client architecture where the master server is called the NGMR Master Node and the Client Nodes are called the NGMR Client Node or NGMR Cluster Units. These units communicate with each other with REST architecture.

The NGMR Master Node maintains several data structures for managing various tasks. For storage, the Master Node keeps the metadata information in a JSON file for each of the large files stored in the cluster. For Map-Reduce execution, it maintains the information, like *Idle State*, *In-progress state*, *Completed State* of each job. The NGMR Master Node also stores the NGMR Client Node information like names, hostnames, total usable space, total available space and total used space in NGMR file system, last heartbeat message receiving time from each node, dead node and live node status etc. The Master Node also stores the details of the intermediate file regions produced by each of the Map tasks. The information is pushed to the Client Node that has Map-Reduce task in progress. Client Node also maintain two types of data structures, one is to maintain the metadata for each data block of each file and one is for storing the data block itself.

NGMR file system maintains two levels of indexing. The Master Node level indexing deals with the names of the files/documents, respective cluster IDs and data unit level indexing references.

The raw indexing file structure of the Master Node is shown in Fig. 2 below.

DocumentName = < FileName^{*+} >
KeyID = < DistributionAttributeKey^{+}>
Distribution = < KeyID^{+} >
DistributionAttributeKey = < DocumentContent >
ClusterID = < ClusterIdentifierIDs^{+} >
ClusterIndexRef = < ClusterID^{+}, ClusterFileName^{+} >

Fig. 2. Indexing file structure of master node

The Client Node level indexing stores the information of all blocks, their sizes, locations, and root file names. Similarly the raw structure of the cluster unit indexing file is shown in Fig. 3,

DocumentName = < ClusterIndexRef^{+} >
Distribution = < BlockID^{+}, ClusterID^{+}, KeyID^{+}, FragmentID^{+}>

Fig. 3. Indexing file structure of cluster node

3.1 Data Fragmentation

NGMR file system is designed for distributed storage and parallel processing systems. The file system gives the end user full flexibility to fragment the big data file according to any policy, such as vertical fragmentation or horizontal fragmentation or both, based on the selected single or multiple key attributes from the data file. A single or a set of key attributes from the big data file are chosen as a distribution key to fragment and distribute the data to the NGMR cluster nodes.

NGMR fragments the data by reading the records one by one, buffering them in a temporary memory created on each distinct occurrences of records based on the chosen attributes. Buffering of data takes place on the NGMR Master Node and when the buffer size reaches a particular size, the data is sent to a NGMR Client Node. By default, the NGMR Master Node creates a buffer of 64 MB size for each value of the key (or set of keys) attribute. Size of the buffer is same as the size of the data block and can be changed by the user. When the buffer is full, the data is transferred from the NGMR Master Node to the Client Node as a 64 MB data block for storing.

For each such transfer action of the buffered data to the NGMR Client Node, a unique identification for that fragmented file is recorded in a JSON file in the NGMR Master Node that is stored as master metadata. Another JSON file is stored on the Client Node that is called child metadata. The metadata at the Master Node not only guides the query execution on the desired Client Node, but it also plays a vital role in failure recovery.

3.2 Fault Tolerance by Data Distribution Using Zone Division

NGMR file system is designed in such a way that the recovery is achieved smoothly. A single attribute or set of attributes are input to fragment the big data file. Thus every set of fragments based on a key can act as a replica of another set of fragments based on a different key.

The above policy addresses another challenge related to recovery of the fragmented blocks stored in different cluster units, in case of failure of some nodes. This may also help in case of mobile nodes when some nodes are not within the communication range. So, the main idea is not to replicate the exact copy of the fragments to other locations, but to rearrange every copy based on the chosen key attributes and to place them optimally, so that no data is lost during node failure or when node is not reachable in the cluster. The set of Client Nodes containing all the data blocks fragmented on the basis of a particular key attribute is termed as a *Zone* [1]. Similarly, other Zones may be created based on another key. Every Zone must contain all the fragments based on the same key, so that even if a unit of a particular Zone fails, the fragments stored in any of the remaining Zones can be accumulated to recover the big data file and that particular key based fragments of the failed node are regenerated by the algorithm itself ensuring failure recovery without data loss.

A simple example may be given using our data-model. The data blocks fragmented based on the "location" attribute as key are stored in Client nodes one, two and three. In this case, the data blocks fragmented based on the diseases will be stored in Client nodes other than one, two or three (for example in four and five). So if any one or all of the Client nodes one, two and three are down or not reachable, the data can be recovered from nodes four and five and re-fragmented based on the "location" attribute. The main idea is to distribute all the fragments across the nodes in such a way, that if we take the union of all blocks from these nodes, it will return the original data file.

4 Setting Up an NGMR Cluster

The huge amount of health data cannot be stored in single conventional storage as the size of the big data file is too large to fit in a single system. It should be distributed for fast processing of user-given queries. For the experiment with the big data, large Datacenters are required along with large capacity servers to build the infrastructure. Significant capital is required to produce the replica of such infrastructure which may sometime be a limitation for research.

Based on the discussion presented in Sect. 4, a Raspberry Pi based cluster has been deployed for this work. The advantage of this deployment is that, while experimenting with the NGMR file system, we have also observed the effect of mobility on such systems.

4.1 Raspberry Pi - Node.js Cluster

Raspberry pi is a low-cost single board computer which is very efficient for making storage cluster [2]. We have made a 16 node storage cluster having one Master Node with 32 GB storage each. The configuration of the Master Node is higher than the Client Nodes. It has a Quad-core Cortex-A72 (ARM v8) 64-bit SoC @ 1.5 GHz. processor with 4 GB LPDDR4-3200 SDRAM. and the Client Nodes have ARM Cortex-A53 1.4 GHz processor with 1 GB SRAM on each node.

The nodes are connected in a self-forming, self-healing, and self-organization ad hoc mesh network using BATMAN-adv (Better Approach To Mobile Ad-hoc Networking [7]) routing protocol, designed and developed to deal with the networks that are based on unreliable links. As the nodes are movable sometime the routes can be broken and need to reestablish the new route. All the cluster nodes will maintain a list of neighbors to the other nodes in the network. A node will select the next hop based on some factors like less packet loss, high signal strength, transmit quality (TQ) etc. Each node perceives and maintains only the information about the best next hop towards all other nodes. The nodes are setup using static IP and connected with 5.2 GHz integrated WLAN to gain access from Master Node.

4.2 Data Storage Using Node.js

Fig. 4. Replica of mobile node

Handling large files is not new to JavaScript. In fact, in the core functionality of Node.js, there are several standard solutions for reading from and writing to files. The most straight forward solution is to stream the data in (and out). This option fails because the large set of data is not expected to be static data ("data at rest").

Hence event streaming is chosen for this work. The event streaming functions effectively for the situation where there is a constant flow of data ("data in motion") and the action needs to be taken in real-time as soon as possible (Figs. 4 and 5).

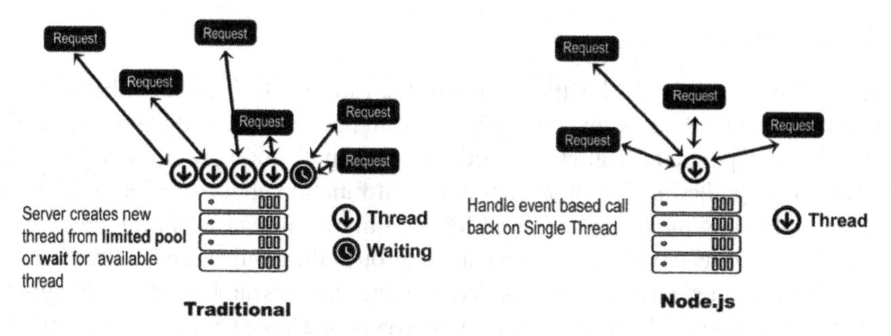

Fig. 5. Node.js request handling thread structure

5 Experimental Setup

In this section, we discuss the experimental setup for comparing our proposed NGMR file system with HDFS. Before discussing the experiments, we discuss the data model used in this work and the data generation technique for carrying out the experiments.

5.1 Data Model

In this paper, we use the data-model presented in [5]. The used data model is shown in Fig. 6. However, data members are not shown here. A *Person* can be a *Doctor*, or a *Patient*. The patient can register one or more *complaints* regarding the illness. A complaint is treated within one or more *treatment episodes* by doctors. A treatment episode may consist of one or more *visits*. In a *visit*, the doctor prescribes one or more *investigations*. The investigation is of two types, *continuous monitoring* and *discrete monitoring*. Diagnosis can be concluded in a visit.

Fig. 6. Used data model

5.2 Data Generation

We have set up kiosk based health centres in the rural areas. The kiosk based health centres are driven by the medical professionals with the help of cloud based application KiORH (Kiosk Operated Rural Healthcare) [4]. The application provides an integrated environment for gathering symptoms and other information about a patient visiting the kiosk, and uploading the details to cloud for real-time treatment by a remotely located doctor. The data collected from the rural patient through the application is stored in cloud storage. This collected data volume has been used for this research work. Although, the data volume was not very large, it has been used to generate large volume of synthetic data by duplicating, restructuring and reusing. The synthetic data set is produced following the above-mentioned data model.

We started with ten patients having 805 diagnosis records and gradually increased the data size by adding locations, permuting doctors and increasing diagnosis records. The data count and sizes are given in Table 2.

Table 2. Data Size and count

Records	Patients	Diagnosis records	Size (in GB)
4 K	4000	2,66,979	0.0528
12 K	12,000	8,00,937	0.1584
20 K	20,000	13,34,895	0.2656
60 K	60,000	40,04,685	0.7969
100 K	100,000	66,74,475	1.3
1.4 Lakhs	1,40,000	93,44,265	1.8
1.8 Lakhs	1,80,000	1,20,14,055	2.25
2.20 Lakhs	2,20,000	1,46,83,845	2.92
2.60 Lakhs	2,60,000	1,73,53,635	3.45

5.2.1 Setting up Hadoop Cluster

In Sect. 6, we discussed about NGMR cluster. As part of our experiment, we have also built a Hadoop cluster with the same set of Raspberry Pi nodes.

For storing the data and executing the user queries using Hadoop, we have installed Hadoop in all 17 nodes (Raspberry pi). One node is made a Master Node, which has a higher configuration and other nodes are setup as DataNodes. The nodes are connected using adhoc network. We have used the latest version of Hadoop (V 3.3.0). For the execution of user-given queries and to support Hadoop execution, Java 8 is used as a programming language and for JVM.

In the next two sections, we present two sets of experiments conducted on the Hadoop cluster and on the NGMR cluster and compare the results.

6 Experimental Results

In this section, the results of the experiments carried out on the NGMR cluster are presented and compared with the results obtained from the experiments carried out on HDFS cluster.

6.1 Data Uploading Time

In our model, health data is primarily stored in cloud storage. At the first step, this large amount of unstructured health-data is fragmented and distributed onto the cluster nodes. Next, the user queries are executed on the distributed data stored on the cluster nodes.

The first experiment is made by fragmenting the data with "City" as a key. The data consists of total 53 number of cities. The system primarily makes an equal distribution of cities for each cluster node. Table 3 shows the distribution of city in each cluster nodes.

Table 3. Cities in each cluster node

NGMR Client node count in cluster	City in each node
1	53
2	27
4	14
8	7
16	4

We have measured the data uploading time for HDFS, as well as for NGMR file system. We have gradually increased the file size from 10 to 2.60 Lakhs records and cluster node count from 1 to 16. When number of patient count is low, the uploading time in NGMR file system is slightly higher than HDFS, as key-wise division takes place before data distribution starts. But when the data size and number of nodes increase, both require nearly the same amount of time. The data upload times are shown in Figs. 7, 8, 9, 10, 11 and 12.

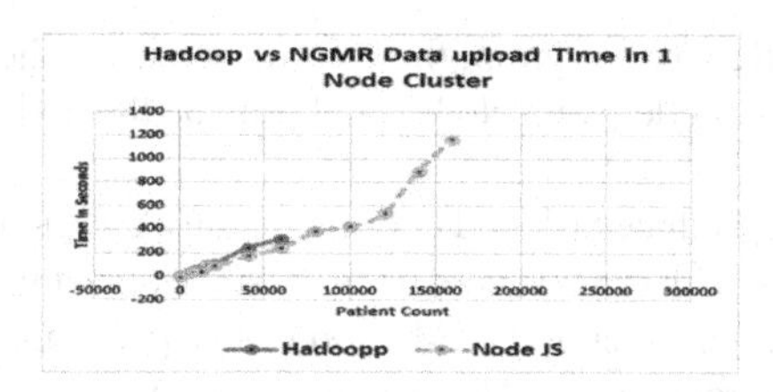

Fig. 7. Data upload time for 1 node cluster

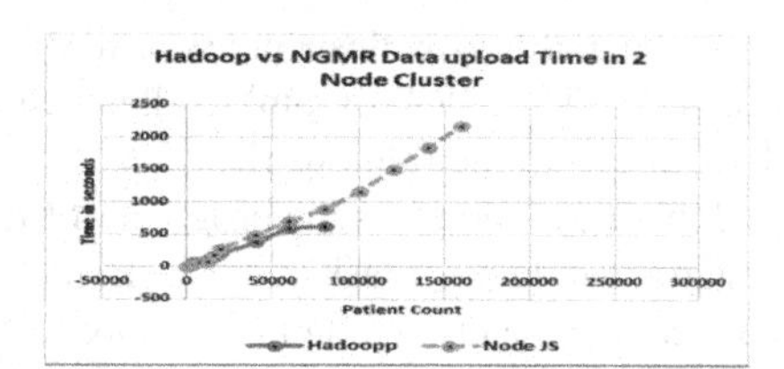

Fig. 8. Data upload time for 2 nodes cluster

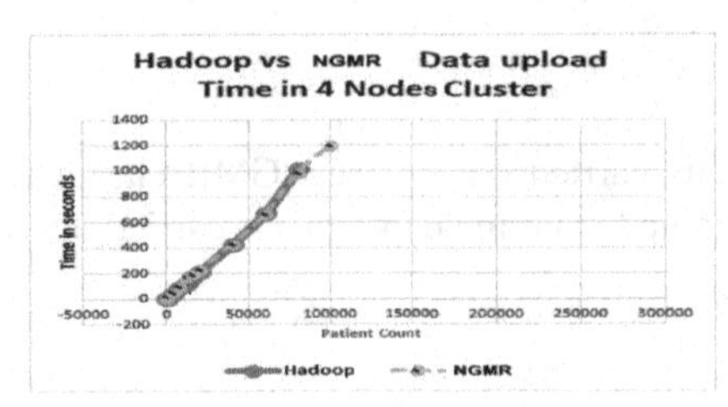

Fig. 9. Data upload time for 4 nodes cluster

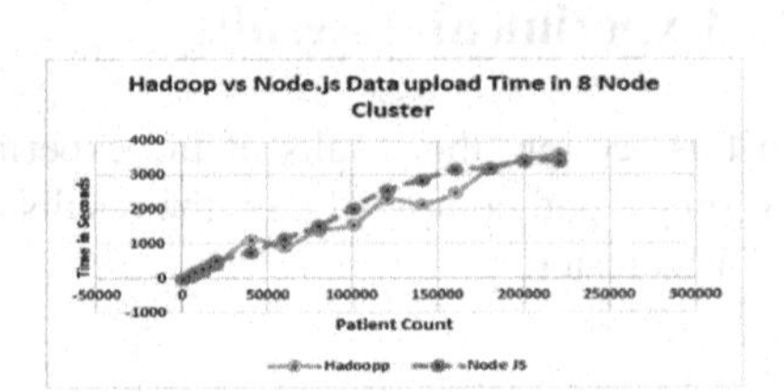

Fig. 10. Data upload time for 8 nodes cluster

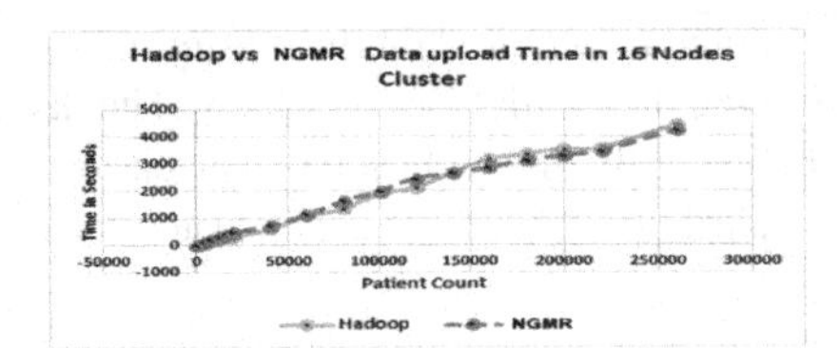

Fig. 11. Data upload time for 16 node cluster

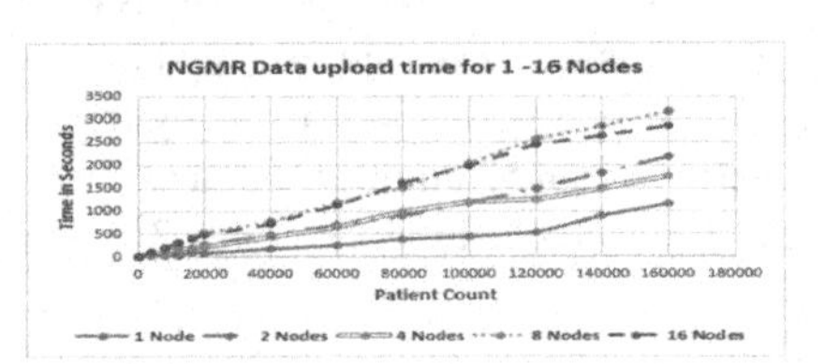

Fig. 12. NGMR data upload time 1–16 nodes cluster

6.2 Execution of User Defined Queries

Hadoop Query Execution: Hadoop framework performs parallel computation across the large cluster having a single Master Node (JobTracker) and multiple Client Nodes (TaskTrackers) using map-reduce architecture. Map-reduce has three basic components that perform distinct operations on the data, i.e. Mapper, Combiner and Reducer.

- **Mapper:** Mapper is applied on input key-value pairs of data, which runs on every block of data on each nodes. It generates intermediate key value pairs for combiner.
- **Combiner:** Combiner can produce any number of key value pairs. But all key-value pairs must be in same type of the mapper output. Partitioners are responsible for dividing the intermediate key space and assigning intermediate key value pairs to reducer.
- **Reducer:** Reducer gathers all the output of reducer from all nodes in key-value format, then shuffle and sort them and stored in HDFS and can be extracted from Master Node.

When a user provides a query as a Map-Reduce Program, the JobTracker sends the user query (The MR Program) to all the cluster node (Data Node), and provide the node block details to the framework.

Job tracker monitors the progress of Map-Reduce task and coordinate the execution of map and reduce the task of the TaskTrackers with run on every node. It can happen that the user given query will not return any data after execution of MR job from a particular node, still, the TaskTracker will run and initiate reduce task. The task returns zero reduced set of data. This process increases the total execution time of the Map-Reduce task.

We take an example of executing health-related queries using the above-mentioned dataset. The dataset is distributed based on a key named as 'city'. As Hadoop distributes the dataset in the Hadoop cluster, the JobTracker initiates the Map-Reduce task on all cluster nodes with the help of TaskTracker, and each TaskTracker runs the MR code to all nodes. There may be some nodes where no data is stored related to the user given key values. TaskTracker runs the query on those node blocks where it returns zero key-value pair data.

In Fig. 13, the end-user needs the details for a key-value *k1*. In this case, the Map-Reduce engine will run on all blocks on all nodes as shown in the figure.

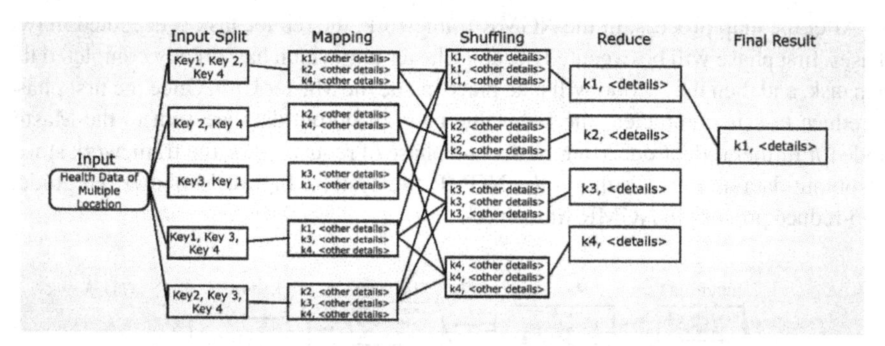

Fig. 13. Hadoop MR process on health records of multiple cities

The user given query may be "Find the details of the patients who belong to Kolkata and suffering from tuberculosis". As our data-model does not store both information in the same file, we have to run two MR jobs. At first, Hadoop runs an MR job to find the details of the patients who belong to the respective cities. These results are stored in the HDFS first as an output and then the patient data is sorted and output is filtered by the diseases The overall execution time is the sum of the two MR jobs executed in Hadoop. However, the entire process requires longer time due to the overhead as mentioned earlier. Therefore, a better strategy will be to discard the records containing key values other than those mentioned in the query either in the Map phase or in the Reduce phase.

Node.js Query Execution: In NGMR, the stored data are processed in parallel both in the Cluster units and inside the Cluster units depending upon their processor cores. In the NGMR framework, instead of running the MR analytics code in all the Cluster units, the code runs in a small subset of the Cluster units. The NGMR system guides the MR code to run in a specific set of selected nodes.

- **Guided Map Reduce:** When a user places a query in the NGMR framework, the NGMR Master Node checks which of the blocks on which nodes are really necessary for execution of the given MR query. Then framework follows the split-apply-combine strategy in the same way as a typical map-reduce programming paradigm on the data blocks distributed in the cluster. The user provides the code written in Node.js. The framework reads the query and extracts the key attributes and their values which are required for executing the query. The Master Node then reads the master meta-data stored for the file referred by the user. From the master meta-data, Master Node gets the reference of the Cluster units along with the meta-data stored in the Cluster units. The framework reads the Cluster unit meta-data referred to by the Master Node and gets the references of the data-blocks which need to be processed for execution of the user given query. The framework sends the MR code to the Cluster units where the requested data-blocks are stored. The map task is then executed in the same way as the conventional Map-Reduce programming model, however, it works only on the selected Cluster unit blocks.

After the map process, in the NGMR framework, the reduce task is executed in two phases, first phase will be executed in each Cluster unit which has already completed the map task, and then the second will take place inside the Master Unit. Once the first phase of reduce task is completed, the key-value pairs of output data are sent to the Master Node for further reduction. After the second phase of reduce tasks, the framework stores the output data as final results in the NGMR file system. Figure 14 shows the guided Map-reduce process in NGMR framework.

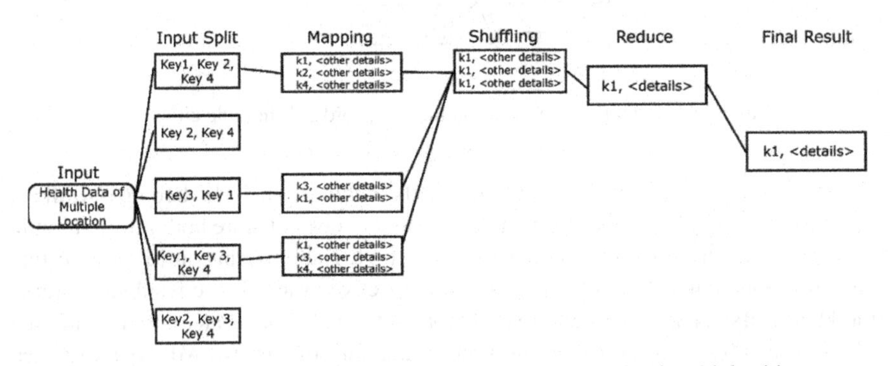

Fig. 14. NGMR framework MR process on health records of multiple cities

- **Parallel execution in processor core:** NGMR uses an NPM package named as 'mapred'. This package follows the Google map-reduce specification. The input should be a key-value paired array-like map(key1, value1), so that a set of intermediate key-value lists(key2, value2) are created for reduce. The reduce program accepts the intermediate key and a set of values for that key like reduce(key2, list(value2)) and produces a smaller set of values like list(value2). Mapred package has the provision to use all cores of the processor or user can fix how many cores should be used by the system to process the MR code.

The OS runs the I/O in parallel and sends the data to single-threaded JS. Our code consists of small portions of synchronous blocks that run fast and pass the data to files and streams. So our JavaScript code does not block the execution of other pieces of JavaScript. A lot more time is spent waiting for I/O events to happen than JavaScript code being executed. NGMR Node.js framework just invokes the functions and does not block the execution of other pieces of code. It will get notified through the callback when the previous code execution is over, and the system will receive the result. Node.js does not evaluate the next code block in the event queue until the previous one finishes executing. So, we split our code into smaller synchronous code blocks and call the callback function to inform Node.js that previous code execution is over and that it can continue executing pending things that are in the queue.

6.2.1 Discussion

As the experiments have been conducted starting from low volume of data, in a single-node cluster, the node stops responding when we start increasing the data volume. In case of Hadoop / HDFS, a Cluster unit gets down when number of patient records crosses 40 thousand (size 531 MB). But when a node cluster is tested with NGMR file-system, it works with more than 2 Lakhs 20 thousand (2.92 GB) data as shown Fig. 15. In case of two node cluster, with HDFS we have been able to execute the query with 80 thousand (1.1 GB) records, but with NGMR file system, the patient count can be increased upto more than 2 Lakhs Fig. 16. This way, it has been observed that our developed NGMR file system has always been able to upload data in higher volume compared to Hadoop HDFS within the same time duration. The results are shown in Figs. 17, 18 and 19. Hadoop distributes the data to all its cluster nodes, but NGMR not only distributes, but also tracks all the block locations using two layer indexing process (Fig. 20).

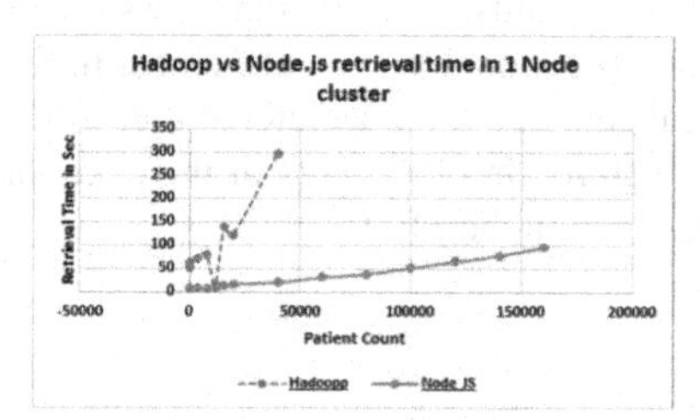

Fig. 15. Execution time of the user given queries with 1 node cluster

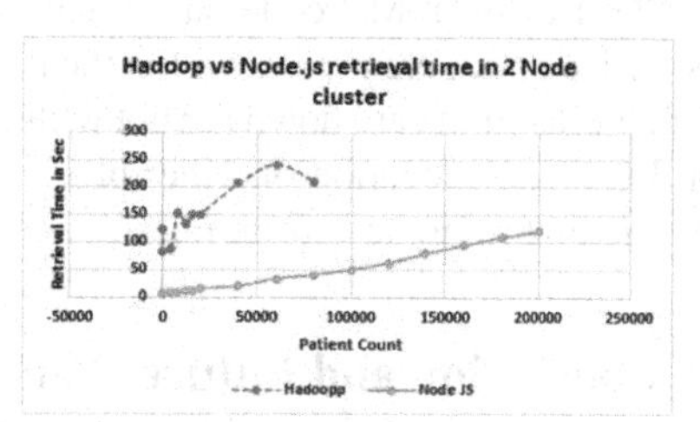

Fig. 16. Execution time of the user given queries with 2 nodes cluster

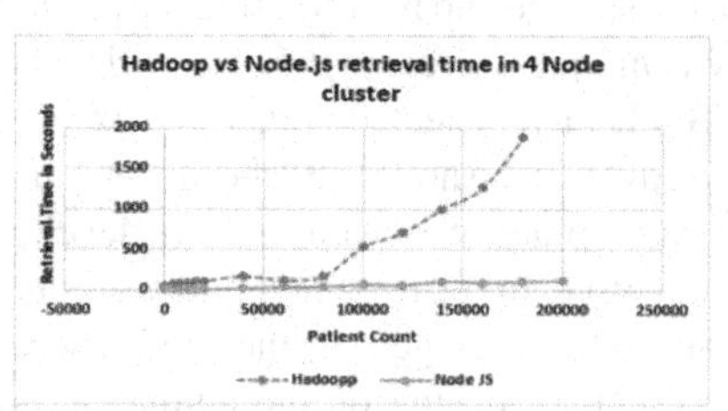

Fig. 17. Execution time of the user given queries with 4 nodes cluster

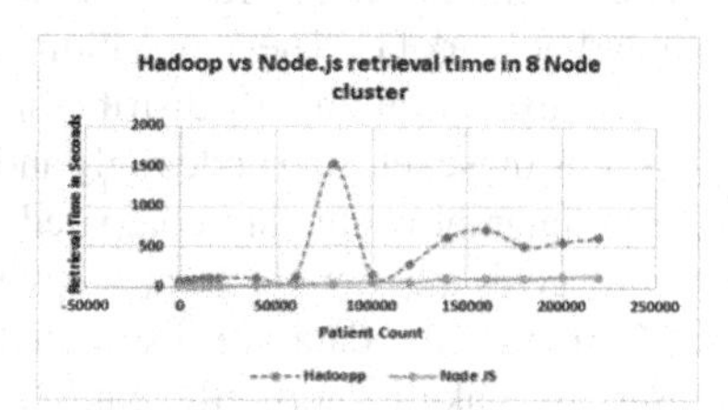

Fig. 18. Execution time of the user given queries with 8 nodes cluster

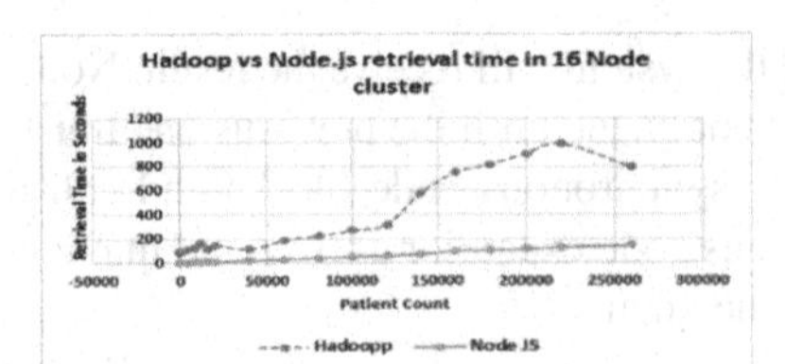

Fig. 19. Execution time of the user given queries with 16 nodes cluster

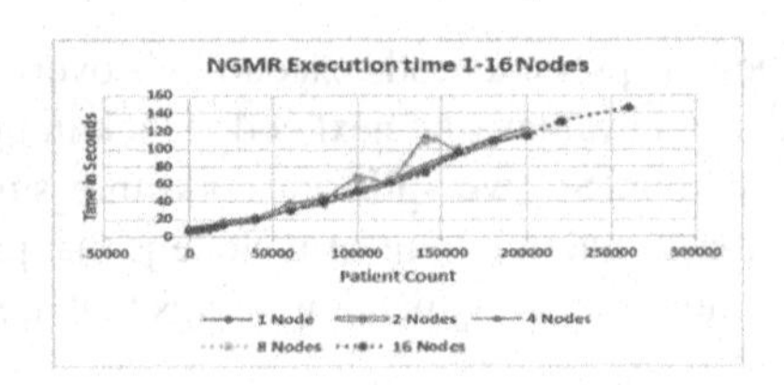

Fig. 20. NGMR execution time 1–16 Nodes

At the time of execution of MR code, Hadoop execution time increases drastically when data size is increased. The reason behind this behavior is that Hadoop runs its MR code on all nodes. In case of NGMR, MR code is run on a small set of Cluster units. So execution time is much less than the previous case.

We have carried out our experiment with mobile nodes. So the time required for data uploading and execution of MR code is always higher in case of Hadoop, because Hadoop framework waits if the node is down or not responding. If Hadoop does not get any response within a certain time, it checks the replication factor and sends the MR code to another node where the same block of data is stored. Figure 18 shows that at data count 60k, the Hadoop framework waits for a longer time as nodes are not within the range. But in case of NGMR, it waits for 3 s and checks for another node (unit). As NGMR runs it MR code on a small set of Cluster units, it is generally possible that respective nodes are always within the range and execution does not take longer time. If all requested units are down (or not within the range), then it immediately (after waiting for 3 s) starts execution on other blocks of data stored in other units of the next zone after getting references from master meta-data stored in the Master Node.

7 Conclusion and Future Work

In this paper, the implementation of Node Guided Map-Reduce technique has been discussed. This research work is done based on the storage and retrieval scheme proposed in our earlier work [1]. The application of this technique decreases the data retrieval and user query execution time by distributing the Big Data to selected nodes, and executing the query on these selected nodes only and not to cover all the nodes attached in the cluster. The experimental results are compared with Hadoop and it is seen that the developed technique gives a better performance with respect to data retrieval and query execution time. Currently, the data are distributed to the nodes after getting the preferences of the user. To make it more efficient, we are presently working on the automatic node selection algorithm based on the key attributes of the Big data file depending on the user given queries. The testbed which has been used is a raspberry pi mobile cluster. Our next endeavor is to deploy a high-end cluster with movable nodes. In future, we aim to carry out our experiment with a large, high-end cluster with mobile nodes and implement automatic cluster node selection algorithm for execution of the user queries for more efficient data retrieval.

Acknowledgement. We would like to acknowledge our supervisor Dr Nandini Mukherjee, Professor, Department of Computer Science and Engineering, Jadavpur University, for her guidance, assistance and timely suggestion which helps us to complete this research work.

References

1. Ray, H.S., Mukherjee, S., Mukherjee, N.: Performance enhancement in big data handling. In: 2020 International Conference on Contemporary Computing and Applications (IC3A), Lucknow, India, pp. 17–22 (2020). https://doi.org/10.1109/IC3A48958.2020.233261
2. Kim, C., Son, S.: A study on big data cluster in smart factory using raspberry-pi. In: 2018 IEEE International Conference on Big Data (Big Data), Seattle, WA, USA, pp. 5360–5362 (2018). https://doi.org/10.1109/BigData.2018.8622539
3. Asb, Andrews, Liz.: Introducing turbo mode: up to 50\% more performance for free, 19 September 2012. https://www.raspberrypi.org/blog/introducing-turbo-mode-up-to-50-more-performance-for-free/. Accessed 23 Aug 2020
4. Mukhopadhyay, P., Roy, H.S., Mukherjee, N.: E-healthcare delivery solution. In: 2019 11th International Conference on Communication Systems and Networks (COMSNETS), Bengaluru, India, pp. 595–600 (2019). https://doi.org/10.1109/COMSNETS.2019.8711429
5. Ray, H.S., Naguri, K., Sen, P.S., Mukherjee, N.: Comparative Study of Query Performance in a Remote Health Framework using Cassandra and Hadoop. HEALTHINF (2016)
6. Liu, L., Liu, J., Qian, H., Zhu, J.: Performance evaluation of BATMAN-Adv wireless mesh network routing algorithms. In: 2018 5th IEEE International Conference on Cyber Security and Cloud Computing (CSCloud)/2018 4th IEEE International Conference on Edge Computing and Scalable Cloud (EdgeCom), Shanghai, pp. 122–127 (2018). https://doi.org/10.1109/CSCloud/EdgeCom.2018.00030
7. Mesh. (n.d.). https://www.open-mesh.org/projects/open-mesh/wiki/BATMANConcept. Accessed 27 Aug 2020
8. Anisha Gnana Vincy, V.G., Karthija, T., Sunil, J.: Understanding hadoop framework through single-node cluster installation. In: 2019 International Conference on Recent Advances in Energy-efficient Computing and Communication (ICRAECC), Nagercoil, India, pp. 1–4 (2019). https://doi.org/10.1109/ICRAECC43874.2019.8995060
9. Qing, W., Yue, Y., Yi, Y., Liang, W.: A method of pre-sentence text based on Map/Reduce storage and indexing classification. In: 2014 IEEE 5th International Conference on Software Engineering and Service Science, Beijing, pp. 195–199 (2014). https://doi.org/10.1109/ICSESS.2014.6933543
10. Yuzuk, S., Aktas, M.G., Aktas, M.S.:On the performance analysis of map-reduce programming model on in-memory NoSQL storage platforms: a case study. In: 2018 International Congress on Big Data, Deep Learning and Fighting Cyber Terrorism (IBIGDELFT), ANKARA, Turkey, pp. 45–50 (2018). https://doi.org/10.1109/IBIGDELFT.2018.8625300

Multipath TCP Security Issues, Challenges and Solutions

Khushi Popat(✉) and Viral Vinod Kapadia

Department of Computer Science and Engineering, Faculty of Technology and Engineering, The Maharaja Sayajirao University of Baroda, Vadodara, India
{khushi.popat-cse,viral.kapadia-cse}@msubaroda.ac.in

Abstract. Multipath TCP (MPTCP) is a bidirectional byte stream transport layer protocol introduced by Internet Engineering Task force (IETF) which provides numerous benefits such as higher throughput, reliability, fault tolerance, backward compatibility and load balancing by supporting multi-homing that allows use of multiple paths for data transfer over single network connection still it is vulnerable to many security intrusions such as Denial of Service, session hijacking, SYN Flooding etc. In this paper, the vulnerabilities of MPTCP leading to some potential attacks and their available solutions are focused. Currently many solutions are available but some of them increase the overhead of MPTCP, some are vulnerable to time-shifted attack and rests are not tested properly. The implementation and experiment details of ADD_ADDR and Eavesdropper in initial handshake attacks performed on MPTCP version0 and version1, detailed analysis of all the available solutions, limitations of these solutions and direction to offer the solution over these limitations are covered in this paper.

Keywords: MPTCP · Session hijacking · Man-in-the-middle attack · ADD_ADDR vulnerability · Time-shifted attack · Hash based solution · Chained based solution

1 Introduction

Transmission Control Protocol (TCP) [1] is the widely used transport layer protocol designed for providing process to process delivery of packet between communicating host over IP network. TCP binds the IP addresses of communicating hosts over connection as shown in Fig. 1 (b) which leads to connection drop if failure occurs during communication. If the host tries to switch to another network interface during on-going communication with the server, the current TCP connection will be dropped as device will get connected to another network and new IP address will be assigned to it. TCP doesn't utilize more than single network interfaces over single connection. For example, Mobile devices have Wi-Fi and LTE network interfaces but at a time one can be connected only to one interface. To overcome these restrictions of TCP, MPTCP [2, 3] was standardized by IETF with the goal to increase the utilization of network resources by supporting routing of packets over multiple disjoint paths for single connection as shown

M. Bhattacharya et al. (Eds.): ICICCT 2021, CCIS 1417, pp. 18–32, 2021.
https://doi.org/10.1007/978-3-030-88378-2_2

in Fig. 1 (a), increases reliability by providing backup paths and improves performance by combining bandwidths of different paths. Each MPTCP path (sub-flow) behaves like single TCP connection at network layer and uses regular TCP Socket APIs at application layer [2]. The mobile devices, laptops, servers etc. with multi-homing facilities can be equipped with MPTCP to offer the higher bandwidth and reliability. Many mobile devices companies like APPLE, Samsung, LG etc. have started using MPTCP with their smartphone in various applications to provide Quality of Service (QoS). Datacenter is another use case of MPTCP in which TCP can be replaced to manage the load through multiple disjoint paths.

Despite MPTCP provides many attractive benefits over TCP such as higher robustness, higher availability, higher performance and backward compatibility with current Internet applications, its multi-homing (connected to more than one networks) support using the various options in TCP header are vulnerable to many suspicious attacks such as session hijacking, flooding, Denial of services etc. [4]. ADD_ADDR, one of the options used in TCP header to inform the peer host about the available IP, is vulnerable to session hijacking attack. ADD_ADDR option plays crucial role in wireless networks as due to the mobility of devices, it is required to inform the peer host about the change of network IP frequently. In version 1 of MPTCP, the Hash based Message Authentication (HMAC) is used to authenticate the peer to add the new address with ADD_ADDR option but the keys used to calculate the HMAC are exchanged in clear form during the initial 3-way handshake of MPTCP. However, it secures the addition of subflows, it doesn't offer the protection for key exchanges during initial handshake which again opens the doors for attackers.

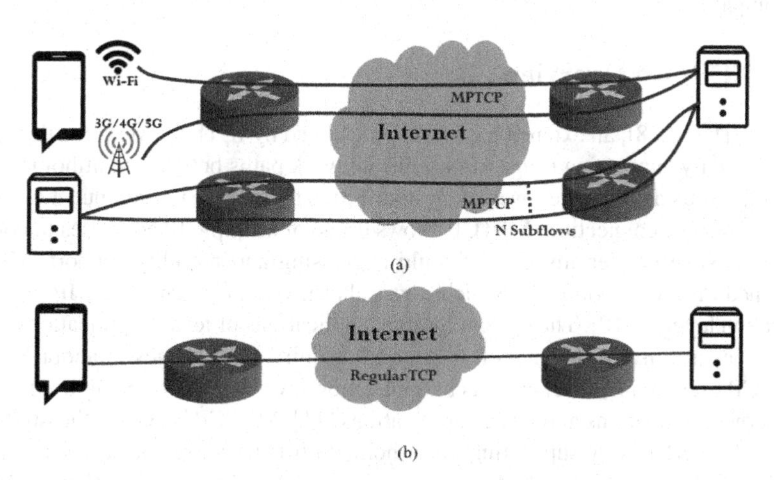

Fig. 1. (a) Multipath TCP connection Scenario (b) Regular TCP connection scenario

MPTCP increases the network performance by transferring data across multiple available paths but this feature allows attackers to transfer malicious codes across different paths. Thus, on different paths different Signature Based Intrusion Detection Systems will be deployed but none of them will have access of full malicious code which prevents them to identify the Signature Based Intrusion [5].

The research in this domain either focus on security issues of MPTCP or deploying MPTCP for providing network security [6, 7]. Many researchers have also proposed the various solutions to overcome the security flaws but either it increases the overhead of the connection, consumes too much energy or leads to another attack. However, security of MPTCP is still an open issue to be resolved for increasing adoption of MPTCP in real life applications. IETF has already identified and listed major security issues essential to be resolved to accelerate deployment of MPTCP in various areas.

In this paper, the following contributions are added to the current research scenario in the area of security of MPTCP:

- Comparison of MPTCP options in MPTCP version0 (v0) and MPTCP version1 (v1).
- MPTCP threat model (Security challenges with MPTCP).
- Experimental setup to perform Session Hijacking using ADD_ADDR and eavesdropper in initial handshake attack on ADD_ADDR2.
- Comparative analysis of available security solutions to overcome vulnerability of ADD_ADDR.

This paper is organized as follow: Sect. 2 covers detailed overview regarding connection establishment, advertisement of new address and adding new sub-flow over connection in MPTCP v0 and v1. The next sections discuss about various vulnerabilities in current MPTCP implementation, session hijacking by using man-in-the-middle attack, experimental analysis of attacks perform by using ADD_ADDR vulnerability, available solutions to overcome ADD_ADDR vulnerability and their limitations with their comparison.

2 MPTCP: An Overview

Multipath TCP [2, 8], an extension of TCP, is designed by IETF to overcome the limitations of TCP by supporting usage of multiple network paths between multihomed hosts for simultaneous data transfer over single connection to improve throughput, availability and robustness of connection. MPTCP allows usage of multiple IP addresses to connect to single host offers certain level of multi-addressing and mobility support. MPTCP is designed on the top of TCP in such a way that it is transparent to the IP layer and Application layer [9, 10]. The current Internet applications or legacy applications which uses TCP as a transport layer protocol doesn't require any changes to adopt MPTCP as MPTCP uses regular Socket APIs at application layer which are being used by current Internet applications and legacy applications [11]. MPTCP improves the utilization of network interfaces by supporting multi-homed hosts to communicate over multiple individual TCP connections which constitute single MPTCP connection. Here each individual TCP connection is known as sub-flow that can be added/removed throughout the MPTCP lifecycle. To fulfil the backward compatibility of MPTCP with applications which uses TCP at transport layer, developers decided to use TCP header options to include data required for MPTCP. The TCP header option number "Kind" is assigned to MPTCP.

MPTCP establishes a connection to the peer host by using 3-way handshake same as TCP, but includes MP_CAPABLE option in TCP packet header to announce that host is

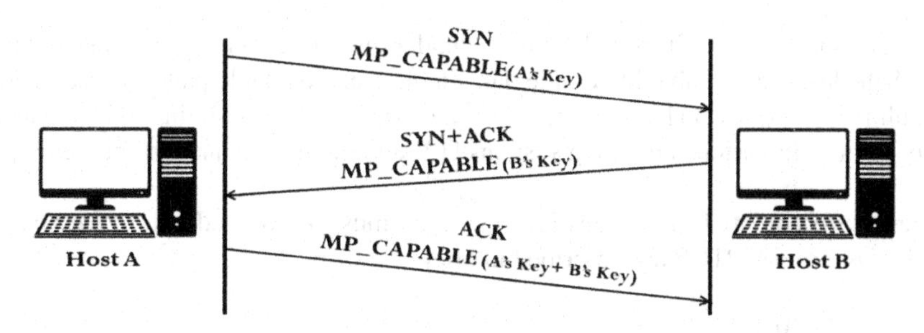

(a) MPTCP connection initiation (3-way handshake)

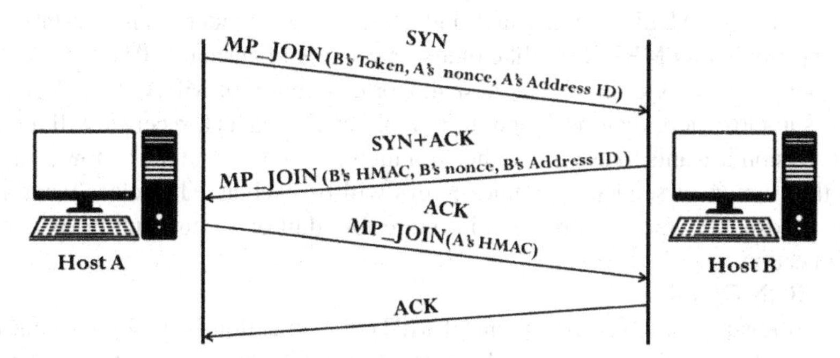

(b) MP_JOIN option in TCP header to add new subflow

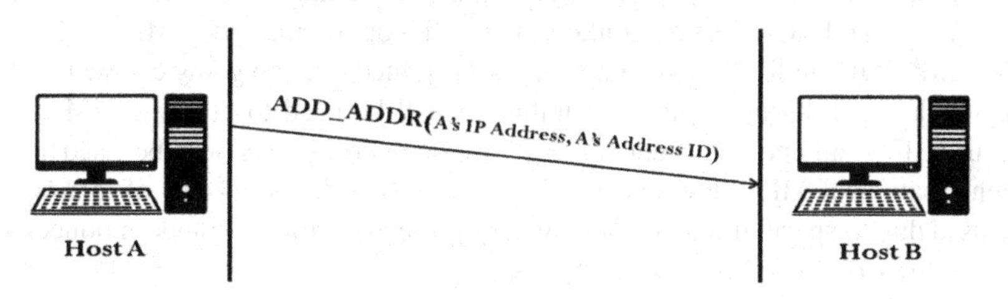

(c) ADD_ADDR option in TCP header to advertise new address

Fig. 2. MPTCP options in Version 0

MPTCP supported as shown in Fig. 2 (a). MPTCP supported host will acknowledge by setting MP_CAPABLE option in SYN+ACK packet while if peer entity is not MPTCP supported will ignore the MP_CAPABLE option and the connection will be treated as normal TCP connection.

In the similar way, options in TCP header will be used for the initiation of new connection (MP_JOIN), advertisement of new address (ADD_ADDR), removal of an address (REMOVE_ADDR) etc. to make the routine TCP work as Multipath TCP as shown in Fig. 2 [11].

MPTCP supports various congestion control algorithms like lia, olia, wVegas etc. as well as various path management techniques such as default, fullmesh etc. those can be configured by using various commands. It also provides various kinds of scheduling techniques such as default, round-robin, redundant that one can select as per requirement. Schedulers are enabled to divide data on different available paths. These various scheduling techniques [11] based on path characteristics are very similar to load balancing techniques in various operating systems [12] and memory management techniques [13].

Here different MPTCP options [2, 14] which must be focused before we analyze security issues of MPTCP are described in brief.

1. MP_CAPABLE
   MPTCP connection initiation takes place with three-way handshake same as TCP by exchanging SYN, SYN+ ACK and ACK packets but each packet is equipped with MP_CAPABLE option which indicates that the sender want to establish the connection using MPTCP. Unlike older version v0, in current MPTCP version (v1) the sender will advertise the highest supported version of MPTCP in SYN packet if it supports more than one version. In reply of this packet, receiver will advertise the version it wants to use for further communication which must be lower or equal to the version of sender. The unique keys will be exchanged in clear form during the handshake which will be used in future for addition of new sub flows over this connection.
2. MP_JOIN Option
   Once the successful establishment of MPTCP connection is done, new sub flows can be added to this connection by using MP_JOIN option. Initiation of new sub flow follows the same steps which are required for connection initiation but SYN, SYN+ACK and ACK packets contains MP_JOIN option instead of MP_CAPABLE. Here MP_JOIN indicates the connection to be joined over on going connection. The key exchanged during connection initiation will be used to generate HMAC will be used for authentication and tokens generated from keys will be used for the identification of MPTCP connection to join. Only leftmost 64 bits of HMAC will be used due to space limitation. MP_JOIN option also contains random nonce, flags and address ID (refers to source address).
3. ADD_ADDR option
   During the lifecycle of the MPTCP connection, availability of additional address can be advertised at any moment of time by any of the hosts. There are two ways to advertise the additional address. One of them is directly initiating a new sub flow by using the MP_JOIN option (Implicitly) and the other by advertising the available address using ADD_ADDR option (explicitly). In MPTCP v0, authentication information was not included while the truncated HMAC (rightmost 64 bits) is being used for the authentication in current version of Linux kernel implementation of MPTCP (v1). Here unique Address ID is mapped with actual address at connection level so that if NAT or middle-boxes changes the actual address, Address ID can be used to add/remove the new subflow on connection.
4. REMOVE_ADDR option
   During the life cycle of MPTCP if host want to remove the previously added address that can be communicated to its peer using REMOVE_ADDR option. REMOVE_ADDR option will remove the address as well as terminate the sub flow initiated using this address. Here Address ID will be used to remove actual address.

5. MP_PRIO option
   Host can change the priority of any of the sub flow whether it should be used as a regular of backup path by using MP_PRIO option of MPTCP. If before removing any of the address by ADD_ADDR option, one wants to close the data transfer on that address can be stopped by using MP_PRIO by changing priority of the path.

3 MPTCP Threat Model

The intruders can be classified into different categories based on where they resided on the path at the time of initiating an attack: off-path attacker, partial time on-path attacker and on-path attacker [4]. The attackers can also be classified in two other categories based on their action or impact: active attacker and passive attacker. Thus, different combinations of mentioned categories can be used to categorize different attacks. The significant threats on MPTCP Linux kernel implementation figured out by IETF in their draft [4, 15] are ADD_ADDR attack, DoS attack on MP_JOIN, SYN Flooding Amplification, Eavesdropper in initial handshake, SYN/JOIN Attacks. Other than these attacks, traffic diversion attacks are another category of attacks which can be implemented by exploiting vulnerability of MPTCP option MP_PRIO [16]. The main attacks on MPTCP which needs to be focused are as below [4, 15]:

1. Eavesdropper in initial handshake
   In this attack, the attacker present during the initial handshake of the MPTCP can capture the keys exchanged during the handshake that can be applied to calculate authentication parameters to acquire the session in future even after moving away from the path. This attack may lead to ADD_ADDR attack and MP_JOIN attack. This attack falls under the category of partial time on path active attack.
2. ADD_ADDR attack
   In this type of attack, attacker can hijack the session by forging the ADD_ADDR packet using man-in-the-middle attack as ADD_ADDR (v0) doesn't pass any authentication parameter with the packet. It is off-path active attack. In next section whole process to perform the attack is covered in detail.
3. DoS attack on MP_JOIN
   This attack comes under the category of off path active attack. In this attack, attacker sends large number of SYN+MP_JOIN request that can exhaust the server by opening large number of half open sessions that will prevent the server from creating new subflow on the existing connection. SYN+MP_JOIN created the state on the receiving host as it contains the 32-bit token and nonce which are not going to be resent on the ACK of the 3-way handshake and these information are used to identify the MPTCP connection and generate the HMAC respectively.
4. SYN Flooding Attack
   By using the SYN message, server can be flooded which prevent it from creating new TCP connections.

The off path active attacker doesn't require any information to initiate any attack as it is the most restricted area. ADD_ADDR vulnerability can be exploited to perform off path active attack as well as partial on time active attack which attracts many researcher to jump in the area of research. Hence, the next section focuses on the ADD_ADDR vulnerability to initiate session hijacking attack.

4 Session Hijacking by Using ADD_ADDR Vulnerability

In session hijacking attack by using ADD_ADDR packet [4, 17, 18], attacker adds its own IP address as an additional address by impersonating identity of a legitimate user and later on establishes the sub-flow by using that address over on-going connection to redirect data on that path with the use of vulnerability of ADD_ADDR packet. This attack falls in the off-path active attack category. Figure 2 (c) shows the information required to advertise new address to the other host in MPTCP version (v0). In the version (v0) of Linux Kernel Implementation of MPTCP, addition of new address by ADD_ADDR packet doesn't include any kind of authentication requirement which leads to MPTCP session hijacking by enabling man-in-the-middle attack. To perform the attack, attacker requires 4-tuple source IP address- port and Destination IP Address-port which identifies the connection. Here destination port can be assumed by identifying the type of service running on the server based on well-known server port i.e. 80 for http, 443 for https etc. Assumption of client port is a challenging task. Different packet manipulation tools such as Scapy, DSniff, Wireshark etc. can be used to sniff the packet to get the prerequisites information such as IP-Port pairs, Sequence number, ACK number etc. SCAPY for MPTCP [3] is available using which 4 tuples and sequence numbers can be obtained to initiate ADD ADDR attack.

Once the attacker gets prerequisite information, following steps [4] can be performed to hijack session:

- Assume that Alice is communicating with Bob and one of the sub-flow is equipped with IP addresses IP-A1 and IP-B1. Here Eve is working at some remote place with IP-E1.
- Now Eve tries to get connection information and succeed to get IP-port pair of both the hosts by capturing packet using the networking tools such as scapy for MPTCP etc.
- Once Eve get the IP-Port pair of both host, he sends spoofed ACK packet with source address as IP-A1, destination address as IP-B1 and ADD_ADDR option with IP address IP-E1 to the Bob. Here ADD_ADDR option also includes Address ID which must be unique so any random bigger number can be selected to prevent collision.
- After receiving ADD_ADDR request, Bob will think that it is coming from Alice so he will initiate the new sub-flow by sending SYN+MP_JOIN request to Eve. After receiving packet, Eve will change the Source address from the packet to IP-E1 and send the same packet to Alice.
- Alice will process this request as a legitimate request by thinking that Bob wants to initiate new sub-flow by using IP-E1 address. All the needed information required for MP_JOIN request will be available to authenticate this packet as it is actually coming from Bob only source address has been changed by Eve. Alice will send SYN/ACK+MP_JOIN to address IP-E1 by adding required parameters.

- Eve will forward this packet to Bob by simply changing source address and create impact of legitimate reply.
- As all the parameters will be correct, Bob will send final ACK+MP_JOIN packet to establish sub-flow towards IP-E1.

In this way, Eve successfully take place as man-in-the- middle between Alice and Bob. Now Eve can change the priority of all the other sub-flows or send RST packet to all the sub flows to fully hijack the connection.

5 Experimental Evaluation of Session Hijacking by Man-in-the-Middle Attack and Eavesdropper in Initial Handshake

The Experimental implementation of the session hijacking by MitM is performed on Linux kernel implementation of MPTCP v0 [19] and Eavesdropper in initial handshake is performed on MPTCP v1. The experimental setup with required topology was created as shown in below Fig. 3 using Oracle virtualBox. The simulation of the attack [17] is performed using two virtual machines with Linux kernel implementation of MPTCP. The use of virtualBox leads to fast experimental setup, reliability of experiments and no risk of damaging or crashing of kernel. To configure this "client server" scenario required compiled kernel and tools are available on official website of MPTCP (http://www.multipath-tcp.org). In order to initiate both the attacks, the attacker host is simulated with Scapy tool which supports MPTCP that is used for capturing and injecting packet on the network. Extended version of scapy specially designed for MPTCP [3] is available by Nicolas Matre on https://github.com/nimai/mptcp-scapy repository. Scapy tool provides inbuilt functions for sniffing, modification, capturing and matching the request-response which is important for initiating an attack.

Here in Fig. 3, Client VM, Server VM and Host Machine are configured with custom MPTCP Linux Kernel. Here client must be equipped with more than one network interfaces to setup MPTCP scenario while server can be equipped with one or more network interface. Here Host Machine is configured with three tap interfaces (virtual network interfaces) to implement multi-homing environment [17].

Session hijacking experiment was performed successful in shown scenario for chat application and file transfer application developed using JAVA socket API on MPTCP Linux kernel v0. Here for gathering prerequisite information such as source-destination IP addresses, SEQ no, ACK no etc. average 2–3 packets need to be captured. Average 6:3 packets are required for client to server to initiate attack successfully. Success rate of session hijacking attack is 77%.

The attack is performed multiple times for capturing different data formats as well as for different file sizes. The above Fig. 4 shows the data lost in percentage (%) while transferring different size of file from client to server. The experiments are performed multiple times and average lost is calculated to represent analysis. From the above graph one can see that as the file size is getting increased, the data lost is increasing. After some threshold, data lost is almost above 90%. The same experiment is also performed multiple times for different data formats and analysis for the same is as shown in Fig. 5.

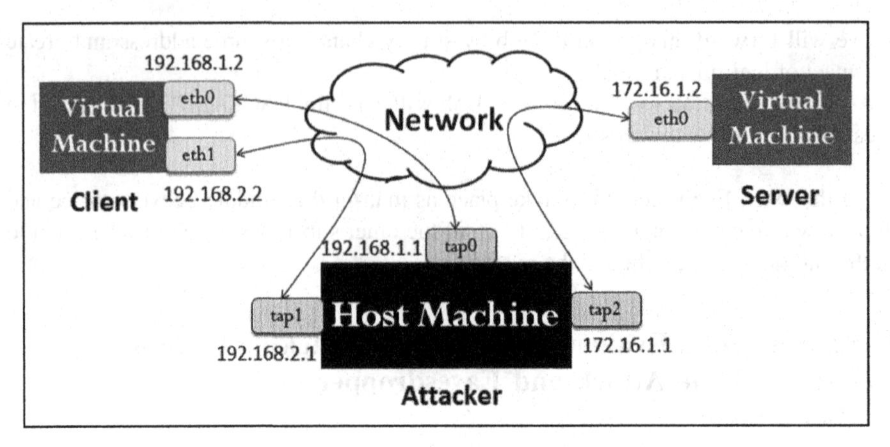

Fig. 3. Network simulations for experiments [16]

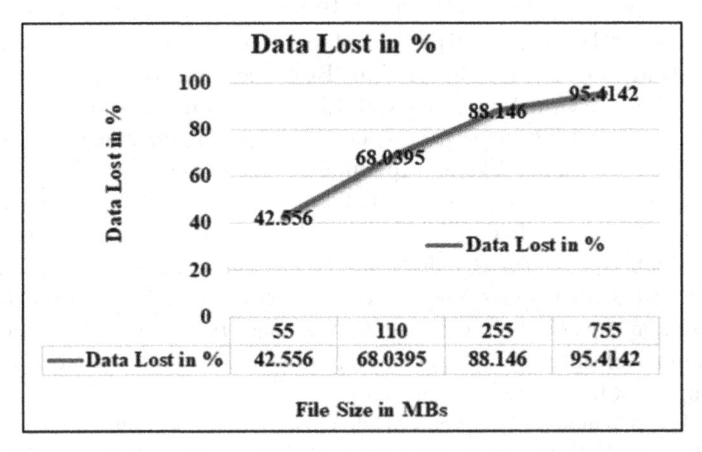

	55	110	255	755
Data Lost in %	42.556	68.0395	88.146	95.4142

Fig. 4. Data lost analysis during different sized file transfer with hijacking attack

The analysis shows that data lost in % is between 95 to 96% for different types of file format for average file size.

To overcome this ADD_ADDR vulnerability, another version ADD_ADDR2 was incorporated with MPTCP Linux kernel v1 but it is vulnerable to Eavesdropper in initial handshake. The Eavesdropper present on the network during the initial handshake can capture the keys which can be used to perform ADD_ADDR Mitm attack on MPTCP v1. Here Eavesdropper in the initial attack has been performed successfully by using Scapy tool to gather client-server keys as shown in Fig. 7 which can be used further to perform ADD_ADD Mitm attack in same manner as mentioned above. Figure 6 is showing the same packet capture in wireshark.

In MPTCP v0, Session hijacking using ADD_ADDR can be implemented as an off path attack by staying away from the ongoing communication and with MPTCP v1 ADD ADDR attack can be implemented as a partial time on path attack by capturing keys by staying on communicating path for short period and then performing session hijacking

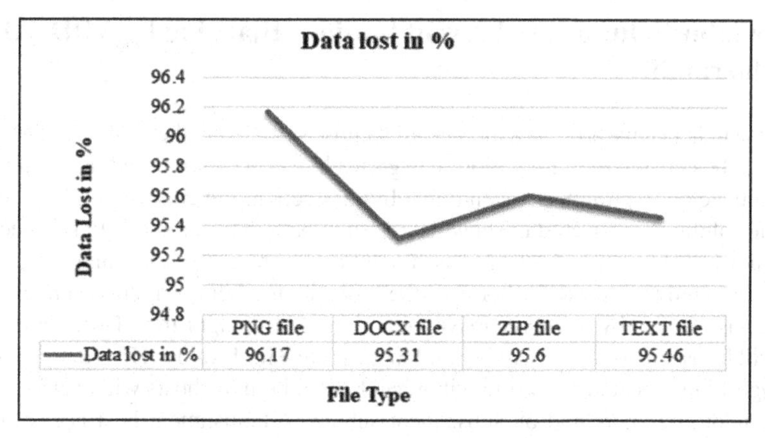

Fig. 5. Data lost analysis in % during file transfer of different format

```
Wireshark · Packet 13 · tap0
▸ TCP Option - No-Operation (NOP)
▸ TCP Option - Window scale: 7 (multiply by 128)
▾ Multipath Transmission Control Protocol: Multipath Capable
    Kind: Multipath TCP (30)
    Length: 12
    0000 .... = Multipath TCP subtype: Multipath Capable (0)
    .... 0001 = Multipath TCP version: 1
  ▸ Multipath TCP flags: 0x81
    Sender's Key: 9857167011537900639
    [Subflow token generated from key: 440892405]
    [Subflow expected IDSN: 1425756494547153003 (64bits version)]
▸ [Timestamps]
▾ [MPTCP analysis]
    [Master flow: Master is tcp stream 0]
    [Stream index: 0]
    [TCP subflow stream id(s): 0]
```

Fig. 6. Wireshark capture for Eavesdropper at initial handshake capturing keys in clear form

```
|   |###[ Multipath TCP capability ]###
|   |  length    = 12
|   |  subtype   = MP_CAPABLE
|   |  version   = 1L
|   |  checksum_req= 1L
|   |  reserved  = 0L
|   |  hmac_sha1 = 1L
|   |  snd_key   = 0x88cbb12be66a545fL
None
9857167011537900639
```

Fig. 7. Extracting keys captured during initial handshake to perform ADD_ADDR attack by using python script

from the off path. From the above analysis one can observe that currently available MPTCP Linux kernel is not up to the mark to overcome ADD ADDR vulnerability in both the versions.

6 Available Solutions to Prevent Session Hijacking by ADD_ADDR Vulnerability

Many researchers have proposed the various solutions to prevent the ADD_ADDR attack which can be categorized into various categories like encryption based, hashing based. Opportunistic encryption based solutions. In this section, the comparison between the available solutions is covered to show the research path in the area of MPTCP security.

Author [20] proposed the hash based solution to prevent session hijacking by the man-in-the –middle attack. Author uses the hash chain scheme which generates list of chained hash values by using recursive execution of hash algorithm. Here some initial value will be assumed which will be used to generate n hash values H0- Hn by applying Hash algorithm repeatedly. During initial handshake, both the hosts will exchange their initial random numbers and on subsequent subflow establishment host needs to send chained values for authentication. But it doesn't provide security against attacker in the initial handshake which can drop the legitimate request of MP_JOIN. Moreover, hash values can be used to authenticate new hosts but ADD_ADDR vulnerability is not covered. Hash chained based solution is also algorithmically costly.

tcpcrypt [21] is an extension of TCP to provide cryptographic protection for session data which helps MPTCP to identify a single TCP sub-flow by using a "session id" which is independent from IP-Port pair. tcpcrypt uses asymmetric cryptosystem to prevent key exposure in initial handshake which requires one additional message for key exchange. However, tcpcrypt is much more efficient in providing security; it is vulnerable man-in-the-middle attack in initial handshake.

In [22, 23], authors implemented TLS with MPTCP. TLS is already implemented and supported by TCP so it is compatible with MPTCP. TLS also uses asymmetric key exchange to prevent key exposure in initial handshake. TLS is better solution than tcpcrypt as tcpcrypt is not well tested with real time applications. TLS and SSH protocols handle application security very well but still they are not able to handle the vulnerabilities of TCP and MPTCP [24]. Moreover TLS provides security against packet injection by falling back MPTCP connection to TCP at time of attack detection which will slow down the communication. To support TLS with Multipath TCP, some modification in MPTCP are required such as MPTCP must support message mode service instead of byte stream, keys generated by TLS must be used with MAC algorithm to authenticate message as well as new subflow etc.

Tcpcrypt and TLS both use asymmetric cryptosystem for key exchange which generates overhead in initial handshake that degrades the performance of TCP when addition of subflow is for the short time. To minimize the overhead of initial handshake, current version of MPTCP (v1) exchanges keys in clear form during initial connection establishment which is vulnerable to eavesdropper in initial handshake.

In [24], authors proposed MPTCPsec which is an extension of MPTCP that assimilates authentication and encryption with the protocol. MPTCPsec authenticates every packet option to prevent DoS attacks. Unlike TLS, MPTCPsec can prevent packet injection attack by avoiding the use of effected sub-flow. MPTCPsec uses authentication and encryption of TCP options to prevent redirection of data on the compromised path by changing the priority of subflows.

Table 1. Comparative analysis of various security solution for ADD_ADDR vulnerability in MPTCP

Paper	Year	Solution	Type of solution	Limitations
[20]	2011	Hash chain based solution	Hash based encryption	Vulnerable to eavesdropper in initial handshake
[21]	2014	tcpcrypt	Opportunistic security protocol	Only offers security against passive attacks still vulnerable to active attacks
[22, 23]	2016, 2015	Transport layer Security (TLS)	Uses third party CA authenticator	Modifications are required in MPTCP, slow down the connection due to overhead on initial handshake
[18]	2016	4-way handshake based solution	Asymmetric cryptography (Elliptic Curve Cryptography)	The point values to plot the elliptic curve are exchanged in clear form during initial handshake which is vulnerable to time-shifted attack
[25]	2017	Sum-chained based solution	Hash based Encryption	Vulnerable to partial time on path active attack
[14]	2018	ADD_ADDR2	HMAC based authentication	Vulnerable to partial time on path active attack
[26]	2019	Key exchange through SDN	Key distribution mechanism using third party SDN	If centralized controller is compromised, whole connection can be compromised
[27]	2019	ADD_ADDR packet confirmation	ADD_ADDR packet conformation	Vulnerable to man-in-the-middle attack
[28]	2020	Secure connection Multipath TCP (SCMTCP)	Uses CA for the authentication of security parameters used for session key generation and exchanged during the initial handshake	Increases the connection overhead by encrypting parameters like ADD_ID and nonce in every future connection establishment

In order to decrease the overhead in initial handshake and prevent attacks by eavesdropper in initial handshake, authors of [18] proposed the solution based on elliptic curve cryptography. Moreover, the public keys are shared during initial 4-way handshake but used in computation at the time of initiating new sub-flows which decreases the computational overhead over network. This scheme is specifically designed for MPTCP which not only decreases computational overhead but also decreases space and time overhead but still it is vulnerable to time-shifted attack.

In order to overcome limitations of hash chained based solution [20], authors of [25] proposed sum chained based solution which generates new values by applying mathematical equation. The proposed solution is suitable for low memory but it is vulnerable to integrity time shifted attack. The maximum length of TCP option field is 40 bytes which is the major limitation for providing strong encryption based security in MPTCP. Different solutions such as hash based solution, sum chained hash based solution, solution with use of elliptic curve for exchanging keys, use of TLS/SSH etc. has been proposed and implemented but still they were not up to the mark and vulnerable to some other attacks.

As discussed in previous section, to overcome ADD_ADDR vulnerability, the new format for ADD_ADDR [14] has been implemented with the current version (v1) of Linux kernel implementation of MPTCP which provides security against man-in-the-middle attack by exchanging HMAC calculated from the keys exchanged during the initial handshake but it is still vulnerable to time-shifted attack which falls under the category of partial-time on path attack.

Table 1 shows the comparative analysis of available security solutions to overcome the ADD_ADDR attack along with the limitations of available solutions which give directions to the researcher in the area o adoption of MPTCP.

7 Conclusion

The traditional transport layer protocol TCP doesn't fulfill the requirements of current network scenario so multipath TCP is the best suitable transport layer protocol which can be deployed easily due to its backward compatibility but security is one the most crucial requirement for today's era. In this paper the focus was on the security challenges in MPTCP Linux kernel implementation version0 and version1 and available solutions for the same with their limitations. By analyzing the current scenario of MPTCP security, it can be concluded that the off path active attackers and Partial On path Active attackers are the most crucial model where attacker can attempt an attack from the remote place without having any information about the communication. Here MiTM is implemented by using ADD ADDR vulnerability of MPTCP (v0) over different applications like chat application and file transfer application developed in JAVA by using Socket application. Currently ADD ADDR2 is also available which authenticate the user before adding new IP address explicitly but still it is vulnerable to eavesdropper present during the initial handshake. Here, available solutions are analyzed deeply and limitations are figured out which shows that the solution must be designed by keeping in mind few things such as space limitation, performance, no worse than TCP and initial overhead and the solution should not open the doors for other threats. The future scope of the research is to design the key exchange mechanism which can secure the MPTCP from off path active attacks by taking care of TCP header size limitation and performance benefit of MPTCP.

References

1. Postel, J.: Transmission Control Protocol, RFC 793, September 1981
2. Ford, A., Raiciu, C., Handley, M., Bonaventure, O.: RFC 6824 TCP Extensions for Multipath Operation with Multiple Addresses (2013)
3. Bonaventure, O.: Multipath TCP: an annotated bibliography, ICTEAM, UCL (2015)
4. Bagnulo, M., Paasch, C., Gont, F., Bonaventure, O., Raiciu, C.: Analysis of residual threats and possible fixes for multipath TCP (MPTCP), (No. RFC 7430) (2015)
5. Ma, J., Le, F., Russo, A., Lobo, J.: Detecting distributed signature-based intrusion: the case of multi-path routing attacks. In: IEEE Conference on Computer Communications (INFOCOM) (2015)
6. Pearce, C., Zeadally, S.: Ancillary impacts of multipath TCP on current and future network security. IEEE Internet Comput. **19**(5), 58–65 (2015)
7. Phung, C.D., Secci, S., Felix, B., Nogueira, M.: Can MPTCP secure internet communications from man-in-the-middle attacks? In: 13th International Conference on Network and Service Management (CNSM), Tokyo, Japan (2017)
8. Bonaventure, O., Handley, M., Raiciu, C.: An overview of Multipath TCP. Login **37**(5), 17–23 (2012)
9. Popat, K., Kapadia, D.V.: Recent trends in security threats in multi-homing transport layer solutions. Int. J. Adv. Sci. Technol. **29**(5), 5641–5648 (2020)
10. Ford, A., Raiciu, C., Handley, M., Bonaventure, O.: Architectural Guidelines for Multipath TCP Development (RFC6182), Internet Engineering Task Force (IETF), March 2011
11. Popat, K.J., Raval, J., Johnson, S., Patel, B.: Experimental evaluation of multipath TCP with MPI. In: Proceedings of the Third International Symposium on Women in Computing and Informatics (2015)
12. Shah, V., Kapadia, V.: Load balancing by process migration in distributed operating system. Int. J. Soft Comput. Eng. (IJSCE) **2**(1), 361–363 (2012)
13. Kapadia, V.V., Thakar, V.K.: Combinatorial system design for high performance memory management. In: 2013 15th International Conference on Advanced Computing Technologies (ICACT) (2013)
14. Ford, A., Raiciu, C., Handley, M., Bonaventure, O., Paasch, C.: TCP Extensions for Multipath Operation with Multiple Addresses RFC6824 (if approved), draft-ietf-mptcp-rfc6824bis-12 (2018)
15. Bagnulo, M.: Threat Analysis for TCP Extensions for Multi-path Operation with Multiple Addresses, draft-ietf-mptcp-threat-08 (2011)
16. Munir, A., Qian, Z., Shafiq, Z., Liu, A., Le, F.: Multipath TCP traffic diversion attacks and countermeasures. In: IEEE 25th International Conference on Network Protocols (ICNP), Toronto, ON, Canada (2017)
17. Demaria, F.: Security Evaluation of Multipath TCP, Analyzing and fixing Multipath TCP vulnerabilities, contributing to the Linux Kernel implementation of the new version of the protocol (Ph.D. Thesis), March 2016
18. Kim, D.-Y., Choi, H.-K.: Efficient design for secure multipath TCP against eavesdropper in initial handshake. In: International Conference on Information and Communication Technology Convergence (ICTC), Jeju, South Korea (2016)

19. Barré, S., Paasch, C., Bonaventure, O.: Multipath TCP: from theory to practice. In: Domingo-Pascual, J., Manzoni, P., Palazzo, S., Pont, A., Scoglio, C. (eds.) Networking 2011. LNCS, vol. 6640, pp. 444–457. Springer, Heidelberg (2011). https://doi.org/10.1007/978-3-642-20757-0_35
20. Díez, J., Bagnulo, M., Valera, F., Vidal, I.: Security for multipath TCP: a constructive approach. Int. J. Internet Protoc. Technol. **6**(3), 146–155 (2011)
21. Bittau, A., Boneh, D., Hamburg, M., Handley, M., Mazieres, D., Slack, Q.: Cryptographic protection of TCP Streams (tcpcrypt), Internet-Draft draft-ietf-tcpinc-tcpcrypt-03 (2014)
22. Hamza, A., Lali, M.I., Javid, F., Din, M.U.: Study of MPTCP with transport layer security. In: Proceedings of the 3rd International Conference on Engineering & Emerging Technologies (ICEET), Superior University, Lahore, PK, 7–8 April 2016 (2016)
23. Bonaventure, O.: MPTLS: Making TLS and Multipath TCP stronger together (2015)
24. Jadin, M., Tihon, G., Pereira, O., Bonaventure, O.: Securing multipath TCP: design & implementation. In: IEEE INFOCOM 2017 - IEEE Conference on Computer Communications (2017)
25. Krishnan, A., Amritha, P.P., Sethumadhavan, M.: Sum chain based approach against session hijacking in MPTCP. In: 7th International Conference on Advances in Computing and Communications, ICACC-2017, Cochin, India (2017)
26. Melki, R., Hussein, A., Chehab, A.: Enhancing multipath TCP security through software defined networking. In: 2019 Sixth International Conference on Software Defined Systems (SDS) (2019)
27. Noh, G., Park, H., Roh, H., Lee, W.: Secure and lightweight subflow establishment of multipath-TCP. IEEE Access **7**, 177438–177448 (2019)
28. Chaturvedi, R.K., Chand, S.: Multipath TCP security over different attacks. Trans. Emerg. Telecommun. Technol. **31**(9), 4081 (2020)

Challenge-Response Based Data Integrity Verification (DIV) and Proof of Ownership (PoW) Protocol for Cloud Data

Basappa B. Kodada[1] and Demian Antony D'Mello[2(✉)]

[1] A J Institute of Engineering and Technology Mangalore, VTU, Belagavi, India
[2] Canara Engineering College Mangalore, VTU, Belagavi, India

Abstract. Cloud Technology enables the users (Smartphone users, Enterprises, individual users) to access on-demand services, and the storage service is the most service used by them to outsource their data and also expect safety of data from cloud vendors w.r.t. confidentiality, integrity and access control. Nowadays, the massive amount data is being generated and outsourced from multiple sources to cloud, this data growth introduces many security issues and challenges. Because data owner (DO) loose physical control on data once it is outsourced to cloud and is fully controlled and managed by cloud service provider (CSP). There is a possibility of data breaching by inner or outer adversaries where employees of CSP can exploit the same or help the intruder to gain access to the sensitive data. It is essential to provide strong cryptography mechanisms to protect the data from unauthorized modification and prevent the unauthorized access by inner and outer adversaries. Hence, this paper presents a challenge-response hybrid-based protocol for data integrity verification (DIV) and Proof of Ownership (PoW) to protect the data from unauthorized users and prevent the data from unauthorized access. This paper also describes the security analysis on proposed approach and results are the proof of concept.

Keywords: Data security · Data integrity · PoW · DIV · Proof of ownership · Data integrity verification · Challenge-response protocol · Cloud security · Cryptography · Information security

1 Introduction

In the big data era, vast volume and different data types are being generated rapidly from different media sources at an unprecedented rate. According to IDC report, by 2025, data growth will be from 61% to 175 zettabytes out of which 42 zettabytes of data will be shipped by industries and 49% of data will be stored in the public cloud environment [1]. The cloud provides unlimited storage service to the cloud users to outsource their data at a lower cost. This will leads to an effective way of managing and maintaining a vast volume of data without establishing and managing local storage infrastructure. With this unlimited storage service from the cloud, more and more enterprises and individuals are interested in outsourcing their data to the cloud. There is a possibility that, out-sourced

M. Bhattacharya et al. (Eds.): ICICCT 2021, CCIS 1417, pp. 33–43, 2021.
https://doi.org/10.1007/978-3-030-88378-2_3

data might compromised during its transmission by outer adversaries and compromised at storage server by inner adversaries. That is why, many enterprises are worried about the security of their outsourced data, because they loose physical control on their own data once it is outsourced to cloud and it is controlled and managed by CSP. Hence there is chance of modifying the data by any employees of CSP. It is essential to provide integrity service to cloud data by using strong cryptographic algorithms.

Data integrity is a type of security service that protects data from modification attack at cloud storage server (CSS). There is the possibility that CSP might not say the truth about the corruption of data for the sake of keeping a good reputation, so CSP cannot be trusted in this regard. In 2018, the incident took place where the organization called "QCloud" lost huge amount of users outsourced data that was exploited by their employees which causes $19 million loss to the organization [2]. One more possibility that CSP might be influenced financially or politically to destroy some users data and attempt to convince the DO that data is still in maintenance. So the integrity of data cannot be secured by CSP and fails to provide normal service to the cloud users form unauthorized data modification by inner and outer adversaries.

The message digest and hashing techniques are mostly used schemes in earlier days to achieve integrity of users data which is outsourced to CSS by generating unique authentication code and is compared when data is downloaded. The most efficient and effective schemes to ensure data integrity are data integrity verification (DIV) techniques where data integrity is verified periodically in CSS and notify to DO if inner or outer adversaries compromise any data modification or integrity of data. That means to say; the DO must be ensured periodically about proof of stored (PoS) data in CSS.

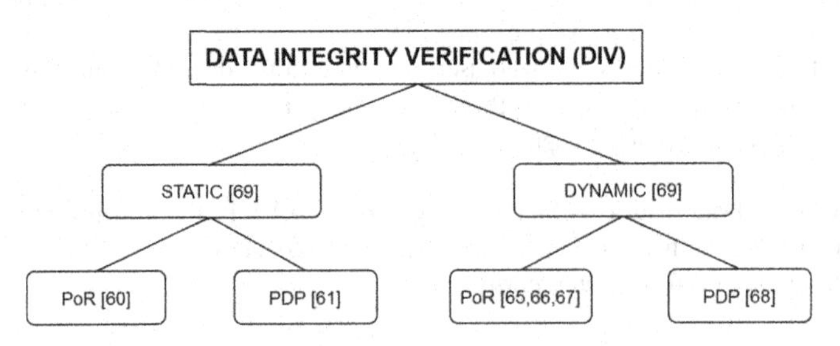

Fig. 1. Taxonomy of data integrity verification schemes

The concept of PoS is a mechanism to ensure the DO regarding the correctness of their outsourced data stored in CSS. This mechanism is achieved by two cryptographic schemes; one is proof-of-retrievability (PoR) [3] and other one is provable-data-possession (PDP) [4]. Both schemes depend on the challenge-response protocol where DO challenges the CSS to prove the reliability of data stored in the server. The server responds with proof, and it proved only when DO verifies the unique authentication code and considered that data is safe and maintained as it is if an authentication code is matched with computed code. The verification is done in two ways by using PoR or PDP schemes: Firstly, the data integrity is verified by DO, and in a second way, the data integrity is verified by a trusted third party (TTP).

The verification process will be carried out between two entities: DO or TTP who has been assigned the responsibility of integrity verification by DO and the CSS. The verification is done on static data which does not have any updation periodically by any cloud user and dynamic data which has frequent updation by the cloud users who has access permission provided by DO. Figure 1 presents the taxonomy of DIV schemes proposed by researchers on static and dynamic data for verifying integrity on outsourced data. Consider the CSA scenario as a motivating example who would like to adopt storage service with a lower cost to outsource their data and expect to maintain the confidentiality, integrity, and access control of data during its transmission on the secure or insecure channel and at a storage location. This inspired us to secure the data from Man-inW protocol to ensure the integrity and access control on data stored in cloud storage server. The main contribution of this paper is briefed as follows:

- Insight into the preliminary concept of hashing algorithms which are used to generate authentication code and integrity verification code
- Propose Challenge-Response hybrid protocol to ensure ownership, the integrity of data from unauthorized access and modification at storage server
- Propose authentication code generation and integrity verification algorithms to ensure the existence of redundant data, DIV and access control mechanism
- The security analysis theoretically proves the strength of the proposed protocol
- The performance analysis and comparison of simulation results as proof of concept

Rest of the paper is organized as follows: The Sect. 2 discuss the preliminary concept used in the proposed protocol and also presented corresponding algorithms, this section also presents the detailed process of protocol communication model along with security analysis. Section 3 presents implementation and result discussion and related work is discussed in Sect. 4. Finally, the Sect. 5 conclude the paper.

2 Challenge-Response Model for DIV

2.1 Preliminaries

One-Way Hash Function. One-way hash function generates unique code from message digest and that can not be reversed to original message. That means, Hash function is defined as H: (0, 1)* (0, 1)* which is computationally infeasible to determine two distinct messages a and b such that $H(a) = H(b)$.

The *SHA*256 hash algorithm is used to generate unique code in the proposed protocol.

Keyed-Message Authentication Code. It is also one-way hash function that generates unique authentication code for message with help of key. The keyed hash function is defined as $HMAC_{key}(P_{text})$: $(0, 1)^* \times (0, 1)^* \rightarrow (0, 1)^*$ and is deterministic that generates unique code for each message P_{text}. It is computationally difficult to determine original message P_{text} from unique authentication code. The $HMAC_{key}(P_{text})$ algorithm is used to generate Tag and is used for verifying PoW of data in the cloud storage.

Authentication Code Tag Generation. The authentication code Tag is generated to verify the existence of redundant data in CSS. This code is generated by using keyed-HMAC hashing algorithm in which secrete key is generated from the file contents and then $HMAC()$ will generate intermediate code t. the keyed- HMAC algorithm is double hashing technique that uses the intermediate code t and $Sha1$ hashing algorithm to generate the final Tag that will be used for proving legitimate owner at CSS. The step-by-step process of generating Tag is presented in the Algorithm 1.

Algorithm 1: Algorithm to generate Authentication code Tag

Input : Plaintext, Key P_{text}, K_s
Output: Tag - Authentication code
GenTag(P_{text}, K_s):;

$$\begin{aligned} t &= HMAC(P_{text}, K_s) \\ Tag &= HMAC - Sha1(t) \end{aligned} \tag{1}$$

Return Tag;

Integrity Verification Code H_{code} Generation. The integrity verification code H_{code} is generated to verify the integrity of data outsourced by CSA to the CSS. The H_{code} is generated mainly by using $SHA256()$ hashing algorithm where P_{text} will be encrypted first by using FSA based encryption algorithm, and then H_{code} will be generated from C_{text}. This H_{code} will also be used in PoW challenge-response protocol initiated by CSS to prove the ownership on requested file as shown in Algorithm 2.

Algorithm 2: Algorithm to generate Hash code H_{code}

Input : Plaintext, Ciphertext P_{text}, C_{text}
Output: H_{code} - Hash Code
GenHcode(P_{text}):;

$$\begin{aligned} A &= (Q, \Sigma, O, \delta, \lambda) \\ C_{text} &= Enc(A, P_{text}) \\ &= \lambda(S, P_{text}) \\ H_{code} &= SHA256(C_{text}) \end{aligned} \tag{2}$$

Return H_{code}

Challenge-Response Model for PoW. Proof of ownership (PoW) is a type of mechanism to demonstrate ownership of file by exchanging some sort of authentication codes between client and server. The PoW protocol protects from fake keying attack and fake ownership claiming attack (discussed in Sects. 1 and 2). In this regard, we design PoW challenge protocol between client and server that is used to identify legitimate owner of file where server assign a challenging task to the client to prove the ownership on requested file τ. Client responds with two authentication codes ν_1 and ν_2 that are hash value of plain text and cipher text of file contents respectively. Server computes $\nu = \nu_1 \oplus \nu_2$ on received authentication codes, if the codes are valid and legitimate, then server learns that user is legitimate and authorized, so CSA will be given an access to the requested file. The same process is depicted in Fig. 1.

Client		*Server*
		$Identify(\tau)\ for\ Client$
	$\xleftarrow{Challenge\_PoW(\tau)}$	
$Get(\nu_1, \nu_2)$		
	$\xrightarrow{Challenge_r esp(\nu_1, \nu_2)}$	
		$Find\ \nu\ of\ \tau$
		$is\ \nu == \nu_1 \oplus \nu_2?$
		$On\ True$
	$\xleftarrow{GiveAcces(\tau)}$	
		$On\ False$
	$\xleftarrow{TermConn(\tau)}$	

Fig. 2. PoW Protocol

Challenge-Response Model for DIV. Data Integrity Verification (DIV) is a type of mechanism to verify whether data is modified during its transmission and at-rest by exchanging some authentication codes between client and server. The process of communication is between client and server is same as Challenge- response model shown in Fig. 1, but in reverse manner. In this model, client will initiate the challenge-response protocol to verify the integrity of data, to this, server has to reply with three codes *Tag*, H_{code}, T_{stamp} to cloud service adopter (CSA). The CSA receive the same and compute $Tag \oplus H_{\text{code}} \oplus T_{\text{stamp}}$ and is compared with existing DIV_{code}. If the comparison is true, then CSA assume that, there is no data modification is done on the outsourced data.

2.2 DIV and PoW for Cloud Computing

In order to streamline the functionality of proposed protocol for DIV and PoW is divided into two phases: data outsourcing phase and data downloading phase.

Data Outsourcing Phase: In this phase, C_{Adopt} outsource their data to cloud by maintaining data confidentiality and data integrity. To achieve confidentiality and integrity of data, C_{Adopt} encrypts data using encryption algorithm and generates hash code using *SHA*256 algorithm. C_{Adopt} also generates authentication code Tag using *HMAC* algorithm that is used for verifying the integrity of data same is depicted in Fig. 2 In the initial data outsourcing phase, C_{Adopt} gets encryption key A from TPC to encrypt the data *GetKey*$(A, U_{id}) \leftarrow TPC$. C_{Adopt} also compute authentication code Tag by using *HMAC*() algorithm that consider hash value of P_{text} as a key $HMAC(Hash(P_{text}), P_{text})$ as shown in Algorithm 1, and generates hash code H_{code} of C_{text} using *ShHA256*() algorithm for integrity check $H_{code} = SHA256(C_{text})$ as shown in Algorithm 2. Finally C_{Adopt} outsources the data that includes C_{text}, *Tag*, H_{code} to CSS $Upload(C_{text}, H_{code}, Tag, U_{id}) \rightarrow CSS$. The CSS verifies C_{Adopt} credentials like U_{id}, S_{sixe}, T_{val} and if U_{id} is valid then CSS generates file id F_{id} for corresponding data and upload into CSS. The CSS also generate time stamp T_{stamp} automatically when data is uploaded to the cloud device that will be used for integrity check and send these information to C_{Adopt} i.e. $Send(F_{id}, T_{stamp}) \rightarrow CSA$. The CSS computes Proof of ownership code by generating hash of $Tag \oplus H_{code}$ so that it can be used for comparing C_{Adopt} response to Challenging task of PoW protocol during subsequent data outsourcing or downloading phase. At the same time C_{Adopt} also compute $DIV_{code} = (Tag \oplus H_{code} \oplus T_{stamp})$ used for verifying the integrity of data when it is downloaded as shown in Fig. 2.

Data Downloading Phase: In this phase, C_{Adopt} downloads data from CSS and decrypt it and also perform integrity verification. Before downloading data from CSS, C_{Adopt} needs to clear PoW protocol challenge by sending authentication code Tag and Hash code H_{code} of cipher text to CSS. The detailed step by step process is depicted in Fig. 3 C_{Adopt} gets decryption key K_s and other meta data like *Tag*, H_{code}, T_{stamp} from *TPC* before downloading it from CSS $GetFileInfo(F_{id}, K_s, U_{id}) \rightarrow TPC$. Once meta data of file is received from *TPC*, C_{Adopt} sends a download request to CSS by providing their credentials i.e. U_{id}, F_{id}. The CSS verifies T_{val} for U_{id} and U_{id} *ACL*. If C_{Adopt} is legitimate and has access permission to the F_{id}, then CSS give PoW protocol challenge to C_{Adopt} to prove ownership on data file F_{id} "*Challenge PoW* $(F_{id}) \rightarrow CSA$": C_{Adopt} accepts challenge and give a response to the challenge by sending Tag and H_{code} of F_{id} to prove ownership. The CSS computes hash value of $Tag \oplus H_{code}$ and compare with PoW_{code} which has been computed by CSS during initial outsourcing phase. If the comparison is true, then CSS sends C_{text}, H_{code}, *and* T_{stamp} to CSA $Send(C_{text}, H_{code}, T_{stamp}) \rightarrow CSA$. C_{Adopt} receive cipher text C_{text}, hash code H_{code} and time stamp T_{stamp} from CSS and check for integrity of data by computing and comparing $DIV_{code} = (Tag \oplus H_{code} \oplus T_{stamp})$. If all comparisons are true, then decrypt cipher text C_{text} by using decryption algorithm, otherwise it is assumed that, data is compromised and modified by either inner adversaries or outer adversaries. The complete step-by-step process of data downloading phase is presented in Algorithm 8.

2.3 Security Analysis

The proposed protocol protects data against modification attack from inner and outer adversaries. In the proposed system, C_{Adopt} generates integrity verification codes $H_{code} = SHA256(C_{text})$ before outsourcing into CSS for verifying the data modification is carried out by inner or outer adversaries when data resides in storage device. C_{Adopt} outsource their data pack to CSS by including C_{text}, H_{code}, Tag to prove data possession, and CSS responds with time stamp T_{stamp} to C_{Adopt} that has been generated while

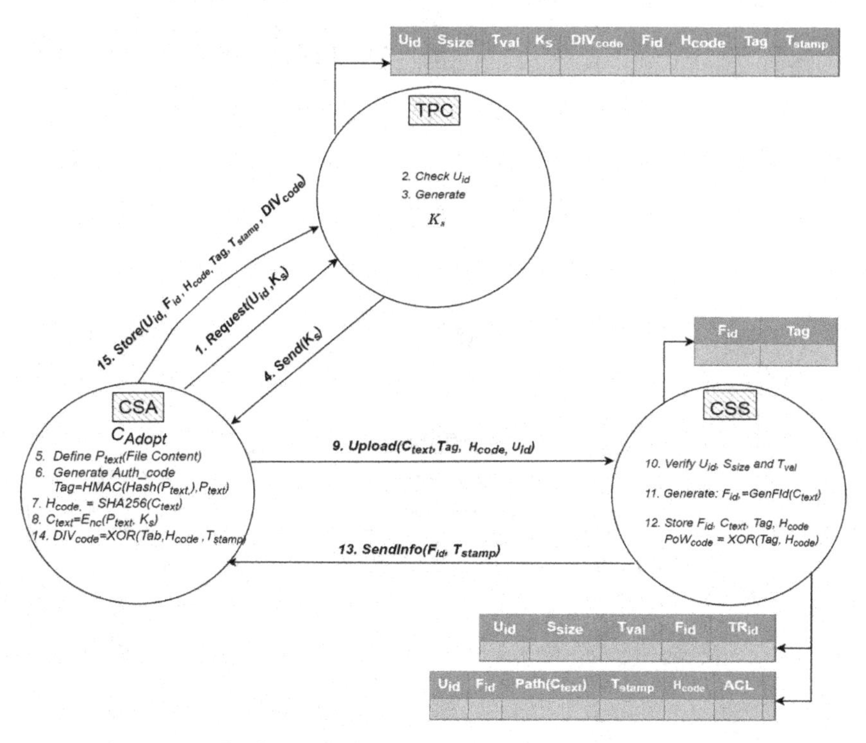

Fig. 3. Initial data outsourcing phase with CSS

uploading into CSS. C_{Adopt} computes $DIV_{\text{code}} = (Tag \oplus H_{\text{code}} \oplus T_{\text{stamp}})$ used for verifying integrity of data when data is downloaded from CSS. C_{Adopt} receives data pack that includes $C_{\text{text}}, H_{\text{code}}, and\ T_{\text{stamp}}$ and C_{Adopt} compares DIV_{code} with received data, if comparison is true, then it proves that, data is not compromised.

3 Implementation and Results Discussion

We presents performance evaluation of our proposed protocol to measure the performance of DIV and PoW. We consider following factors to evaluate our proposed system: 1. Authentication code generation time 2. Integrity verification code generation time 3. PoW protocol 4. DIV Protocol are essential to achieve data data integrity and access control in secure data outsourcing process. The propose protocol is implemented in python3.8 with crypto, hashlib and hmac cryptographic libraries to implement the scheme on Intel 32bit machine having i3 processor with 4GB RAM running in windows 7 basics.

The Fig. 4 demonstrates computation time of authentication code generation algorithm Tag and the integrity verification code H_{code} generation algorithm, the average time taken by these algorithms are 0.0010 s to 0.0421 s and 0.0015 to 0.0067 s respectively. The same is depicted in Table 1 and Fig. 4. The Table 1 and Fig. 4 also shows the time taken for the process of PoW protocol and verifying the integrity of data.

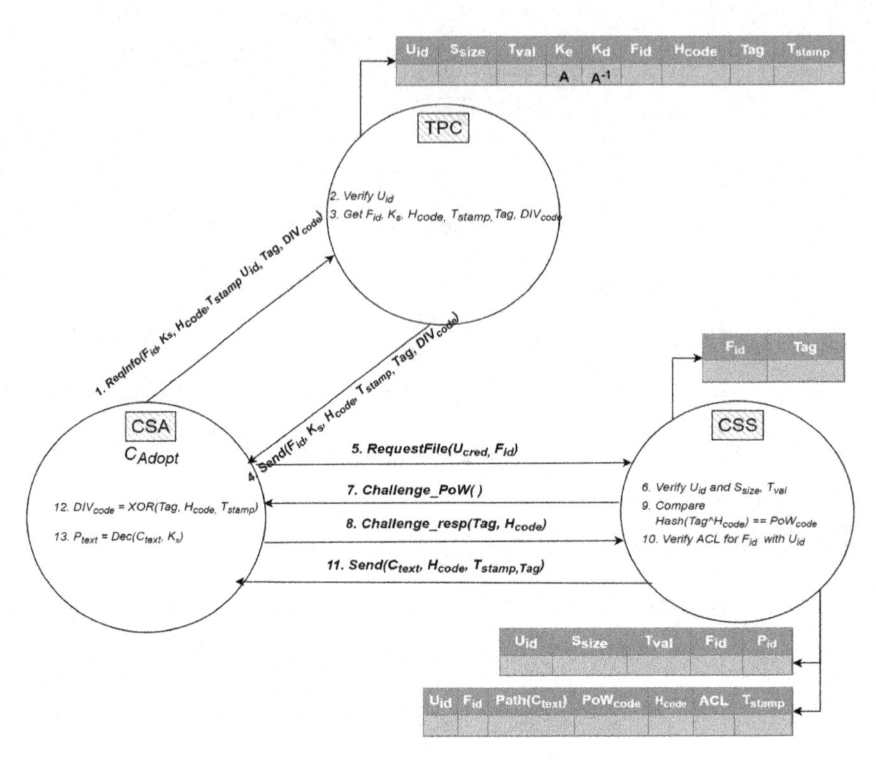

Fig. 4. Data downloading from CSS

Table 1. The time evaluation of FSACE protocol during data outsourcing and downloading phase

File size in KB	Time in sec			
	H_{code}	PoW	*Tag*	Integrity
32	0.0015	0.0010	0.0010	0.0015
64	0.0019	0.0013	0.0069	0.0020
128	0.0026	0.0019	0.0130	0.0027
192	0.0031	0.0026	0.0153	0.0035
256	0.0040	0.0032	0.0211	0.0041
320	0.0047	0.0041	0.0297	0.0053
384	0.0054	0.0049	0.0349	0.0061
448	0.0062	0.0055	0.0385	0.0072
512	0.0067	0.0062	0.0421	0.0079

4 Related Work

The CSP must ensure the confidentiality and privacy of data stored in CSS and ensure the secure integrity service. The author of [5] proposed an efficient scheme for data integrity for cloud computing. This scheme reduces the computational cost of cloud user during the auditing process. The security of this scheme depends on Diffie-Helman problem. Dong Dong and Yue et al. proposed Markle tree and block-chain based integrity verification scheme in [6]. The scheme uses random numbers with Markle tree as challenging tasks for analyzing and optimizing the system performance. The scheme adopts decentralized scheme for DIV and auditing mechanism in CSS.

Due to the enormous computing and storage service provided by different CSP, the health organizations are also moving towards outsourcing electronic health records (EHR) and the compromising these records in the cloud becomes very critical security and integrity issue. So the author of [7] proposed operation log auditing scheme based on HE based hash function. In this scheme, the operation logs are grouped based on the verification tags and type of operations carried out. The author claims that the scheme is secure and efficient (Fig. 5).

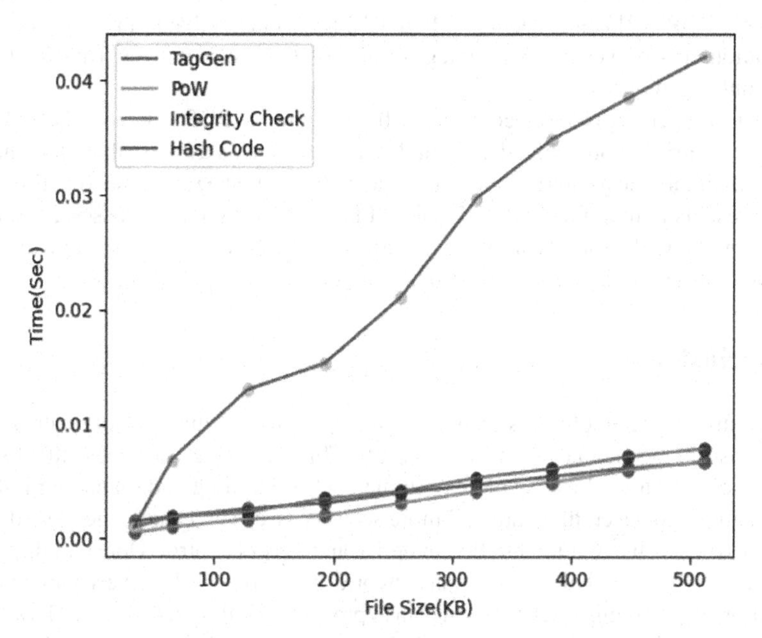

Fig. 5. The time evaluation of proposed algorithms for different file size

Nowadays, society is moving towards using smart applications by interconnecting more IoT devices, and data which are generated from these smart applications through IoT devices are stored in the CSS. This requires a very efficient DIV scheme for auditing cloud data. The author of [8] propose Hyperledger Fabric-based decentralized auditing scheme and use more than on TTP which can be selected dynamically for auditing task.

This scheme is designed to improve efficiency and scalability of TTP based on bilinear pairing.

Many researchers have proposed hybrid models for providing data integrity service to the cloud. Initially, data integrity is verified locally by downloading entire content from CSS or remotely with the help of challenge-response protocol model [9] called static and dynamic DIV techniques [10]. The author of [3] introduced the static DIV scheme called PoR, and author of [11–13] proposed dynamic DIV schemes to reduce the issues and computational and communication cost at DO side. The author of [4] has proposed a type of static based DIV scheme called PDP where DO verifies whether outsourced data is possessed or not. Later, the author of [12] has proposed a dynamic PDP based DIV scheme.

The multiple replicas based static PDP is proposed by Curtomala et al. in [13] called MR-PDP, where DO download the various files at a time for DIV. The computation cost will be increased on both user and CSS due to multiple downloads. So in 2013, the Wang et al. proposed proxy-based static PDP for DIV where the proxy server will verify the integrity of data on behalf of user [14]. This scheme is called the PPDP. The static PDP scheme suffers from computation overhead due to data downloading from CSS and verify the integrity of data locally. The author Ateniese et al. proposed an improved version of dynamic PDP [12] called scalable - PDP (S-PDP) scheme that improves storage and computation cost by verifying the integrity of data remotely on CSS. But this fails to support unlimited queries.

In 2012 Zhu et al. proposed cooperative based dynamic PDP scheme (CPDP) [15] to verify the integrity of outsourced data. In this scheme, the data is distributed to multiple CSS and maintain the hash tree index of metadata for DIV at a remote server. The author of [16] presents a multicopy based dynamic PDP to verify the data possession on CSS and prevent the CSS from cheating the cloud user. The further static and dynamic PoR schemes are found in [17] for DIV on both user and CSS.

5 Conclusion

Cloud computing technology is an emerging service model that grabs the attention of many industries, academia and individual users. This technology provides attractive and low cost services to end users compare to the cost of building and managing in-house infrastructure. However, there are still more security issues and challenges for adopting the cloud services like confidentiality, integrity and access control. Hence in this paper, we proposed a simple challenge-response hybrid-based protocol for verifying integrity of data as well as proving ownership on data to prevent unauthorized access. The analysis on obtained results shows the strength of proposed approach. The proposed approach takes 0.0015 s to 0.0079 s for PoW protocol execution and 0.0010 s to 0.0060 s for integrity verification.

References

1. Reinsel, D., Gantz, J., Rydning, J.: The digitization of the world from edge to core. International Data Corporation, Framingham (2018)

2. Xinhuanet: ten million yuan data loss tencent cloud lost data torture cloud security (2020). http://www.xinhuanet.com/finance/2018-08/13/c129931885.htm. Accessed 24 July 2020
3. Juels, A., Kaliski, B.S.: PORS: proofs of retrievability for large files. In: Proceedings of the 14th ACM Conference on Computer and Communications Security, pp. 584–597 (2007)
4. Ateniese, G., et al.: Provable data possession at untrusted stores. In: Proceedings of the 14th ACM Conference on Computer and Communications Security, pp. 598–609 (2007)
5. Garg, N., Bawa, S., Kumar, N.: An efficient data integrity auditing protocol for cloud computing. Future Gener. Comput. Syst. **109**, 306–316 (2020)
6. Yue, D., Li, R., Zhang, Y., Tian, W., Huang, Y.: Blockchain-based verification framework for data integrity in edge-cloud storage. J. Parallel Distrib. Comput. **146**, 1–14 (2020)
7. Tian, J., Jing, X.: Cloud data integrity verification scheme for associated tags. Comput. Secur. **95**, 101847 (2020)
8. Lu, N., Zhang, Y., Shi, W., Kumari, S., Choo, K.K.R.: A secure and scalable data integrity auditing scheme based on hyperledger fabric. Comput. Secur. **92**, 101741 (2020)
9. Gudeme, J.R., Pasupuleti, S.K., Kandukuri, R.: Review of remote data integrity auditing schemes in cloud computing: taxon omy, analysis, and open issues. Int. J. Cloud Comput. **8**(1), 20–49 (2019)
10. Martin, J.A.: What is access control? A key component of data security (2020). https://www.csoonline.com/article/3251714/what-is-access-control-a-key-component-of-data-security.html. Accessed 24 July 2020
11. Erway, C.C., Ku¨pc¸u¨, A., Papamanthou, C., Tamassia, R.: Dynamic provable data possession. ACM Trans. Inf. Syst. Secur. (TISSEC) 17(4), 1–29 (2015)
12. Cash, D., Ku¨p¸cu¨, A., Wichs, D.: Dynamic proofs of retrievability via oblivious ram. *J. Cryptol.* 30(1), 22–57 (2017). https://doi.org/10.1007/s00145-015-9216-2
13. Curtmola, R., Khan, O., Burns, R., Ateniese, G.: MR- PDP: multiple-replica provable data possession. In: 2008 the 28th International Conference on Distributed Computing Systems, pp. 411–420. IEEE (2008)
14. Wang, C., Chow, S.S., Wang, Q., Ren, K., Lou, W.: Privacy-preserving public auditing for secure cloud storage. IEEE Trans. Comput. **62**(2), 362–375 (2011)
15. Zhu, Y., Hongxin, H., Ahn, G.-J., Mengyang, Y.: Cooperative provable data possession for integrity verification in multicloud storage. IEEE Trans. Parallel Distrib. Syst. **23**(12), 2231–2244 (2012)
16. Barsoum, A.F., Hasan, M.A.: Provable multicopy dynamic data possession in cloud computing systems. IEEE Trans. Inf. Forensics Secur. **10**(3), 485–497 (2014)
17. Bowers, K.D., Juels, A., Oprea, A.: Proofs of retrievability: theory and implementation. In: Proceedings of the 2009 ACM Workshop on Cloud Computing Security, pp. 43–54 (2009)

Multi-layer Parallelization in Transportation Management Software

Anton Ivaschenko[1](✉), Sergey Maslennikov[1], Anastasia Stolbova[2], and Oleg Golovnin[2]

[1] Samara State Technical University, Molodogvardeyskaya 244, Samara, Russia
[2] Samara National Research University, Moskovskoye shosse 34, Samara, Russia

Abstract. In this paper there is introduced a comparative analysis of multi-layer data model used for parallelization of transportation network objects processing in the process of logistics optimization. A concept of multi-layer parallelization is based on distribution of the objects and subjects of a complex transportation logistics system including the road network, pick-up and delivery points, drivers, tracks and trailers, transportation instructions and order flows to the logical layers established considering the requirements of parallel computing. This paper continues this research by study of parallel computing characteristics comparing to traditional approach. The proposed data model allows increasing the performance of parallel computing application in transportation logistics problem domain. The considered performance metrics include execution time, total parallel overhead, speedup, and efficiency. Research results are implemented and applied in practice as a part of a transportation management software solution.

Keywords: Parallel computing · Transportation logistics · Routing · Multi-layer model

1 Introduction

Transportation logistics industry faces a number of challenges nowadays including the focus on digitization-driven challenges to established innovative business models. Transportation management software implementation provides efficient resources management and thus allows optimization of business performance. In order to meet these goals new software solutions are developed that provide a possibility to simulate, compare and analyze multiple alternative options of transportation resources utilization. Their implementation requires new algorithms of parallel computing capable of providing acceptable performance while processing multiple allocation options in real time.

To solve this problem, there was proposed a new multi-layer data model [1]. The main idea is to support the parallel algorithm of resources allocation by decomposing the knowledge of the problem domain considering the requirements of processing efficiency of modern computer systems. This paper continues this research by study of parallel computing characteristics comparing to traditional approach.

M. Bhattacharya et al. (Eds.): ICICCT 2021, CCIS 1417, pp. 44–51, 2021.
https://doi.org/10.1007/978-3-030-88378-2_4

2 State of the Art

At the moment, automated resource optimization using modern information and intelligent technologies has become an important service for transport companies [2, 3]. This situation affects our lives, especially in urban areas, while people need to move more quickly between different places. To meet the new requirements of an increased efficiency and service level the transportation management software start implementing new algorithms that consider an extended set of problem parameters and solution options.

The state of the transport system is influenced by a large number of factors, such as the intensity of vehicle traffic, weather conditions, traffic rules, the processing of which takes a long time and complicates the calculation of simple metrics, such as the cost of the path. Transportation logistics industry targets at optimization of key performance indicators such as efficiency, optimization, speed and timing. New opportunities to improve these parameters are concerned with digital transformation affected by the next revolution of Industry 4.0 [4, 5].

Taking into account the volume, variety and velocity of these options the problem of transportation logistics optimization belongs to a Big Data scope [6, 7]. Application of data science to provide transportation resources management in real time allows considering multiple options of problem solving and thus improves the quality of decision-making support.

Positive examples of Big Data algorithms application in transportation logistics [8–10] illustrate new capabilities and benefits of their practical use. Modern transport systems routinely use parallel computing to process large amounts of data generated by traffic monitoring systems [11, 12]. Parallel computing is the execution of many operations at the same time. This methodology is used to accelerate and distribute the load on server hardware.

It is important to emphasize that the efficiency of parallel computing technologies application essentially depends on the way of their application. It is important to consider the semantics of data processed and relations between the objects and subjects in one transportation system. For example, when solving a problem of combination of inbound and outbound delivery [13] distribution of data between the layers can be performed in two different ways: by type of data or by belonging to the route or delivery time.

Proper decision on the type of data classification and pre-processing can significantly increase the effect of parallelization [14]. In this paper we present the results of multi-layer model application to distribute the data objects processed by optimization parallel algorithm. Efficiency analysis of parallel computing algorithms is performed using specific metrics like execution time, total parallel overhead, speedup, efficiency, and cost [15, 16]. The main research goal is to evaluate a contribution of proper data classification and pre-processing before optimization using parallel algorithms.

3 Method Description

3.1 A Problem of Transport System Processing Performance

Let us express the logistic network as an undirected connected graph. Vertices represent cities, and edges represent paths between them using the roads. Each edge of this graph has a weight that denotes the distance that the transport must travel in order to get from one city to another. An example is shown in Fig. 1.

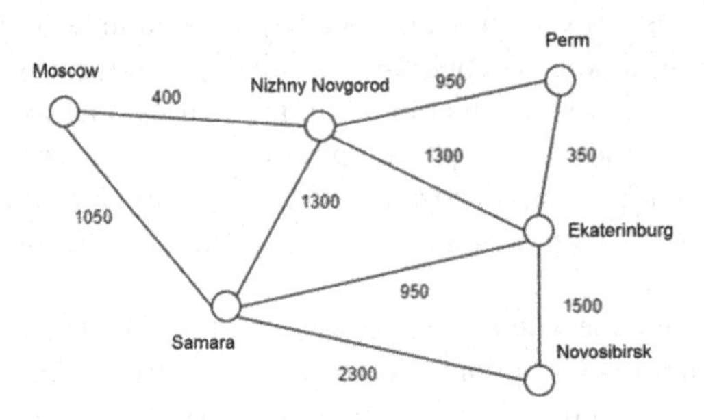

Fig. 1. Undirected connected road graph between cities

When calculating the optimal path between cities in the form of graphs, you can also display many factors affecting the route, such as weather conditions, the quality of the road surface, traffic congestion, and the rules for the delivery of goods provided by the companies. These factors can change over time and make the routing problem computationally intensive.

The problem of transportation logistics optimization usually contains a number of challenges including efficient routing, cargoes consolidations, shipment management, truck and trailer allocation, scheduling, etc. Decision-making support requires solving them multiple times while choosing the best option in the field of various constraints. In addition to this, the citation permanently changes, which triggers new optimization and corresponding calculations.

The main problem of Big Data in the transport system is the inability to process the data that comes to the input of the algorithm for analysing the movement of vehicles in an acceptable amount of time. The simplest solution of the problem of optimizing the movement of transport units is to analyse and enumerate a variety of options for system states.

3.2 Multi-layer Data Model

Multiple iterations, in which an enumeration of possible states is carried out, always require time for calculations, and the process of predicting profitable routes becomes more expensive. Therefore it becomes impractical to maintain powerful servers for calculating the optimal traffic flow.

Implementation of parallel algorithms can solve this problem by efficient distribution of computing jobs between the servers. The principles of routing algorithms parallelization can vary depending of the problem domain specifics.

The proposed approach is described in general by Fig. 2. Considering the specifics of the routing problem the objects are distributed by location proximity rather than type (i.e. processing all vehicles in one layer and transportation instructions in another). The procedure of data classification can be done either manually or automatically based on the analysis of experimental data taken for a test period of time.

For example, combination of inbound and outbound deliveries of a distribution center can be mixed up considering the regular and occasional orders. Routing can also be rearranged considering long distances and regular backhauls. These logical patterns that are currently used in the form of heuristics in scheduling algorithms can also be applied at the stage of parallelization to reduce interaction between the layers and improve calculations efficiency.

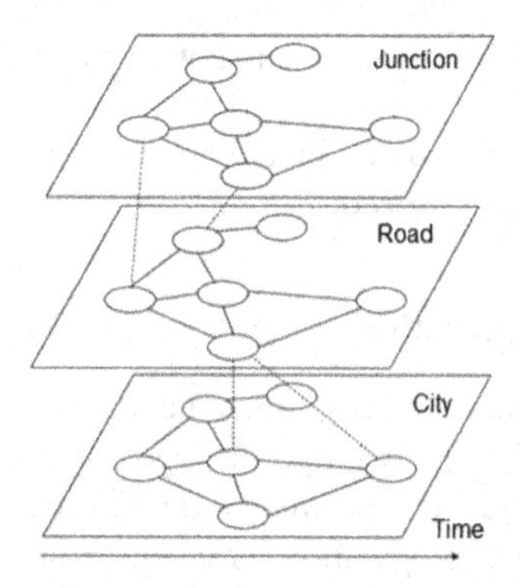

Fig. 2. Multi-layer computing oriented modification

Dividing the data into layers makes it easier to parallelize computations. And with the correct distribution of the layers over which the calculations will be performed by the parallel algorithm, it is possible to achieve a significant increase in the speed of execution by combining various layers of the transport system at the stage of preparation for computation.

3.3 Sequential Routing Algorithm

Let us consider a typical routing algorithm as a basis. Input parameters include a list of transport system correlations that represent the weights of the point-to-point transition, such as the quality of the pavement, the weather on the sections, and the road accident rates. The algorithm contains the following steps:

1. Constructively iterate over the weights of each criterion. Normalization is performed using the reference weights of the graph vertices;
2. Weigh the graph and distribute the weights along the edges;
3. Find the optimal path using Dijkstra's algorithm.

Result is the optimal path from point A to B.

3.4 Parallel Routing Algorithm

Parallel version of the routing algorithms contains the following steps:

1. Divide the data by criteria;
2. Assign weights to the edges of each graph according to a given criterion (for example, fare, length of the path);
3. Combine the predefined layers into one graph (in our case, this is the weather, the quality of the road and the accident rate – these are correlated criteria that affect the road to the destination);
4. Use several streams to find the shortest paths from a given point to the final one using Dijkstra's algorithm;
5. Combine routes, using a reference value for each route according to a given criterion, and choose the most suitable, in accordance with expert judgment.

Result is a number of consolidated links between transport arteries indicating the impact of road quality on vehicle speed.

3.5 Multi-layer Routing Algorithm

The parallel routing algorithms can be optimized considering the features of combined data processing. Step 3 of the parallel algorithm greatly facilitates data processing by reducing the number of connections between cities. Combining different layers of the transport system at the stage of preparation for calculations allows to:

- Reduce the total number of threads created to calculate Dijkstra's algorithms;
- Reduce the number of operations for combining calculation results;
- Increase the accuracy of calculations.

The described algorithms were compared using the performance metrics specific for parallel computing technology.

4 Performance Metrics Analysis

4.1 Test Case Introduction

The sample of data for testing solutions covers 21 settlements in the Netherlands, connected by 137 weighted connections: weather, distance, road surface quality, and path costs. This stage of optimization of the parallel algorithm can be improved by identifying special patterns for constructing routes. For example, let's take the construction of the route between Hoogeveen and Den Haag, (see Fig. 3).

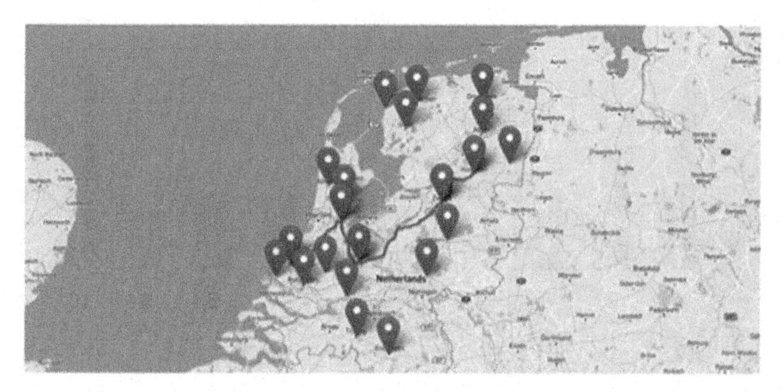

Fig. 3. Optimal route between Hoogeveen and Den Haag

This route can be divided into two large clusters: the vicinity of the city of Hoogeveen within a radius of 100 km and the vicinity of the city of Den Haag within a radius of 150 km, since in fact there is one road between these clusters.

Therefore if the clusters are processed separately, it becomes possible to improve the performance of the parallel algorithm.

5 Performance Metrics Analysis

Performance metrics calculated for the introduced case are presented in Figs. 4, 5, 6 and 7. The proposed modification presents lower execution time, which is the most common metric that illustrates the benefits of parallel computing, and lower parallelization overhead, which is conversely used to describe extra costs caused by algorithms modification. Therefore based on the calculated parameters, the improved multi-layer parallel algorithm is optimal for finding the path on n processing elements.

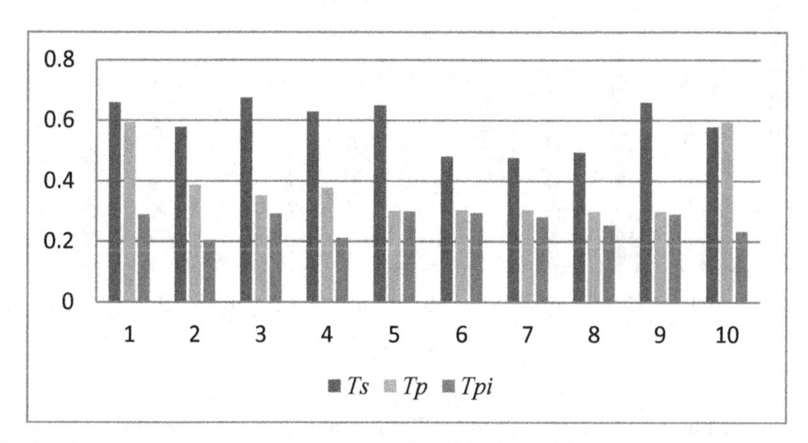

Fig. 4. Execution time of sequential *Ts*, parallel *Tp* and multi-layer *Tpi* routing algorithms

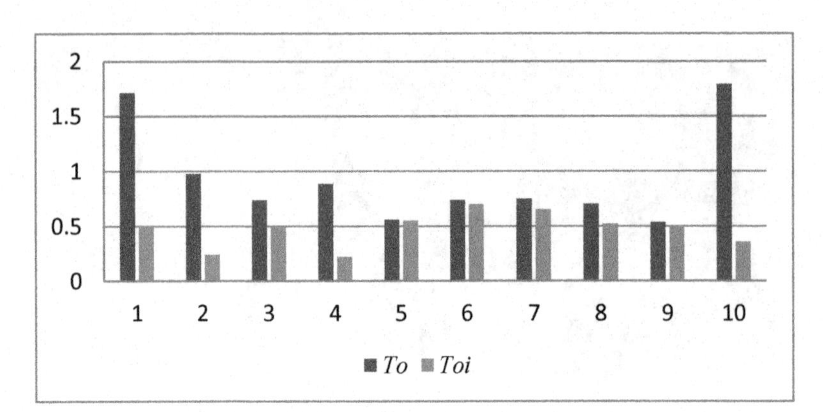

Fig. 5. Parallel concurrent overhead for parallel *To* and multi-layer *Toi* routing algorithms

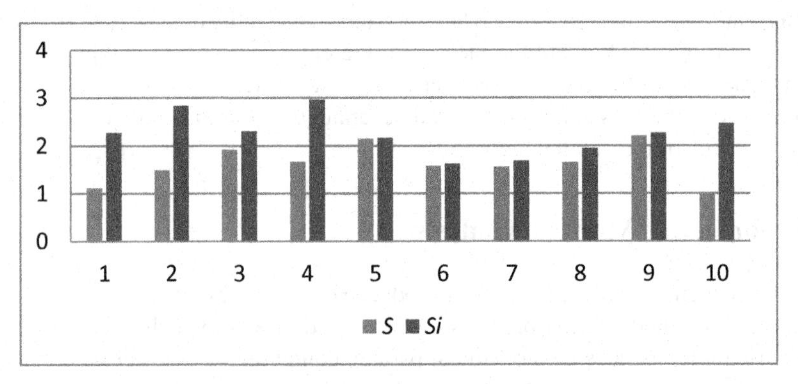

Fig. 6. Acceleration of parallel (*S*) and multi-layer (*Si*) routing algorithms

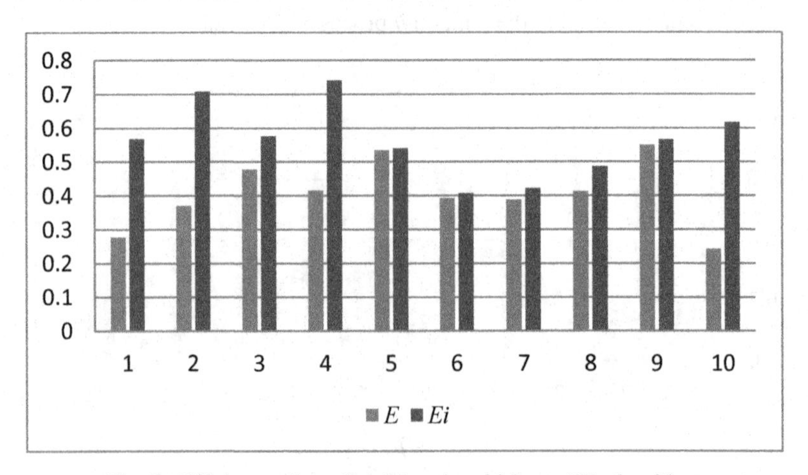

Fig. 7. Efficiency of parallel (*E*) and multi-layer (*Ei*) algorithms

6 Conclusion

In order to improve the quality of parallel computing in transportation logistics it is recommended to implement a multi-layer model of parallelization. Experimental results demonstrate an improvement of algorithm performance metrics including execution time, total parallel overhead, speedup, and efficiency. Next steps are concerned with automated construction of layers using the intelligent algorithms.

References

1. Ivaschenko, A., Maslennikov, S., Stolbova, A., Golovnin, O.: Multi-layer data model for transportation logistics solutions. In: Proceedings of the 26th Conference of Open Innovations Association FRUCT, Helsinki, Finland, pp. 124–129 (2020)
2. Griffis, S., Goldsby, T.: Transportation management systems: an exploration of progress and future prospects. Transportation **18**, 18–33 (2007). https://doi.org/10.22237/jotm/1175385780
3. Surnin, O., Sitnikov, P., Suprun, A., Ivaschenko, A., Stolbova, A., Golovnin, O.: Urban public transport digital planning based on an intelligent transportation system. In: Proceedings of the FRUCT'25, pp. 292–298 (2019)
4. Kagermann, H., Wahlster, W., Helbig, J.: Recommendations for implementing the strategic initiative Industrie 4.0. Final report of the Industrie 4.0 Working Group, p. 82 (2013)
5. Lasi, H., Fettke, P., Feld, T., Hoffmann, M.: Industry 4.0. Bus. Inf. Syst. Eng. **4**(6), 239–242 (2014)
6. Baesens, B.: Analytics in a Big Data World: The Essential Guide to Data Science and its Applications, p. 232. Wiley, Hoboken (2014)
7. Batty, M.: Big data, smart cities and city planning. Dialogues Hum. Geogr. **3**(3), 274–279 (2013)
8. Zhu, L., Yu, F.R., Wang, Y., Ning, B., Tang, T.: Big data analytics in intelligent transportation systems: a survey. IEEE Trans. Intell. Transp. Syst. **20**(1), 383–398 (2018)
9. Kim, T.-H., Kim, S.-J., Ok, H.: A study on the cargo vehicle traffic patterns analysis using Big Data. In: ACM International Conference Proceedings Series, pp. 55–59 (2017)
10. Javed, M.A., Zeadally, S., Hamida, E.B.: Data analytics for cooperative intelligent transport systems. Veh. Commun. **15**, 63–72 (2019)
11. Lee, K.-H., Lee, Y.-J., Choi, H., Chung, Y.D., Moon, B.: Parallel data processing with MapReduce: a survey. SIGMOD Rec. **40**(4), 11–20 (2011)
12. Zhang, D., Shou, Y., Xu, J.: A mapreduce-based approach for shortest path problem in road networks. J. Ambient. Intell. Human Comput. 1–9 (2018).https://doi.org/10.1007/s12652-018-0693-7
13. Falsafi, M., Marchiori, I., Fornasiero, R.: Managing disruptions in inbound logistics of the automotive sector. IFAC-PapersOnLine **51**, 376–381 (2018)
14. Golovnin, O.K.: Data-driven profiling of traffic flow with varying road conditions. In: CEUR Workshop Proceedings, vol. 2416, pp. 149–157 (2019)
15. Kumar, V., Grama, A., Gupta, A., Karypis, G.: Introduction to Parallel Computing: Design and Analysis of Algorithms, p. 597. Benjaming/Cummings, San Francisco (1994)
16. Sahni, S., Thanvantri, V.: Parallel Computing: Performance Metrics and Models, p. 42 (2002)

A Post Quantum Signature Scheme for Secure User Certification System

Swati Rawal(✉) and Sahadeo Padhye(✉)

Department of Mathematics, Motilal Nehru National Institute of Technology Allahabad, Prayagraj 211004, India

Abstract. Enhancement of technology requires better systems for user certifications, which are resistant to quantum attacks. A nominative signature is a better candidate for the user certification system. Most of the existing nominative signature schemes are based on bilinear pairing. The desire to expand the hardness assumption beyond factoring or discrete logarithm problem on which digital signatures can rely on. Lattice-based assumptions have seen a certain rush in recent years, as they are quantum resistant. In this article, we proposed the quantum secure user certification schemes signature scheme. The security of proposed signature scheme relies on Shortest-Integer problem, which can withstand the quantum attacks. Moreover, we prove that the proposed scheme satisfies unforgeability and other security aspects (invisibility, non-repudiation and impersonation) under the standard security model.

Keywords: Lattice cryptography · Digital signature · User certification schemes

1 Introduction

In a user certification system a user ask a certification authority for its signature on its certificate (e.g. diver licence, passport, digital marksheets etc.). In 2007 Dennis et al. [8] showed the application of nominative signature in user certification/identification system which was first introduced in 1993 by Kim, Park and Won [6]. It has several many other practical applications such as in insurance, banking, mobile communication etc. Nominative signature scheme works with three parties: *Nominator* A, *nominee* B and *verifier* V. A and B jointly involving in generation of a signature s on message m known as nominative signature. Only B (nominee) can verify the validity of s, if s is valid then B proves the verifier V the authenticity of s with a *confirmation protocol.* Otherwise, B proves V the invalidity of s with *disavowal protocol.* Security notions of nominative signature involves unforgeability, invisibility, security against impersonation and non-repudiation.

Before, 2007 it was believed that universal designated verifier signature (UDVS) suits best for certification systems. The following scenario clearly explains that nominative works better than UDVS for certification.

M. Bhattacharya et al. (Eds.): ICICCT 2021, CCIS 1417, pp. 52–62, 2021.
https://doi.org/10.1007/978-3-030-88378-2_5

For instance, Suppose an organization say a driving licence authority signs and issues the signature on the driving licence who applies for it. Above security notions makes nominative signature best suited cryptographic primitive by generating the mutual consent between the licence authority and the customer. Unforgeability of the scheme ensures that the signing authority cannot make false claims on the user and the other way round. Only customer can verify the whether the details included in the issued licence (signature) is correct or not as permitted by invisibility of the signature. Due to impersonation, the customer can prove his possession of a licence to the licence authority. However, due to non-repudiation, the customer can't fool the authority by possessing a fake licence.

Kim et al. [6] introduced the *Nominative Signature* in 1996 using Schnorr's scheme under the hardness of DLP (discrete logarithm problem). But, Huang et al. [4] found a fault in the latter scheme and introduced the concept of convertible nominative signature. But this scheme was also proved to be flawed and insecure [5].

Later, in 2007 Liu et al. [8] gave the formal structure of nominative signature using the undeniable signature under the hardness of computational and decisional Diffie-Hellman problem. But this scheme require a multiple rounds between the nominee and certification authority for generating a signature, thus Liu et el. [7] proposed a nominative signature with one round of communication using a ring signature under the hardness of DLP. But both these schemes were have weak invisibility as certification authority is not involved in some of valid signature generation. To tackle this, nominative signature with strong invisibility were proposed by Huang et al. [3]. Moreover, the construction have a strong notion for unforgeability as well with only one round of communication. Then in 2011, Schuldt et al. [10] introduced another nominative signature based on bilinear pairing. These are the only schemes secure in the classical setting. But now, due to a threat of quantum computers we need schemes to be quantum secure. Scheme based on hash function, codes, lattices and multivariate functions are resistant to quantum computers. Lattice based schemes have an upper-hand over other areas as the security of these systems can be reduced to worst-case problems, and most of the schemes require simple computations to compute signatures instead of standard modular exponentiation.

Taking this as a motivation, in this paper we introduce quantum resistant signature scheme based on lattices for user certification systems. Lattice cryptography is a promising candidate for post quantum cryptography as it involves simpler computations like matrix multiplication, vector additions etc. unlike modular exponentiation in number theoretic schemes. Moreover, here the security is guaranteed through the worst case hardness of lattice problems unlike the average case hardness of number theoretic problem.

The paper first introduces some ground work on lattice background and preliminaries required. Then, gives the formal structure and security requirements best suited for a secure user certification system in Sect. 2. Then Sect. 3 and 3.2

discuss the proposed scheme and its security analysis. Last section concludes the paper.

2 Lattice Background

Definition 1. ***Lattice-*** *Its is viewed as a discrete additive subgroup of* $\mathbb{R}^n$. *Due to its discrete structure and closure property under addition, it is also defined as a vector space, where instead of taking any linear combination, integer linear combination of basis vectors forms its members. Formally,*

A Lattice $\mathcal{L}$ *in* $\mathbb{R}^m$ *is an integer linear combination of vectors* $\mathbf{b}_1, \mathbf{b}_2, \ldots, \mathbf{b}_n$ *in* $\mathbb{R}^m$ *where* $n \leq m$,

$$\mathcal{L}(\mathbf{B}) = \mathcal{L}(\mathbf{b}_1, \mathbf{b}_2, \ldots, \mathbf{b}_n) = \{\Sigma_{i=1}^{n} \mathbf{x}_i \mathbf{b}_i | \mathbf{x}_i \in \mathbb{Z}\} = \{\mathbf{B}^T \mathbf{x} | \mathbf{x} \in \mathbb{Z}^n\}$$

$\mathbf{B} = \{\mathbf{b}_1, \mathbf{b}_2, \ldots, \mathbf{b}_n\}$ is known as the basis of the lattice. The integers n and m represents the rank and dimension of the lattice. If $n = m$, then it is a full rank lattice. Ajtai [1] introduced certain families of lattices, that have been used in various scheme based on latices.

Definition 2. ***q-ary Lattice:*** [1] *A Lattice* $\mathcal{L}$ *is known as q-ary lattice if* $q\mathbb{Z}^n \subseteq \mathcal{L} \subseteq \mathbb{Z}^n$ *holds for some integer* q. *For matrix* $\mathbf{A} \in \mathbb{Z}_q^{n \times m}$, *we can define two q-ary lattices of dimension* m,

1. $\mathcal{L}^{\perp}(\mathbf{A}) = \{\mathbf{x} \in \mathbb{Z}^m | \mathbf{A}\mathbf{x} = 0 \mod q\}$
2. $\mathcal{L}_y^{\perp}(\mathbf{A}) = \{\mathbf{x} \in \mathbb{Z}^m | \mathbf{A}^T \mathbf{y} = \mathbf{x} \mod q \text{ for } y \in \mathbb{Z}^n\}$

Lattice based scheme utilizes guassian distributions, but since their structure is discrete, hence they involve a slight variant of guassians distributions knowns as *discrete guassians distributions.*

Definition 3. ***Gaussian Function:*** *For* $\sigma > 0$ *a gaussian function* $\rho : \mathbb{R}^n \longrightarrow \mathbb{R}^+$ *centred at c is defined as*

$$\rho_{\sigma,c}(x) = exp(-\pi \|x - c\|^2 / \sigma^2), \forall x \in \mathbb{R}^n$$

Definition 4. ***Discrete Gaussian Distribution:*** *For some* $c \in \mathbb{R}^n$, $\sigma > 0$ *and an n-dimensional lattice* $\mathcal{L}$, *discrete gaussian distribution can be defined as*

$$D_{\mathcal{L},\sigma,c}(\mathbf{x}) = \frac{\rho_{\sigma,c}(\mathbf{x})}{\rho_{\sigma,c}(\mathcal{L})}, \forall \mathbf{x} \in \mathcal{L}$$

where, $\rho_{\sigma,c}(\mathcal{L}) = \sum_{\mathbf{y} \in \mathcal{L}} \rho_{\sigma,c}(\mathbf{y})$.

Note that if σ and c are 1 and 0 respectively then we use $D_{\mathcal{L}}(\mathbf{x})$ instead of $D_{\mathcal{L},\sigma,c}(\mathbf{x})$.

Now we define Shortest Integer Solution Problem (SIS) introduced by Ajtai [1] and used as hardness assumption by Zhang et al. in their scheme.

Definition 5. ***Shortest Integer Solution Problem*** (SIS)**:** *For an integer* q, *a real* β *and a matrix* $\mathbf{A} \in \mathbb{Z}_q^{n \times m}$, *determine a non-zero integer vector* $\mathbf{s} \in \mathbb{Z}^m$ *such that* $\mathbf{A}\mathbf{s} = 0 \mod q$ *and* $\|\mathbf{s}\| \leq \beta$.

2.1 Formal Structure of Signature Scheme for Certification System and Security Requirements

It works with three PPT (probabilistic polynomial time) algorithms (*Setup*, $KeyGen, Verify_{user}$) and protocols ($SignGen, Confirmation/Disavowal$). For a security parameter n, it works as follows:

- *Setup*: For security parameter n, it returns the system parameters $params$.
- *KeyGen*: It takes system parameters $params$ as input and output the keys for certification authority A (pk_A, sk_A) and user B (pk_B, sk_B).
- *SignGen*: It involves both A and B with common inputs $params$ and the message msg to be signed. A, additionally inputs pk_B specifying B as a user and pk_A is an additional input from B specifying A as certification authority. At last, either A or B output the signature σ on msg.
- $Verfiy_{user}$: This algorithm takes input $params$, (σ, msg), public keys (pk_A, pk_B) and sk_B and returns valid or invalid. Since it involves sk_B, the algorithm works only for B.
- $Confirmation/Disavowal$: If signature turns out to valid/invalid then B runs Confirmation/Disavowal protocol to assure the validity/invalidity of the signature to verifier V.

Security notions of the signature involves unforgeability, invisibility, security against impersonation and non-repudiation outlined as follows

Definition 6. ***Unforgeability-*** *Any adversary shouldn't able to forge any valid signature under chosen message attack.*

Definition 7. ***Invisibility-*** *It states that solely B (user) will confirm whether or not the signature is valid or invalid.*

Definition 8. ***Security against impersonation-*** *It is also known as non-transferability, this property prevents the proving capability of B being transferred to anyone else.*

Definition 9. ***Non-repudiation-*** *Invisibility property empowers the user, that on one except him can run the verification. Thus, non-repudiation is required against the user B.*

The proposed construction is based on GPV [2] signature scheme, the scheme uses trapdoor for lattices for signature generation. Therefore, before presenting the proposed construction, GPV signature scheme [2] and trapdoor functions are discussed.

2.2 GPV Signature Scheme

Before discussing the GPV signature scheme [2], we describe the construction of one-way preimage sampleable functions (PSF) which is the heart of the GPV signature scheme. The construction of collection of PSF's is as follows.

For the security parameter n, the system parameters q, m, L are such that $q = poly(n)$, $m \geq 5n \log q$ and $\|L\| = m^{2.5}$. The collection is also parametrized by a guassian parameter $s \geq L\omega(\sqrt{\log m})$ and it satisfies the following:

- *TrapGen*(1^n): This generates a pair $(\mathbf{A}, \mathbf{T})$, where $\mathbf{A} \in \mathbb{Z}_q^{n\times m}$ which defines a function $f_{\mathbf{A}} : \{\mathbf{e} \in \mathbb{Z}^m : \|\mathbf{e}\| \leq s\sqrt{m}\} \longrightarrow \mathbb{Z}_q^n$ such that $f_{\mathbf{A}}$ maps $\mathbf{e}$ to $\mathbf{Ae}$ mod q. Trapdoor $\mathbf{T}$ is a basis for $\mathcal{L}^{\perp}(\mathbf{A})$.
- *SampleDom* : This samples the domain $D_{\mathbb{Z}^m,s}$.
- *SamplePre*: For every $\mathbf{y} \in \mathbb{Z}_q^n$, *SampPre* samples $\mathbf{e}$ with conditional distribution, such that $f_{\mathbf{A}}(\mathbf{e}) = \mathbf{y}$.

From the above preimage sampleable function, Gentry et al. [2] constructed a signature scheme as follows

- *KeyGen*(1^n): Obtain $(\mathbf{A}, \mathbf{T}) \leftarrow$ *TrapSamp*(1^n), such that verification key $\mathbf{A}$ defines a function $f_{\mathbf{A}}$ and signing key $\mathbf{T}$ is its trapdoor.
- *SignGen*: To sign a message msg run *SamplePre*$(\mathbf{T}, H(msg))$ (where H is the hash function) to get a preimage σ as a signature on msg.
- *Verify*: If $\sigma \in \{\mathbf{e} \in \mathbb{Z}^m : \|\mathbf{e}\| \leq s\sqrt{m}\}$ and $f_{\mathbf{A}}(\sigma) = H(msg)$, signature is accepted as valid. Else, reject it as invalid.

3 Proposed Construction

Our construction is based on trapdoor for lattices [2] and an strongly unforgeable scheme. For a security parameter n the system parameters are m, q, where $q = poly(n)$, $m \geq 2n\log(q)$ with guassian parameter $\beta > \|\tilde{\mathbf{T}}\|\omega\sqrt{\log n}$. It also works with an hash function $H : \{0,1\}^{\star} \longrightarrow \mathbb{Z}_q^n$. The proposed scheme works as follows:

- *KeyGen*: for the security parameter n, run *TrapGen*(1^n) (trapdoor generation algorithm from [2]) to get the public key $(\mathbf{P}_A, \mathbf{P}_B)$ and secret key $(\mathbf{T}_A, \mathbf{T}_B)$ for A and B. Here, $\mathbf{P}_A, \mathbf{P}_B \in \mathbb{Z}_q^{m\times n}$ and $\mathbf{T}_A, \mathbf{T}_B \in \mathbb{Z}^{m\times m}$ such that $\mathbf{T}_A, \mathbf{T}_B$ forms a basis for the lattice $\mathcal{L}^{\perp}(\mathbf{P}_A)$, $\mathcal{L}^{\perp}(\mathbf{P}_B)$ respectively.
- *SignGen*: For any message $msg \in \{0,1\}^{\star}$, A and B proceeds as follows:
  - **certification authority** A**:** The certification authority A computes $\mathbf{u} = H(msg\|\mathbf{P}_A\|\mathbf{P}_B)$. Then samples $\sigma \leftarrow$ *SamplePre*$(\mathbf{P}_A, \mathbf{T}_A, \mathbf{u}, \beta)$ (pre-image sampling algorithm from [2]). A sends (σ, msg) as his signature to B.
  - **user** B**:** On receiving (σ, msg) from A, B checks:
    1. $\|\sigma\| \leq \beta\sqrt{m}$
    2. $\mathbf{P}_A\sigma = H(msg\|\mathbf{P}_A\|\mathbf{P}_B) \mod q$.

    Then if above holds he proceeds otherwise aborts.
  - First, B chooses an $\mathbf{e} \in D_{\mathbb{Z}_q^m,\beta}$ (secret) and compute $\mathbf{S}_B = \mathbf{P}_B\mathbf{e} \mod q$ and make $\mathbf{S}_B$ public. Then he compute $\mathbf{M} = H'(msg)$ and $\sigma_2 = \mathbf{Me}$ mod q, where hash function H' maps the input to $\mathbb{Z}_q^{n\times m}$.
    Finally computes $\sigma_3 =$ *SamplePre*$(\mathbf{P}_B, \mathbf{T}_B, H(msg\|\sigma_2, \beta))$ and return the signature $(\sigma_1 = \sigma, \sigma_2, \sigma_3)$ on msg.
- *Verify*$_{user}$: Since B has already checked the validity of σ_1, he verifies whether $\sigma_1 \stackrel{?}{=} \mathbf{P}_B\mathbf{e} \mod q$ and $\mathbf{P}_B\sigma_3 \stackrel{?}{=} H(msg\|\sigma_2, \beta)$. Since it involve the secret $\mathbf{e}$ it can only be verified by B.

3.1 Confirmation/Disavowal Protocol

This protocol is an interactive protocol between B and the verifier V which proceeds as follows:

- B chooses a permutation $\pi \in \{1, 2 \ldots, m\}$, $\mathbf{u} \in \mathbb{Z}_q^m$ and computes the commitments.
  - $C_1 = H(\pi \| \mathbf{P}_B\mathbf{u} \mod q)$
  - $C_2 = H(\pi(\mathbf{u}))$
  - $C_3 = H(\pi(\mathbf{u} + \mathbf{e}))$
  - $C_4 = H(\pi \| \mathbf{M}\mathbf{u} \mod q)$

  Then, B sends (C_1, C_2, C_3) to the verifier.
- Verifier V chooses a challenge $c \in \{0, 1, 2\}$ and sends it to B.
- B responds as follows:
  - If $c = 0$, B sends π, $\mathbf{u}$.
  - If $c = 1$, B sends π, $\mathbf{u} + \mathbf{e}$.
  - If $c = 2$, B sends $\pi(\mathbf{u})$, $\pi(\mathbf{e})$.
- For $c = 0$, V verifies that C_1, C_2 and C_4 are worked honestly.
- For $c = 1$, V checks validity of C_1, C_3 and C_4

$$\text{if } C_4 = H(\pi \| \mathbf{M}(\mathbf{u} + \mathbf{e}) - \sigma_2 \mod q), \text{ it is confirmation.}$$
$$\text{if } C_4 \neq H(\pi \| \mathbf{M}(\mathbf{u} + \mathbf{e}) - \sigma_2 \mod q), \text{ it is disavowal.}$$

- For $c = 2$, V checks validity of C_2, C_3 and whether $\|\pi(\mathbf{e})\| \leq \beta\sqrt{m}$.

After checking the validity of σ_2, V confirms the validity of σ_1 and σ_3 as follows

- $\|\sigma_1\|$, $\|\sigma_3\| \leq \beta\sqrt{m}$.
- $\mathbf{P}_A\sigma_1 \mod q = H(msg\|\mathbf{P}_A\|\mathbf{P}_B)$
- $\mathbf{P}_B\sigma_3 \mod q = H(msg\|\sigma_2)$

3.2 Security Analysis

We first prove that the verifier V will always accept a valid signature and detect an invalid signature.

Theorem 1. ***Completeness:*** *If the user B and verifier V works honestly then for any message msg and a valid signature $(\sigma_1, \sigma_2, \sigma_3)$,*

$$Prob(Confirmationrunssuccessfully) = 1$$

and if $(\sigma_1, \sigma_2, \sigma_3)$ is invalid

$$Prob(Disavowalrunssuccessfully) = 1.$$

Proof. If $(msg, (\sigma_1, \sigma_2, \sigma_3))$ is a valid pair then

$$\begin{aligned}(\mathbf{P}_B(\mathbf{u}+\mathbf{e}) - \mathbf{S}_B) \mod q &= \mathbf{P}_B(\mathbf{u}) \mod q,\\ (\mathbf{M}(\mathbf{u}+\mathbf{e}) - \sigma_2) \mod q &= \mathbf{M}(\mathbf{u}) \mod q,\\ \pi(\mathbf{e}) + \pi(\mathbf{u}) &= \pi(\mathbf{e}+\mathbf{u}).\end{aligned}$$

And therefore,

$$\begin{aligned}C_1 &= H(\pi' \| \mathbf{P}_B(\mathbf{u}) \mod q) = H(\pi' \| (\mathbf{P}_B(\mathbf{u}+\mathbf{e}) - \mathbf{S}_B) \mod q),\\ C_4 &= H(\pi' \| \mathbf{M}(\mathbf{u}) \mod q) = H(\pi' \| \mathbf{M}((\mathbf{u}+\mathbf{e}) - \sigma_2) \mod q),\\ C_3 &= H(\pi(\mathbf{e}+\mathbf{u})) = H(\pi(\mathbf{e}) + \pi(\mathbf{u}))\end{aligned}$$

Hence, confirmation will always be accepted if $(msg, (\sigma_1, \sigma_2, \sigma_3))$ is a valid pair.

Now if, $(msg, (\sigma_1, \sigma_2, \sigma_3))$ is invalid pair, then σ_2 will be invalid i.e. $\sigma_2 \neq \mathbf{Me} \mod q$. Therefore, $C_4 = H(\pi' \| \mathbf{M}(\mathbf{u}) \mod q) \neq H(\pi' \| (\mathbf{M}(\mathbf{u}+\mathbf{e}) - \sigma_2) \mod q)$, which is a valid check for disavowal protocol. Thus, if signature is invalid, disavowal runs successfully.

We next proof signer B can't fool verifier V into accepting a fraudulent σ_2 as valid, except with a negligible probability.

Theorem 2. ***Soundness:*** *If σ_2 is invalid then probability that confirmation runs successfully is negligible and if σ_2 is valid then disavowal runs successfully is also negligible.*

Proof. If σ_2 is invalid then $\sigma_2 \neq \mathbf{Me} \mod q$, for confirmation to result in acceptance B need to verify all the challenges. Suppose, in response to the challenges, B reply as follows,

- for $c = 0$, B sends ϕ_1 and α_1
- for $c = 1$, he sends ϕ_2 and α_2
- for $c = 2$, he sends α_3 and α_4, where ϕ_1, ϕ_2 are permutations from $(1, 2, \ldots, m)$.

Then, if the confirmation holds it implies

$$\begin{aligned}C_1 &= H(\phi_1 \| \mathbf{P}_B\alpha_1 \mod q) = H(\phi_2 \| (\mathbf{P}_B\alpha_2 - \mathbf{S}_B) \mod q),\\ C_2 &= H(\phi_1(\alpha_1)) = H(\alpha_4),\\ C_3 &= H(\phi_2(\alpha_2)) = H(\alpha_3 + \alpha_4),\\ C_4 &= H(\phi_1 \| \mathbf{M}\alpha_1 \mod q) = H(\phi_2 \| (\mathbf{M}\alpha_2 - \sigma_2) \mod q),\\ \|\alpha_3\| &\leq \beta\sqrt{m}.\end{aligned}$$

Which implies, either S can find a collision in H or following holds

$$\begin{aligned}\phi_1 = \phi_2, \mathbf{P}_B\alpha_1 \mod q &= (\mathbf{P}_B\alpha_2 - \mathbf{S}_B) \mod q,\\ \phi_1(\alpha_1) &= \alpha_4,\\ \phi_2(\alpha_2) &= \alpha_3 + \alpha_4,\\ \mathbf{M}\alpha_1 \mod q &= (\mathbf{M}\alpha_2 - \sigma_2) \mod q.\end{aligned}$$

Which results in $\alpha_2 - \alpha_1 = \phi_1^{-1}(\alpha_3)$ then $\mathbf{S}_B = \mathbf{P}_B\phi_1^{-1}(\alpha_3) \mod q, \therefore \phi_1^{-1}(\alpha_3) = \mathbf{e} \mod q$.

Since $\sigma_2 = \mathbf{M}\phi_1^{-1}(\alpha_3) \mod q = \mathbf{M}\mathbf{e}^{\mathbf{T}} \mod q$, but as σ_2 is invalid $\sigma_2 \neq \mathbf{M}\mathbf{e}^{\mathbf{T}} \mod q$, which leads to a contradiction. Thus S can't answer all three challenges correctly.

Therefore at the best he can answer with probability $\frac{2}{3}$, for $N-rounds$, the probability become $(2/3)^N$ which becomes negligible if N increases.

Now, if the signature is valid i.e. $\sigma_2 = \mathbf{Me} \mod q$ and he wants to run the disavowal protocol successfully. Suppose signer gives the same response as above and in this case following holds

$$
\begin{aligned}
&C_1 = H(\phi_1 \| \mathbf{P}_B\alpha_1 \mod q) = H(\phi_2 \| (\mathbf{P}_B\alpha_2 - \mathbf{S}_B) \mod q),\\
&C_2 = H(\phi_1(\alpha_1)) = H(\alpha_4),\\
&C_3 = H(\phi_2(\alpha_2)) = H(\alpha_3 + \alpha_4),\\
&C_4 = H(\phi_1 \| \mathbf{M}\alpha_1 \mod q) \neq H(\phi_2 \| (\mathbf{M}\alpha_2 - \sigma_2) \mod q),\\
&\|\alpha_3\| \leq \beta\sqrt{m}.
\end{aligned}
$$

Similarly, we get

$$
\begin{aligned}
\phi_1 = \phi_2, \mathbf{P}_B\alpha_1 \mod q &= \mathbf{P}_B\alpha_2 - \mathbf{S}_B \mod q,\\
\phi_1(\alpha_1) &= \alpha_4,\\
\phi_2(\alpha_2) &= \alpha_3 + \alpha_4,\\
\mathbf{M}\alpha_1 \mod q &\neq \mathbf{M}\alpha_2 - \sigma_2 \mod q.
\end{aligned}
$$

Thus, similar to the first case, we get $\phi_1^{-1}(\alpha_3) = \mathbf{e} \mod q$, but we have $\sigma_2 \neq \mathbf{M}\phi_1^{-1}(\alpha_3) \mod q \neq \mathbf{Me} \mod q$, but since signature is valid, we have $\sigma_2 = \mathbf{Me} \mod q$. Therefore we have a contradiction. Hence signer can't get all the checks correct, at best his success probability is $\frac{2}{3}$. For N rounds, probability will be $(2/3)^N$ which becomes negligible as N increases.

Now, we will prove that the proposed scheme is secure under existential forgery. The scheme can have three types of adversaries,

- Type 1: Adversary $\mathcal{A}$ having user's secret key.
- Type 2: Adversary $\mathcal{A}$ having certification authority's secret key.
- Type 3: Adversary $\mathcal{A}$ acting as third party no knowledge of any the secret keys.

Clearly, if we prove it secure against *Type 1* and *Type 2* adversaries then it will be secure against *Type 3*. the security proves runs as follows.

Theorem 3. *The proposed certification authority signature is secure against Type 1 adversary $\mathcal{A}$ with the underlying hardness of SIS problem.*

Proof. Suppose $\mathcal{A}$ can successfully forge the certification authority signature with probability ϵ then we can construct an attacker $\mathcal{S}$ against the SIS problem.

Setup: $\mathcal{S}$ runs $\mathcal{A}$ on public key $\mathbf{P}_A$ and user's keys $(\mathbf{P}_B, \mathbf{T}_B)$.

Since $\mathcal{A}$ knows B's secret key, therefore, forgery of the signature, can be regarded as the forgery on certification authority's part. For certification authority's part, Without loss of generality (WLOG) we assume that $\mathcal{A}$ queries H before making any sign query on every message msg.

HashQuery: For every hash query on any distinct $msg \in \{0,1\}^\star$, $\mathcal{S}$ obtains $s'_{msg} \leftarrow SampleDom(1^n)$ and store (msg, s'_{msg}) then returns $\mathbf{P}_A s'_{msg} \mod q = H(msg\|\mathbf{P}_A\|\mathbf{P}_B)$ to $\mathcal{A}$.

SignQuery: For every sign query on msg, $\mathcal{S}$ looks up the local storage and return s'_{msg} as a signature on msg.

Now, we assume that before any attempted forgery $(m^\star, s^\star)$, $\mathcal{A}$ queries hash H on $m^\star$. When $\mathcal{A}$ outputs a forgery $(m^\star, s^\star)$, then $\mathcal{S}$ looks up the local storage for a pair $(m^\star, s'_{m^\star})$. Since it is a forgery $s'_{m^\star} \neq s^\star$ but we have

$$\begin{aligned} \mathbf{P}_A s^\star &= \mathbf{P}_A s'_{m^\star} \mod q \\ \therefore \mathbf{P}_A(s^\star - s'_{m^\star}) &= 0 \mod q \\ \text{and } \|(s^\star - s'_{m^\star})\| &\leq 2\beta\sqrt{m} \end{aligned}$$

Thus, $s^\star - s'_{m^\star}$ becomes a solution for SIS problem. By pre-image min-entropy property, $s'_{m^\star} \neq s^\star$ except with a negligible probability $2^{-\omega(logn)}$. Hence, with probability $\epsilon(1 - 2^{-\omega(logn)})$, $\mathcal{S}$ can obtain solution to the SIS problem.

Theorem 4. *The Proposed certification authority Signature is secure against Type 2 adversary under the hardness of SIS problem.*

Proof. Similar to above proof, assume that adversary $\mathcal{A}$ can output a valid forgery with probability ϵ then the solver $\mathcal{S}$ can solve the SIS problem.

Setup: $\mathcal{S}$ runs $\mathcal{A}$ on public key $\mathbf{P}_B$ and certification authority's keys $(\mathbf{P}_A, \mathbf{T}_A)$.

Since $\mathcal{A}$ knows certification authority's secret key, therefore, forgery of the signature, can be regarded as the forgery on user's part. For user's part, to generate σ_2, $\mathcal{S}$ follows the signature generation protocol and for σ_3 he works in the following manner.

WLOG, we assume that $\mathcal{A}$ queries H before making any sign query on every message msg.

For every hash query on any distinct $msg \in \{0,1\}^\star$, $\mathcal{S}$ obtains $\sigma_3 \leftarrow SampleDom(1^n)$ and store (msg, σ_3) then returns $\mathbf{P}_B\sigma_3 \mod q = H(msg\|\sigma_2)$ to $\mathcal{A}$. Then for sign query, $\mathcal{S}$ looks up the local storage and returns σ_3.

WLOG, we assume that before returning a valid forgery $(\sigma_2^\star, \sigma_3^\star)$, $\mathcal{A}$ call for hash query. Then on receiving the forgery $(msg^\star\sigma_2^\star, \sigma_3^\star)$, $\mathcal{S}$ looks up the local storage for pair $(msg^\star, \sigma_3)$. Since, it is forgery, $\sigma_3 \neq \sigma_3^\star$, but

$$\begin{aligned} \mathbf{P}_B\sigma_3^\star &= \mathbf{P}_B\sigma_3 \mod q \\ \therefore \mathbf{P}_B(\sigma_3^\star - \sigma_3) &= 0 \mod q \\ \text{and } \|(\sigma_3^\star - \sigma_3)\| &\leq 2\beta\sqrt{m} \end{aligned}$$

Thus, $\sigma_3^\star - \sigma_3$ becomes a solution for SIS problem. By pre-image min-entropy property, $\sigma_3^\star \neq \sigma_3$ except with a negligible probability $2^{-\omega(logn)}$. Hence, with probability $\epsilon(1 - 2^{-\omega(logn)})$, $\mathcal{S}$ can obtains the SIS problem.

In addition to these, the proposed signature scheme satisfies all other security requirements also as described below.

Invisibility- Since the verification involves user's secret key, therefore on one except him can run verification. Thus signature satisfies invisibility.

Security Against Impersonation- Since certification authority and user jointly work together and both uses their secret keys in signature generation individually. Thus, certification authority can't carry out impersonation attack.

Non-repudiation- Since the secret key of user is involved in verification. Therefore he can't deny later of his role in signature generation.

4 Conclusion

The proposed signature based on lattices is the user certification signature scheme secure against quantum attacks. The proposed construction is proved to satisfy all the security requirements for a secure certification scheme.

References

1. Ajtai, M.: Generating hard instances of the short basis problem. In: Proceedings of ICALP, pp. 1–9 (1999)
2. Gentry, C., Peikert, C., Vaikuntanathan, V.: Trapdoors for hard lattices and new cryptographic constructions. In: Proceedings of ACM Symposium on Theory of Computing, pp. 197–206 (2008)
3. Huang, Q., Liu, D.Y., Wong, D.S.: An efficient one-move nominative signature scheme. Int. J. Appl. Cryptogr. **1**(2), 133–143 (2008)
4. Huang, Z., Wang, Y.: Convertible nominative signatures. In: Wang, H., Pieprzyk, J., Varadharajan, V. (eds.) ACISP 2004. LNCS, vol. 3108, pp. 348–357. Springer, Heidelberg (2004). https://doi.org/10.1007/978-3-540-27800-9_30
5. Susilo, W., Mu, Y.: On the security of nominative signatures. In: Boyd, C., González Nieto, J.M. (eds.) ACISP 2005. LNCS, vol. 3574, pp. 329–335. Springer, Heidelberg (2005). https://doi.org/10.1007/11506157_28
6. Kim, S.J., Park, S.J., Won, D.H.: Zero-knowledge nominative signatures. In: Proceedings International Conference on the Theory and Applications of Cryptology 1996, pp. 380–392 (1996)
7. Liu, D.Y.W., Chang, S., Wong, D.S., Mu, Y.: Nominative signature from ring signature. In: Miyaji, A., Kikuchi, H., Rannenberg, K. (eds.) IWSEC 2007. LNCS, vol. 4752, pp. 396–411. Springer, Heidelberg (2007). https://doi.org/10.1007/978-3-540-75651-4_27
8. Liu, D.Y.W., et al.: Formal definition and construction of nominative signature. In: Qing, S., Imai, H., Wang, G. (eds.) ICICS 2007. LNCS, vol. 4861, pp. 57–68. Springer, Heidelberg (2007). https://doi.org/10.1007/978-3-540-77048-0_5

9. Steinfeld, R., Bull, L., Wang, H., Pieprzyk, J.: Universal designated-verifier signatures. In: Laih, C.-S. (ed.) ASIACRYPT 2003. LNCS, vol. 2894, pp. 523–542. Springer, Heidelberg (2003). https://doi.org/10.1007/978-3-540-40061-5_33
10. Schuldt, J.C.N., Hanaoka, G.: Non-transferable user certification secure against authority information leaks and impersonation attacks. In: Lopez, J., Tsudik, G. (eds.) ACNS 2011. LNCS, vol. 6715, pp. 413–430. Springer, Heidelberg (2011). https://doi.org/10.1007/978-3-642-21554-4_24

Comparative Analysis of Serverless Solutions from Public Cloud Providers

Darshan Baid[1](✉), Pallavi Murghai Goel[1], Pragya Bhardwaj[1], Astha Singh[1], and Vishu Tyagi[2]

[1] CSE, Galgotias University, Greater Noida, UP 201308, India
hey@darshanbaid.com

[2] CSE, Graphic Era Deemed to be University, Dehradun, Uttarakhand 248171, India

Abstract. Cloud Providers such as Google Cloud Platform (GCP), Microsoft Azure, Amazon Web Services (AWS) offer have started expanding their serverless platforms, promising higher availability and dynamic auto-scaling while at the same time diminishing operational and maintenance costs. One such is serverless processing, or function-as-a-service (FaaS). Serverless has emerged as a decent contestant for web applications, APIs, backend-services due to their unique event-driven architecture and pay-as-you-go model. In this paper, we discussed the current architecture of serverless functions & compared the strength of serverless offering of two major cloud providers - Google's GCP & Microsoft Azure and conducted a load testing experiment against their node js environment to better understand their capabilities. We conducted 2 different scenarios of user-load testing via the open-source framework, k6. These scenarios included several phases of sudden and spontaneous load increment and decrement, mimicking how it happens in real-world situations. Finally, we observed how quickly and effectively they scale, their cold boot times, and their response times.

Keywords: Serverless · Load testing · Gcp · Azure · Cloud functions

1 Introduction

Serverless computing is one of the cloud computing execution models and is emerging as a compelling paradigm for deployment of cloud applications and has gained increasing attention in the past five years [1].

Apart from Infrastructure-as-a-service (IAAS), Platform-as-a-service (PAAS), serverless begins their own Function-as-a-service model, which offers an extremely cost efficient solution for simpler runtime loads that might not require running full-blown applications. Serverless lowers cost of deploying by charging only for execution time and memory which are directly proportional to number of executions. They generally do not have any standing cost or cost that incur even if no instances are running or active [2].

In this paper we aim to compare the performance of two serverless cloud computing platforms available, GCP's Cloud Functions and Azure Functions. We ran load balancing tests on them, measured their response time and server utilization to compare the two.

M. Bhattacharya et al. (Eds.): ICICCT 2021, CCIS 1417, pp. 63–75, 2021.
https://doi.org/10.1007/978-3-030-88378-2_6

Serverless computing was popularized by AWS during their 2014 re:Invent session "Getting Started with AWS Lambda" [3] Later, they were followed by Google quietly releasing their Google Cloud Function [4] and IBM releasing IBM Cloud Functions [5] in 2016. Microsoft launched it's Azure Functions soon after and Cloudflare also entered the market with their Cloudflare Workers in 2017.

Serverless are generally considered more cost-efficient than assigning a fixed number of compute servers like AWS's EC2 or GCP's Compute Engine, which generally involves significant periods of idle time or underutilization. Due to its bin-packaging, serverless can also be more cost effective than autoscaling groups. Serverless sees immediate cost efficiency for smaller companies and startups.

The entire cloud infrastructure service market has is $111 Billion industry with AWS leading with 33% share followed by Azure and GCP with 18% and 9% respectively [6] Other providers like Alibaba Cloud, IBM Cloud, Salesforce, etc. follow up quickly. Serverless being a newer paradigm, it would be interesting to see which provider is here to take a leap.

The main difference between serverless platforms & servers is the automatic scaling up & down and charging of resources only when the submitted code is executing. That is not considering the free resource utilization, where a limited amount of monthly resources are available by major cloud providers such as GCP & AWS [18].

Tech Industry today tends to favor smaller computation units to make it easier to manage and scale in the cloud. This is where Serverless Computing comes into play. The calculation, which can be interrupted or restarted, now no more depends on a cloud platform to maintain its status.

1.1 Literature Review

Serverless computing can be defined by its nameless thinking (or caring) about servers [21]. Serverless computing is a platform that hides server usage from developers and runs code on-demand automatically scaled and billed only for the time the code is running.

The term 'serverless' does not mean it's easily comprehended definition of not using servers. Of course, there are servers involved in the process of execution but are not the developer's tension to maintain [1].

The currently used method for serverless computing is providing a function that is executed by the cloud computing platform. This change led to the rise of Function-as-a-service (FaaS) platforms, instead of building the entire architecture, focusing on small blocks of code collated as functions that run for a limited amount of maximum time, with executions that are triggered by events (pub-sub) or HTTP requests [17].

In Serverless, the developer currently has management over the code they deploy into the Cloud, although that code needs to be written within the kind of stateless functions [1].

Software-as-a-Service (SaaS) might support the server-side execution of user-provided functions, however, they are running within the environment of the application and thence restricted to the application's env or domain [15].

Seeing the trend Amazon Web Services & Google Cloud Platform launched their own serverless functions around 2014–16 [3, 4]. They were soon followed by IBM & Microsoft [5, 10].

The economical advantages of serverless computing have gained its serious popularity as seen by the increasing rate of the "serverless" search term by Google Trends. Its market size is estimated to grow to 7.72 billion by 2021 [14].

We further studied load testing and how it is performed for serverless functions. Load testing is used to analyze whether the infrastructure used for the application is sufficient or not [12]. We used k6, a developer-centric, open-source load testing tool, which also happened to be free-of-cost [13] to identify the maximum operating capacity of a single instance of each function of AWS & GCP platforms [12].

Compared to other IAAS or PAAS platforms, serverless offers different choices in control, cost & flexibility. Serverless allow the developers to carefully think about the costs when designing the modularity of the application, rather than worrying about its scalability [1].

2 Architecture

The name serverless often creates an impression that servers are not actually needed in a serverless environment. This is a misconception, and servers are still needed [1]. The name serverless is coined as developers do not need to be concerned about setting up and scaling servers, provisioning new ones or decommissioning unused ones.

2.1 Event-Driven Nature

Serverless functions are built upon Service-Oriented Architecture (SOA) or the more commonly used Event-Driven (EDA) [7]. Event-driven implies that the function is run on the occurrence of an event. This event can be an HTTP event or an internal event like a database trigger or storage trigger. CRON events [9] could also be attached in most cloud functions.

2.2 Programming Model

Serverless are built to scale, hence, they have very limited expressiveness. In most scenarios, serverless do not maintain an internal state between execution. [1] They can connect with an external database [8] or state management system. The functions can also access global objects or environment variables.

3 Studied Cloud Platforms

In this paper, we studied 2 major cloud platforms based on their offerings to serverless functions. We analyzed the capabilities they are offering and how well they are suited to sudden changes in load, how they scale up or down and finally their cold boot times.

3.1 Google Cloud Platform (GCP)

Google Cloud Functions are the main offering from Google in the serverless compute paradigm. The product is extended with the Firebase platform offering. Other than Cloud Function, google also has a newer product Cloud Run, a docker based event-driven system.

For GCP customers, Cloud Functions acts as a connection layer to connect logic from various GCP products such as pub-sub, monitoring, etc. services for listening and responding to services [11].

For Firebase customers, Cloud Functions for Firebase acts as an extended version with more event triggers such as Firestore trigger, realtime database triggers, authentication trigger, etc. and integrates with Firebase to provide server-side code to most mobile applications [11].

3.2 Microsoft Azure

Microsoft entered the Cloud Computing market in 2010 with the launch of Azure and has since then dominated the number 2 spot. It's serverless offering, Azure Functions were launched in 2016.

One unique feature about azure functions are their deployment method. Unlike AWS and GCP where you can either edi the functions on the cloud console or deploy via CLI, the main method of deploying Azure Functions are via a VS Code Extension [10], though it also supports CLI deployments.

4 Experiment Method

4.1 Function Configuration

We had several machine settings for testing our cloud functions across different cloud providers. We selected NodeJs 10 environment as it was widely available and had a large developer-base.

The selected functions were run across all memory sizes with timeout set to 3 s. This was done to vary the degree of memory stress on the functions as load increases.

The server code computed a Fibonacci series up to a random integer selected from the interval of *[25, 20]*.

```
function fibonacci_gen(n) {
    if (n <= 1) return 1;
    return n * fibonacci_gen(n-1);
}

exports.loadFn = functions.https.onRequest((request, response) => {
    const rn = (Math.Random * (25 - 20) + 20).toFixed(0);
    response.send(fibonacci_gen(+rn));
})
```

4.2 Client Configuration

For sending varied load to the created functions we used a Compute Engine instance with 8 virtual CPU cores and 32 GB of memory (Table 1).

Table 1. Client side configuration

Machine type	Virtual CPUs	Memory
GCP e2-standard-8	8	32 GB

4.3 Load Testing

Load testing is a non-functional testing against a web application conducted to analyze how well the application functions under a specific expected load [12].

Load testing is conducted by creating an N number of Virtual Users, that mimic the behavior of actual users. This number is gradually or instantaneously increased or decreased, commonly called "rampage".

We used k6.io scripting to send a ramping load to serverless functions. k6 is an open-source load testing tool with several features like scripting scenarios, virtual user (VU) concept among others [13].

k6 docker package was installed on the e2-standard machine along with other necessary tools. 2 scenarios were tested - the first being a 5 min stress test and other being a 17-min long extreme test (Table 2).

Table 2. Load testing scenarios

Scenario	*Phase*	*Target VUs*	*Duration*
S1	Phase 1 (Warmup)	50	1 m
	Phase 2	50	5 m
	Phase 3 (Cooldown)	0	1 m
S2	Phase 1 (Warmup)	50	2 m
	Phase 2	600	5 m
	Phase 3	100	5 m
	Phase 4	1000	3 m
	Phase 5 (Cooldown)	0	2 m

Target VUs represent the target of Virtual User to be reached.

4.4 Platform Configuration

GCP offers Cloud Function which can be extended with Firebase to support events from the firebase platform such as authentication & database triggers. We choose the default available with HTTP trigger for the experiment.

The experiment was run on the basic 128 mb memory size. The timeout was set to 3 s and no maximum limit on the number of instances were set. GCP offers several runtime environments like NodeJS, Go, we choose Nodejs 12 for our experiment. The location was set to *us-central1*.

Microsoft offers serverless functions to run either on Linux or Windows. To maintain uniformity, we went with the Linux System and selected Central US as the function location.

5 Experiment Result

5.1 GCP Scenario 1

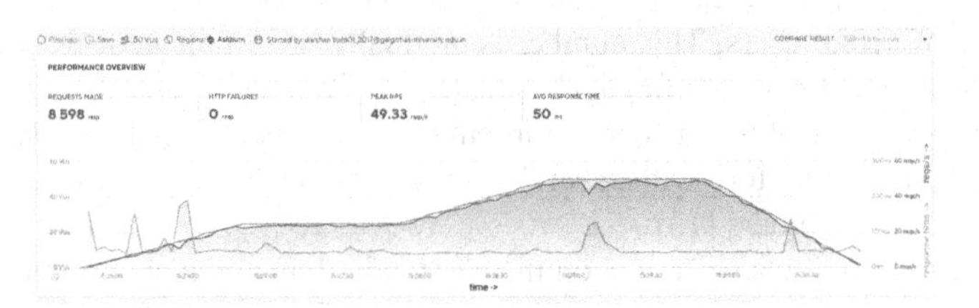

The first thing we noticed was that GCP resulted in 0 downtimes or non-200 status codes. This meant that auto-scaling was properly able to handle the load.

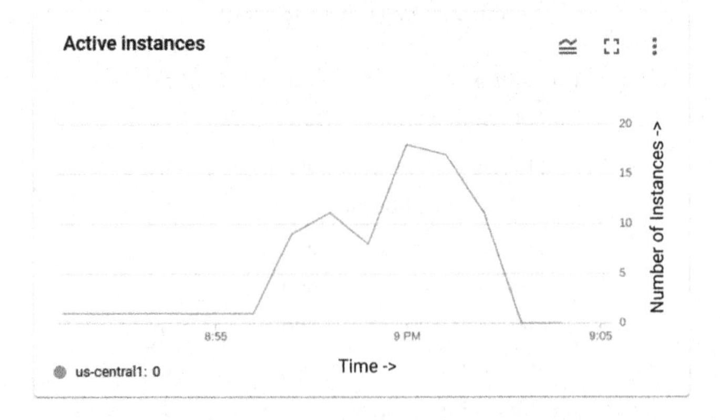

Instead of preferring a one-by-one or gradual increase, the number of active instances quickly went to 9 and then to a maximum of 18 instances after a mid-stage decrease to 8. As the load provided was increasing stress, a reduction in the number of instances was slightly shocking.

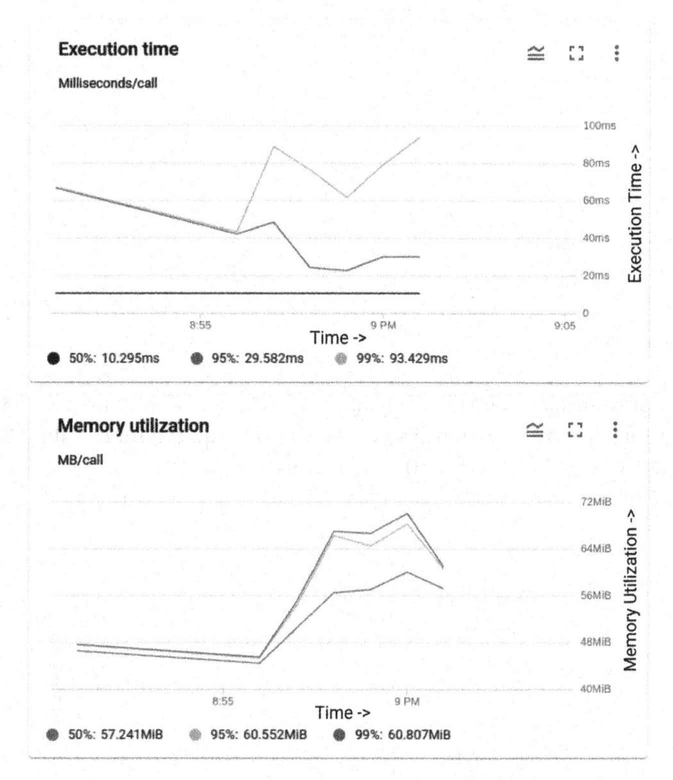

Next, we measured execution time and memory utilization. It was evidently clear that this was an easy scenario for GCP as it didn't even reach it's memory limit and execution time remained well under 100 ms. The 99 percentile result was set at 93.42 ms and memory utilization for 99 percentile at 60.807MiB.

5.2 GCP Scenario 2

The second test showed a similar result. Unsurprisingly, there were no server errors or downtimes. At the peak, there were about 885.3 executions per second.

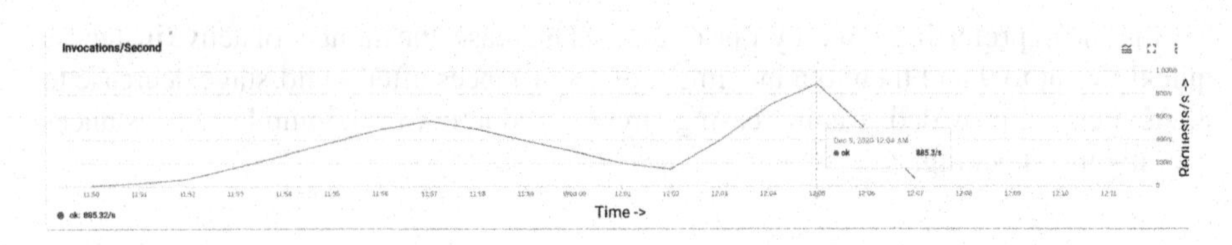

The number of instances reached a peak of 60 with a half-way mark at 38. It can be observed that for the same function, 38 instances were required for handling 500 Virtual Users whereas 60 were enough for 1000 virtual users.

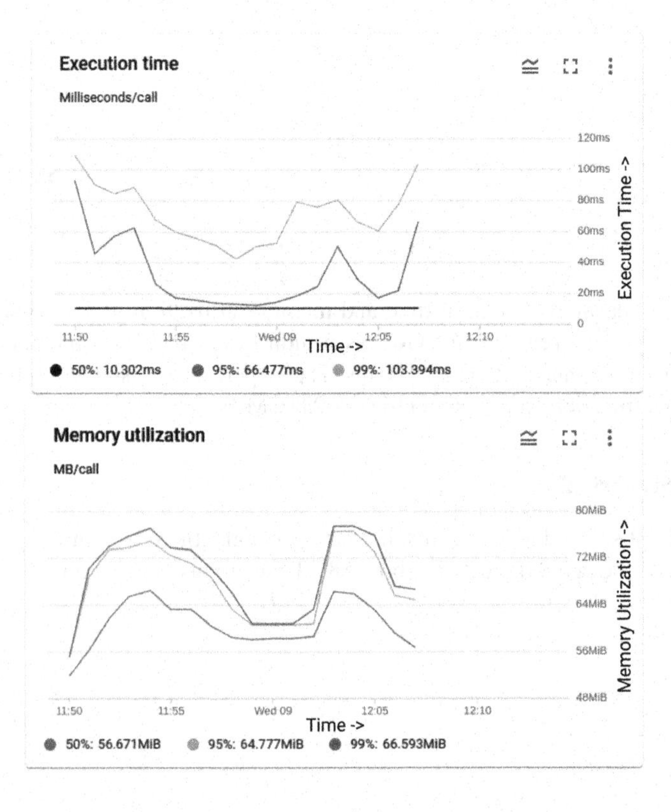

The memory utilization and execution time remained nearly the same. Though, this time execution time crossed the 100 ms mark for the 99th percentile.

http_duration:
avg: 17.21ms
min: 6.98ms
med: 12.23ms
max: 2.63s
p(90): 26.56ms
p(95): 45.41ms

5.3 Azure Scenario 1

Azure had an average of 62 ms response time, minimum being 43 ms and P99 being 515 ms.

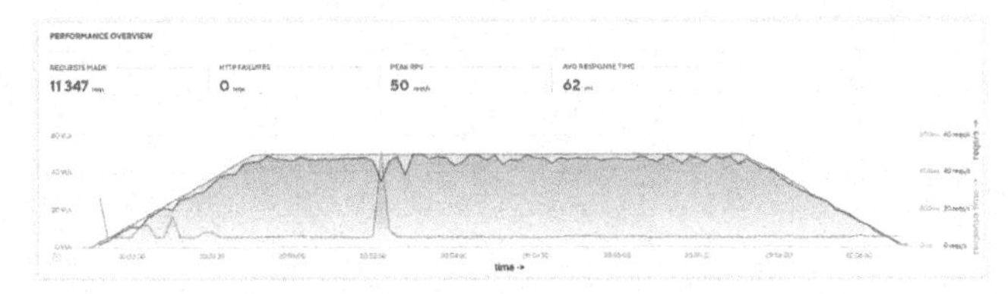

As azure doesn't provide graphically how the instances are scaling up or down, we could only note their maximum number.

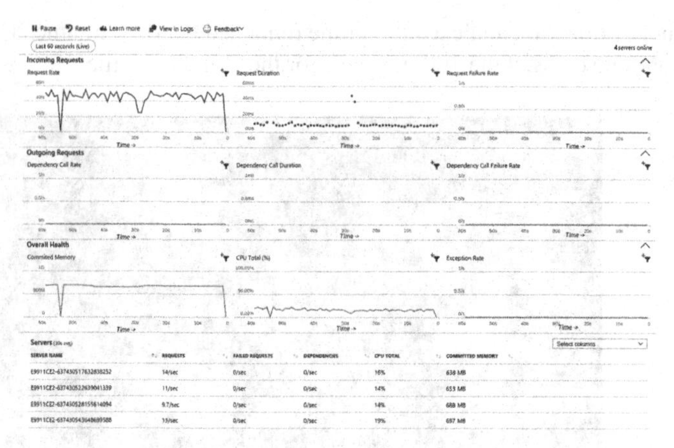

Quite surprisingly, azure was able to handle the entire load with just 4 server instances. Each instance was able to handle an approx of 12–15 requests per second with a max CPU utilization of 20%.

It is important to note that the Committed Memory was set to about 692 MB which is significantly higher than GCP.

Same as GCP, azure resulted in 0 downtime or non-200 status codes. It was able to scale up and down with ease.

5.4 Azure Scenario 2

In the second scenario, Azure started with just a single server instance for about 200 Virtual User mark and then scaled upto 5 server instances. Before scaling, the single server reached a CPU utilization rate of over 120%.

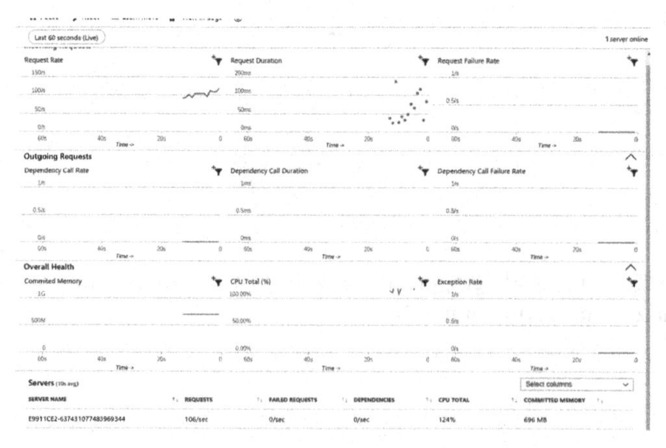

When scaled to 5 instances (at phase-2), each of the servers were handling about 100 requests per second, at about 80–90% CPU Utilization.

During scaling down (phase-3), the number of instances didn't decrease. This means Azure didn't automatically scale down for reduced load.

During the final stress phase (phase-5), we observed that all of the 5 servers reached at 120% CPU Utilization peak.

Servers (10s avg)

SERVER NAME	REQUESTS	FAILED REQUESTS	DEPENDENCIES	CPU TOTAL	COMMITTED MEMORY
E9911CE2-637431077483969344	101/sec	0/sec	0/sec	117%	691 MB
E9911CE2-637431079974877324	120/sec	0/sec	0/sec	112%	716 MB
E9911CE2-637431086981118728	94/sec	0/sec	0/sec	123%	674 MB
E9911CE2-637431090485919897	93/sec	0/sec	0/sec	124%	691 MB

This was confirmed by Application Insight Metric which reported a CPU utilization of 133% Max.

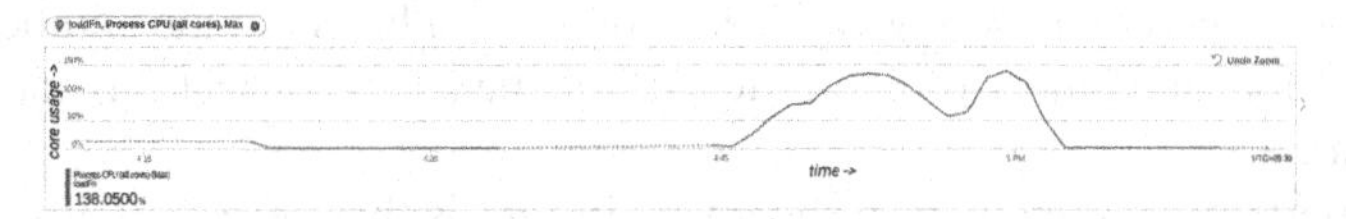

Azure scaled to a maximum of 9 server instances during the final phase with each of the server having a CPU utilization of over 100%.

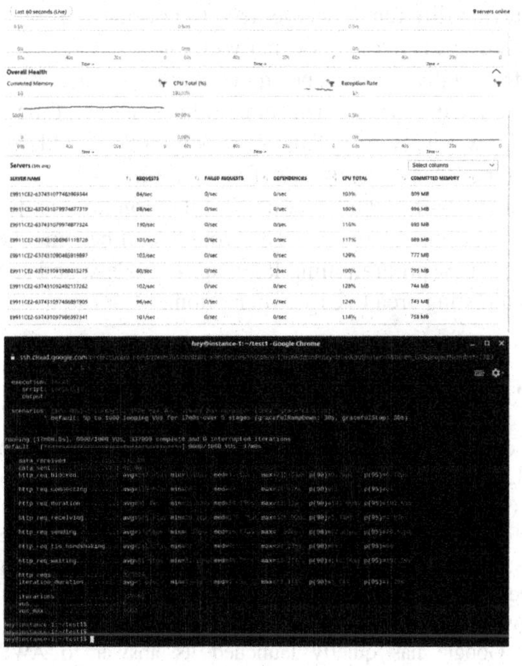

http_duration:
avg: 82.4ms
min: 31.35ms
med: 50.18ms
max: 12.17s
p(90): 141.96ms
p(95): 192.66ms

6 Conclusion

We observed that both the cloud platforms performed well with 0 server errors that could be caused due to scaling issues. This was a good notion to stick to serverless computing. This could mean it is a good time for corporations to switch to cloud functions over preferring server environments like App Engine.

We noticed that GCP needed about 60 instances to handle the peak 1000-VAU load whereas Azure did the same with just 9. This seems to favour the fact that individual GCP instances can handle less load than Azure. But the fact that CPU utilization of Azure instances crossed the 120% mark gives a negative light that the server might fail if the computation is complex as well. A developer is not billed for the number of active instances but for the number of calls. Hence, it is better to connect this fact with if a server might fail.

One important consideration in the above insight is that GCP calls it instances as *"function instances"* whereas Azure calls them *"server instances"*. It is unclear whether they mean the same. This study assumes that both correspond to the same.

The average response time from GCP was at 17.21 ms whereas for Azure was 82.4 ms. This is a significant difference but we need to normalize as the client was also running on the GCP premises, hence the low network costs.

We should not apply the same normalization for the max response time where Azure performed significantly worse. The maximum response time for Azure was 12.17 s whereas for GCP was just 2.63 s. This corresponds to the Cold Boot time of new instances pointing to scaling issues in Azure's infrastructure.

We conclude stating that serverless infrastructure is just getting started and already showing promising results. This could soon become a champion for backend-development and new programming languages, models, and service infrastructures are certainly an interesting area to experimentation.

References

1. Baldini, I., et al.: Serverless computing: current trends and open problems. In: Chaudhary, S., Somani, G., Buyya, R. (eds.) Research Advances in Cloud Computing, pp.1–20. Springer, Singapore (2017). https://doi.org/10.1007/978-981-10-5026-8_1
2. Jamieson, F.: Losing the server? Everybody is talking about serverless architecture, 2 October 2017. https://www.bcs.org/content-hub/losing-the-server/. Accessed 8 Dec 2020
3. Aws re:invent 2014 (mbl202) new launch: getting started with aws lambda. https://www.youtube.com/watch?v=UFj27laTWQA. Accessed 9 Dec 2020
4. Novet, J.: Google has quietly launched its answer to AWS Lambda. VentureBeat, 9 February 2016. https://venturebeat.com/2016/02/09/google-has-quietly-launched-its-answer-to-aws-lambda/. Accessed 9 Dec 2020
5. Zimmerman, M.: IBM Unveils Fast, Open Alternative to Event-Driven Programming, 23 February 2016. https://www-03.ibm.com/press/us/en/pressrelease/49158.wss. Accessed 9 Dec 2020
6. Richter, F.: Amazon Leads $100 Billion Cloud Market, 18 August 2020. https://www.statista.com/chart/18819/worldwide-market-share-of-leading-cloud-infrastructure-service-providers/. Accessed 7 Dec 2020

7. Chou, D.: Event-Driven Serverless Architectures, 15 October 2018. https://dachou.github.io/2018/10/15/event-driven-serverless.html. Accessed 9 Dec 2020
8. Firebase: Extend Cloud Firestore with Cloud Functions. https://firebase.google.com/docs/firestore/extend-with-functions
9. Firebase: Schedule Function. https://firebase.google.com/docs/functions/schedule-functions. Accessed 9 Dec 2020
10. Microsoft: Develop Azure Functions by using Visual Studio Code. https://docs.microsoft.com/en-us/azure/azure-functions/functions-develop-vs-code. Accessed 9 Dec 2020
11. Firebase: Google Cloud Functions and Firebase. https://firebase.google.com/docs/functions/functions-and-firebase
12. TryQA: What is Load Testing in Software Testing? http://tryqa.com/what-is-load-testing-in-software
13. Getting Started with k6. https://k6.io/docs/getting-started/running-k6. Accessed 9 Dec 2020
14. Businesswire: $7.72 billion function-as-a-service market 2017Global forecast to 2021: Increasing shift from DevOps to serverless computing to drive the overall Function-as-a-Service market. https://bwnews.pr/2G3ZzQY
15. CNCF Serverless Landscape. http://s.cncf.io
16. NGINX: NGINX announces results of 2016 future of application development and delivery survey. https://www.nginx.com/press/nginx-announces-results-of-2016-future-of-application-development-and-delivery-survey/
17. Barga, R.S.: Serverless computing: redefining the cloud [Internet]. In: Proceedings of the 1st International Workshop on Serverless Computing, Atlanta, GA, USA, 5 June 2017. http://www.serverlesscomputing.org/wosc17/#keynote
18. Armbrust, M., et al.: A view of cloud computing. Commun. ACM 53(4), 5058 (2010). https://m.cacm.acm.org/magazines/2010/4/81493-a-view-of-cloud-computing/fulltext
19. Sbarski, P., Kroonenburg, S.: Serverless architectures on AWS with examples using AWS Lambda. In Preparation (2016). https://www.manning.com/books/serverless-architectures-on-aws29
20. Sbarski, P., Kroonenburg, S.: Serverless architectures on AWS with examples using AWS Lambda. In preparation (2016). https://www.manning.com/books/serverless-architectures-on-aws29
21. Castro, P., Ishakian, V., Muthusamy, V., Slominski, A.: The rise of serverless computing. Commun. ACM **62**(12), 44–54 (2019). https://doi.org/10.1145/3368454

Comparative Analysis of Security Protocols in IoT

Saif Saffah Badr Alazzawi(✉) and Tamanna Siddiqui

Department of Computer Science, Aligarh Muslim University, Aligarh, India

Abstract. A comprehensive analysis has been carried out in this paper for security issues and uncertainties for the most common IoT protocols. It discusses the possible security risks and attacks outlined in protocol specifications and categorize them in detail. To provide a valuable perspective into the materialization of and effect of safety risks, these challenges were mostly analyzed in a selective manner, which consists of the assessment of various threats and vulnerabilities [1]. This paper further explores and addresses the steps and methodologies implemented in the literature to increase protection and to minimize the risks involved.

Moreover, this paper analyzes different procedures to assess whether they are susceptible to attack against confidentiality and integrity. Furthermore, the other protocols studied, apart from TLS and DTLS protocols that are not generic security procedures. With spreading of M2M communication there are many standardized communication protocol for IOT applications, the performance of these protocols can vary greatly even under the same operating conditions.

Keywords: Security · Internet of Things · Security layers · Security protocol · CoAP · MQTT · XMPP

1 Introduction

A coherent security approach is complicated because of the heterogeneity of systems and protocols within the IoT scenario [2]. Besides, there are questions about privacy because of the large volume of data obtained from IoT products. New smart devices ensure ease of usage and improved livelihood, but the volume and volume of user data gathered, processed, held and stored at all stages of IoT is a weakness that compromises user data security.

Additionally, for the IoT technology containing various identifying layers of IoT frameworks and architectures, several techniques have been adopted. Previous research [3], proposed a 3-tier model comprising Physical layer, Transportation/Network layer and Application layer, in which the perception (physical) layer depicts the sensors and actuators of the physical (layer), for instance, RFID tags that have a direct interface with the physical environment. The application layer allows the IoT clients to interact with intelligent and smart features. The Network exchanges information from different wireless communication systems between the various layers. More recent literature reveals frameworks for cybersecurity that describe additional layers. A 5-layer architecture with

M. Bhattacharya et al. (Eds.): ICICCT 2021, CCIS 1417, pp. 76–86, 2021.
https://doi.org/10.1007/978-3-030-88378-2_7

a Data Link/Element layer that offers a smart interface between the device and network layers was suggested in [4], where information from the Physical layer is processed over wireless communication channels through services that include parallel computing, data mining and cloud computing (Fig. 1).

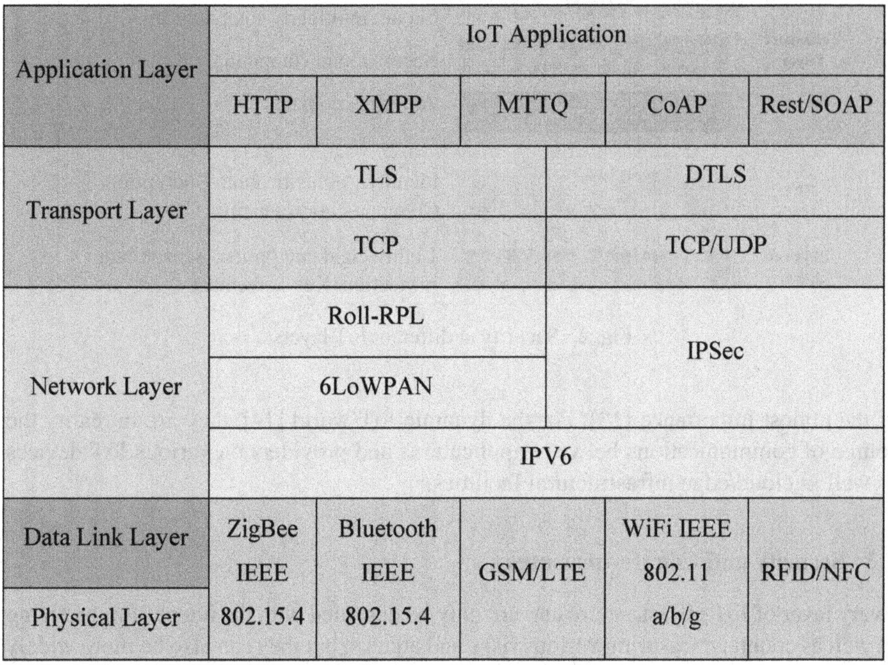

Layer					
Application Layer	IoT Application				
	HTTP	XMPP	MTTQ	CoAP	Rest/SOAP
Transport Layer	TLS			DTLS	
	TCP			TCP/UDP	
Network Layer	Roll-RPL			IPSec	
	6LoWPAN				
	IPV6				
Data Link Layer / Physical Layer	ZigBee IEEE 802.15.4	Bluetooth IEEE 802.15.4	GSM/LTE	WiFi IEEE 802.11 a/b/g	RFID/NFC

Fig. 1. Various layers in IoT architecture

1.1 Challenges in IoT Security

Constraints for security should already be considered and developed during the planning process of new technologies in addition to making operational IoT devices more resistant to cyber-attacks [5]. The broad variability of IoT devices furthermore, prevents the production of well-defined IoT design security protocol [6, 7]. The problem is further exacerbated by extreme limitations in the capability of several IoT systems for capacity, networking, energy, computations as well as memory. These limitations do not allow the utilization of standardized security frameworks in more conventional Internet-connected devices [8] and require new technologies which are not yet established.

IoT systems are based on a range of hardware/software technologies, e.g. software frameworks, repositories, middleware, device constructs, protocols, sensors, including wireless connections, often without the supervision of the agencies or people utilizing these systems [9–11] as well as on various supporting technologies (Fig. 2).

A special account must be taken of future security issues [12], of the heterogeneity of the systems as well as their technologies in practice. Application layer protocols are

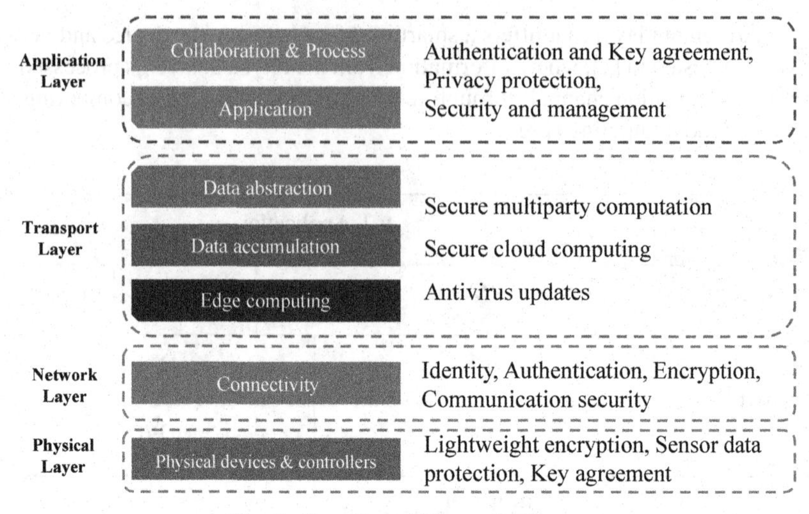

Fig. 2. Security at different IoT layers

of the utmost importance [13]. For the dynamic IoT world [14], they are in reality the source of communications between applications and providers on various IoT devices as well as cloud/edge infrastructural facilities.

1.2 Security and Countermeasures

Every layer of IoT infrastructure can not only be designed with the intent of minimizing as well as counter-measuring various risks and attacks, but they can also be more widely

Layer			
Application Layer	Availability Service	Antivirus Service	Anti-phishing Service
	IDS mechanism	Antivirus mechanism	Spam filtering mechanism
Transport Layer	Availability Service	Access control Service	Auditing log Service
	IDS mechanism	Access control Service	Event monitoring mechanism
Network Layer	Availability Service	Integrity Service	Antivirus Service
	Router filtering mechanism	Data encryption mechanism	Antivirus mechanism
Data Link Layer	Authentication Service	Access control Service	Confidentiality Service
	Digital signature mechanism	Access control mechanism	PKI mechanism

Fig. 3. Services and security mechanisms at IoT layers

perceived through several levels. Few of the commonly recognized mechanism and methods under different IoT layers are given in (Fig. 3).

2 Application Layer Security Protocols

Even though several application layer protocols are present such as MQTT, AES, SNMP, XMPP, etc., the most widely utilized is the Constrained Application Protocol (CoAP). It's also quite lightweight mostly because it operates over UDP (User Datagram Protocol). Another factor, however, is that both unicast and multicast communication mechanism can be handled under the CoAP protocol [15]. Table 1 represents a sample of commonly used security protocols in the application layer.

2.1 CoAP at Application Layer

The Constrained Application Protocol (CoAP) is actualized through the UDP. Based on the UDP IoT application layer decreases the bandwidth requirements as well as facilitates unicast and multi-cast communication channels, the aim of CoAP is to provide infrastructure for resource-restricted devices such as mobile phones. The CoAP protocol offers a conceptual framework of communication for "requests and responses" between the endpoint devices and uses the AES encryption as the cryptographic algorithm for delivering security interface services including integrity, confidentiality and authentication. Additionally, to enable secure transmission CoAP often utilizes universal resource identifiers (URIs) to gain access to the services and resources on a sensor point or interfaces like HTTP. The CoAP protocol's biggest benefit is that the total charge is low since it runs through UDP, and not TCP [16, 17].

3 Encryption and Cryptography

It is the most critical and principal activity during the correspondence to ensure secrecy. The original message (plaintext) can be converted to another message (Ciphertext), using a hash function which can only be reversed quickly knowing a hidden key.

The cryptographic encryption/decryption technology provides an excellent means of exchanging messages among users and without the unwanted threat of losing the messages to unwelcome agents. This type of encryption/cryptographic functions is achieved using a particular series of algorithms in different ways [18–20]. As such cryptographic libraries intended to be used in microprocessors and embedded systems, the required security layers for IoT devices are given as well as the overall IoT infrastructure is shielded.

Nevertheless, data between devices must also be encrypted to prevent abuse and also to protect the safety, integrity and authenticity of information exchanged, whether convergent or asymmetrical encryption is usually debated, but algorithms that use fewer resources are typically favored because of system limitations. A potential unauthorized person can only reach the ciphertext through unauthorized access, but he should not be able to read the message data. It may be encrypted using the symmetrical or asymmetrical method for encryption. In symmetric encryption method, both the sender and the recipient will have to have the same cryptographic key to use to encrypt and decrypt the data for communications.

3.1 Point-to-Point and End-to-End Security

End-to-end (E2E) security mechanism mitigates the possibility of WiFi inter-device connectivity, irrespective of the protocols implemented; although, based on the protocol(s), separate suites should be used in the respective layers [21]. Likewise, point to point networking methods, which could be VPNs, IPSec or MPLS, offer an authentication equivalent to E2E, but have higher energy usage requirements. Also, it was noted that the prevention of tertiary vulnerabilities is a core strength of end-to-end security confidence models. Final to full protection was also suggested as a way to protect data privacy and confidentiality in a P2P communication network for IoT environment, while this type of networking architecture is not necessarily dependent on such protocols.

3.2 Authentication

IoT authentication is due to the number of computers one of the main challenges. It's not easy to authenticate every unit. Many safety protocols have been suggested because of the characteristics of the quick calculation and energy consumption, based on private, key cryptographic primitives [22].

4 Security in Message and Service Discovery

A significant component of this type of IoT environment is perhaps the parameters for communications at the application layer because it encapsulates all the connections between IoT devices as well as between IoT devices including cloud/edge infrastructure [23, 24]. Therefore, messaging and service discovery are the standard tasks introduced by these protocols. Messaging applies in particular to data exchange and sharing within device-to-device communications when exploring technologies and facilities that are implemented.

4.1 Message Queue Telemetry Transport (MQTT)

For the lightweight machine-to-machine telecommunications, IBM's Message Queue Telemetry Transport (MQTT) protocol was designed. It actuates over the TCP protocol as a basis for applying an integrative interaction approach to publish/subscribe [25]. This model has been picked since hardware components will not need upgrades individually, which will potentially minimize resource and energy drainage at the IoT connected nodes.

Table 2 identifies the major characteristic features of various security protocols that have been standardized for utilization in IoT infrastructure, for instance, 5 messaging protocols (i.e. CoAP, AMQP, MQTT, DDS and XMPP), as well as 2 service discovery protocols, have been identified for various standard protocols (i.e., mDNS and SSDP). Figure 4 represent potential use of various security protocols under different layers of IoT framework.

Application Layer	AMQP	ISO	IETF	ITU SG20
	OASIS	MQTT	CoAP	SC&C
Service Layer	Microsoft	Intel	AT&T	Novell
Network Layer	IETF	Thread (Group)	OIC	
	6LoWPAN	Thread	IoTivit	
Data Link Layer	IEEE 802.15.4			

Fig. 4 Various protocols at different levels

Thus, for certain purposes, including various architectures, communication and transport models, these protocols are distinctive. Some protocols are based on unified architectures, including client/server and architectures. For instance, the broker plays the server function and communicates with customers by receiving and transmitting messages for protocols like MQTT and AMQP. Message exchanges are typically carried out following templates for subscribing/publish or request/response mechanism.

5 Service Discovery Protocols

Multi-cast Domain Name Service (mDNS) and Domain Name Service Discovery DNS-SD are common IoT protocols which deal with applications of this type. Multicast DNS is good for locally saved stuff networks since no details must be hunted. There already is data available that is considered necessary and would be available without significant machine resources [5]. mDNS gets the names it needs from a straightforward multi-cast message that is transmitted to the networked system [5].

5.1 Simple Service Discovery Protocol (SSDP)

SSDP is placed under an open protocol that is generally utilized in domestic or small-scale business environments (UPnP Forum) to explore services and also in advertisings. The protocols in the UPnP architecture make possible remove and play devices transparently that don't require any manual setup. The SSDP protocol is very poor as regards stability, close to mDNS, because it doesn't have any incorporated mechanism. Various protection threats, therefore, impact devices that are SSDP-enabled. These threats normally take advantage of the characteristics of operation exploration.

A major challenge faced by SSDP based clusters is a distributed denial-of-service (DDoS) attack by amp/reflection that aims at rendering devices and resources inaccessible. Such attacks target the features of the UDP as well as SSDP protocols also capable of producing even some system configuration issues for devices. More adequately an attacker might use manipulated IP address nodes influenced by attack to generate an M-SEARCH request.

5.2 Extensible Messaging and Presence Protocol (XMPP)

With SASL (Simple Authentication and Security Layer) for the process of authentication and with TLS to ensure data protection, privacy and integrity, XMPP offers solid security services. These facilities are included in the core protocol criteria and are hence allowed by default. The protocol is therefore vulnerable to different forms of threats because of lack of end-to-end encryption protection. An attacker may, for example, change, erase, play or obtain an unwanted server entry. In addition to protocol protection concerns, several bugs concern XMPP-based devices and services. In particular, significantly less than 100 CVEs have been identified in the last few years, primarily concerning to message validation and authentication processes.

6 Transport Layer Security Protocols

The security standard most often used for this layer is the internet Transportation Layer security (TLS) protocol as well as the Datagram Transport Layer Security (DTLS) for IoT devices. TLS is a standard intended to protect stable protocols for transport, such as the TCP. TLS requires two devices to exchange a common key on the Internet and then establish a protected contact channel [7, 12]. In general, TLS allows for two devices. Even though considering various framework implementations, still this protocol is very resource-intensive, CoAP based IoT devices cannot implement this security mechanism. Another cause is that TLS operates on reliable messaging networks, which is unsafe for IoT environments.

DTLS is a protocol that imitates TLS via an insecure channel, such as the UDP protocol. The biggest drawback of the DTLS protocol is nevertheless that this protocol was not exclusively developed for the IoT applications. Thus as a consequence, multi-cast messaging is not supported [26]. Furthermore, it means that CoAP based IoT systems will have to have lengthy authentication keys that have been used for the previous applications utilizing the said protocol. A further limitation of the DTLS protocol is that under various conditions an exhaust attack can occur on a constrained machine under the DTLS handshake mechanism [27].

7 Security in Cloud Infrastructure

7.1 Thingplug

Several security features are supported for secured communications in the LoRa network, including a MAC, AES as well as ADR (Adaptive Data Rate) [10]. Further, Thingplug supports HTTP and MQTT protocols for accessing one machine-to-machine REST definition based API from the outside to acquire these security and confidentiality based services. It still does not, however, accept CoAP since it utilizes its exclusive GMMP, a short as well as low power protocol (global M2M protocol). Moreover, there are not several market-specific CoAP implementations for commercial use, but in different facilities GMMP has been used before actively [11, 28]. TLS and the data encryption system for data privacy and confidentiality was used under the transport layer protocol that secures data with the help of a standard domestically standard encryption algorithms as AES, the ECC (elliptic curve cryptosystem) and ARIA.

7.2 Cryptographic Libraries in IoT

Owing to its limited footprints as well as low execution time contained in ANSI C the cryptographically library "wolfSSL" is intended to be utilized within embedded systems, in RTOS (Real Time Operation System) and in areas where computing resources are restricted [10]. WolfSSL (formerly referred to as CyaSSL) This library facilitates the creation of platform-to-platform algorithms and also includes several other algorithms.

AvrCryptoLib: the cryptographic library "AvrCrypto-Lib" approved under GPLv3 has the application of its 8-bit micro-controller AVR for cryptographic protocols and encryption algorithms. Which is similar to all of the other cryptographic library programs utilized in the IoT area.

WiseLib: The cryptographic library of "Wiselib" has been written in C++ which is intended for use with the embedded network infrastructures.

8 Analysis and Discussion (Comparison of Security Protocols in IoT)

Table 1. Commonly use application layer protocols with security implements

Protocol	Transport	QoS	Communication/Messaging	Security
CoAP	UDP	Yes	Request/Response Publish/Subscribe	DTLS
MQTT	TCP	Yes	Publish/Subscribe	TLS/SSL
XMPP	TCP	No	Request/Response Publish/Subscribe	TLS/SSL

QoS (Quality of Service)

Table 2. Comparison of various protocols under the application layer in IoT

Protocol	Standard	Function		Architectural model		Interaction model		Transport protocol	
		Messaging	Discovery	c/s	Decentralized	Pub/Sub	Req/Resp	TCP	UDP
CoAP	IETF	+	#	+		#	+	#	+
MQTT	OASIS	+		+		+		+	
DDS	OMG	+	#		+	+	#	+	+
XMPP	IETF	+	#	+		+	+	+	
AMQP	OASIS	+		+		+	#	+	
mDNS	IETF		+		+		+		+
SSDP	UPnP		+	+			+		+

+: native features
#: added functions

Table 3. Comparing various security protocols in IoT

Protocol	Transport	Messaging	2G-4G (1000's)	Low Power & Lossy (1000's)	Compute Resources	Security	Implements	Archit
CoAP	UDP	Rqst/Resp	Excellent	Excellent	10Ks/Ram Flash	DTLS Medium-Optional	Utility field area	Tree
DDS	UDP	Rqst/Resp Pub/Subscb	Fair	Poor	100Ks/Ram Flash ++	High-Optional	Military Industries	Bus
MQTT	TCP	Rqst/Resp Pub/Subscb	Excellent	Good	10Ks/Ram Flash	TLS/SSL Medium-Optional	IoT Messaging	Tree
XMPP	TCP	Rqst/Resp Pub/Subscb	Excellent	Good	10Ks/Ram Flash	TLS/SSL High-Optional	Management White Gds	Client-Server
SNMP	UDP	Rqst/Resp	Excellent	Fair	10Ks/Ram Flash	High-Optional	Network Monitoring	Client-Server
Azure-IoT	Https/TCP	Rqst/Resp	Excellent	Good	10-100Ks RAM Flash	TLS/SSL High-Mandatory	Werables	Client-Server
UPnP	UDP	Rqst/Resp Pub/Subscb	Excellent	Good	10Ks/Ram Flash	High-Mandatory	Consumer goods	P2P/Client-Server
HTTP	TCP	Rqst/Resp	Excellent	Fair	10Ks/Ram Flash	Low Optional	Smart Energy Phase2	Client-Server
Thread	UDP	Rqst/Resp	Excellent	Excellent	10Ks/Ram Flash	High-Mandatory	Nest	Mesh
DPWS	TCP	Rqst/Resp	Good	Fair	100Ks/Ram Flash++	High-Optional	Web server	Client-Server

Table 3 briefs the comparison of a few of the commonly utilized security protocols and communication models and transport channels implemented under various cases.

Consequently, performance is a priority when we rely on resource-limited machines. Thus while creating the security protocols, the nodes should be secured by transferring the workload as close to the confidentiality and privacy standards. Moreover, these should be energy efficient and lightweight as possible as it is usually meant to be implemented in a resource-limited IoT device, although, not necessarily requiring frequent updates.

9 Conclusion

For purposes of protection and accessibility, the relationship between the broader Internet and the Internet of Things has remained tenuous. The creation of network-based APIs for devices accessing the Web securely for practical purposes and conceptual implementations of TCP as a standard protocol on transportation layer was opened up, based upon previous implementations in this area. Context-aware, as well as lightweight IoT models, have been presented to protect the privacy and more recently virtualization techniques are implemented to maintain the data confidentiality. New solutions are required to consume limited resources of an IoT for lightweight cryptographic primitives. Besides, with the aid of unified routing on the SDN (Software Defined Networking) controller, the SDN solution offers to introduce lightweight cryptography implementations through IoT. Due to heterogeneous security threats, IoT network failure is correlated with IoT nodes, and the use of the above purpose along with the adoption of various security measures that can be adopted in various communication modes and implementation of intrusion detection systems and schemes will lead to a more secure and reliable IoT infrastructure for wide adoption of its security devices.

References

1. Ammar, M., Russello, G., Crispo, B.: Internet of Things: a survey on the security of IoT frameworks. J. Inf. Secur Appl. **38**, 8–27 (2018)
2. Kalra, S., Sood, S.K.: Secure authentication scheme for IoT and cloud servers. Pervasive Mob. Comput. **24**, 210–223 (2015)
3. Kumar, N.M., Mallick, P.K.: The Internet of Things: insights into the building blocks, component interactions, and architecture layers. Proc. Comput. Sci. **1**(132), 109–117 (2018)
4. Mrabet, H., Belguith, S., Alhomoud, A., Jemai, A.: A survey of IoT security based on a layered architecture of sensing and data analysis. Sensors **20**(13), 3625 (2020)
5. Das, A.K., Zeadally, S., He, D.: Taxonomy and analysis of security protocols for Internet of Things. Future Gener. Comput. Syst. **89**, 110–125 (2018)
6. Sequeiros, J.B., Chimuco, F.T., Samaila, M.G., Freire, M.M., Inácio, P.R.: Attack and system modeling applied to IoT, cloud, and mobile ecosystems: embedding security by design. ACM Comput. Surv. (CSUR) **53**(2), 1–32 (2020)
7. McManus, J.: Security by design: teaching secure software design and development techniques. J. Comput. Sci. Coll. **33**(3), 75–82 (2018)
8. Wu, X.W., Yang, E.H., Wang, J.: Lightweight security protocols for the Internet of Things. In: 2017 IEEE 28th Annual International Symposium on Personal, Indoor, and Mobile Radio Communications (PIMRC), pp. 1–7 (2017)

9. Silva, B.N., Khan, M., Han, K.: Internet of things: A comprehensive review of enabling technologies, architecture, and challenges. IETE Tech. Rev. **35**(2), 205–220 (2018)
10. Dhanvijay, M.M., Patil, S.C.: Internet of Things: a survey of enabling technologies in healthcare and its applications. Comput. Netw. **22**(153), 113–131 (2019)
11. Sankar, S., Srinivasan, P.: Internet of things (IoT): a survey on empowering technologies, research opportunities and applications. Int. J. Pharm. Technol. **8**(4), 26117–26141 (2016)
12. Ojo, M.O., Giordano, S., Procissi, G., Seitanidis, I.N.: A review of low-end, middle-end, and high-end IoT devices. IEEE Access **9**(6), 70528–70554 (2018)
13. Hassan, A.M., Awad, A.I.: Urban transition in the era of the internet of things: social implications and privacy challenges. IEEE Access **18**(6), 36428–36440 (2018)
14. Samaila, M.G., Neto, M., Fernandes, D.A., Freire, M.M., Inácio, P.R.: Challenges of securing Internet of Things devices: a survey. Secur. Priv. **1**(2), e20 (2018)
15. Siddiqui, T., Alazzawi, S.S.B., Khan, N.A: Generalization of IoT applications: systematic review. Int. J. Sci. Res. Comput. Sci. Eng. Inf. Technol. IJSRCSEIT **3**(5), 688–694 (2018). ISSN 2456-3307, (Impact Factor 6.135) UGC Journal No: 64718
16. Keoh, S.L., Kumar, S.S., Tschofenig, H.: Securing the internet of things: a standardization perspective. IEEE Internet Things J. **1**(3), 265–275 (2014)
17. Siddiqui, T., Alazzawi, S.S.B.: Comparative analysis of Internet of Things (IoT) security models. In: Luhach, A.K., Jat, D.S., Bin Ghazali, K.H., Gao, X.-Z., Lingras, P. (eds.) ICAICR 2020. CCIS, vol. 1394, pp. 186–196. Springer, Singapore (2021). https://doi.org/10.1007/978-981-16-3653-0_15
18. Dey, S., Al-Qaheri, H., Sanyal. S.: Embedding secret data in HTML web page. arXiv preprint arXiv:1004.0459 (2010)
19. Caviglione, L., Merlo, A., Migliardi, M.: Covert channels in IoT deployments through data hiding techniques. In: 2018 32nd International Conference on Advanced Information Networking and Applications Workshops (WAINA), pp. 559–563 (2018)
20. Lin, Y.H., Hsia, C.H., Chen, B.Y., Chen, Y.Y.: Visual IoT security: data hiding in AMBTC images using block-wise embedding strategy. Sensors **19**(9), 1974 (2019)
21. Sridhar, S., Smys, S.: Intelligent security framework for IoT devices cryptography based end-to-end security architecture. In: 2017 International Conference on Inventive Systems and Control (ICISC), pp. 1–5 (2017)
22. Cabrera, C., Palade, A., Clarke, S.: An evaluation of service discovery protocols in the internet of things. In: Proceedings of the Symposium on Applied Computing, pp. 469–476 (2017)
23. Dizdarević, J., Carpio, F., Jukan, A., Masip, X.: A survey of communication protocols for internet of things and related challenges of fog and cloud computing integration. ACM Comput. Surv. **51**, 116 (2019)
24. Naik N.: Choice of effective messaging protocols for IoT systems: MQTT, CoAP, AMQP and HTTP. In: 2017 IEEE International Systems Engineering Symposium (ISSE), pp. 1–7 (2017)
25. Karagiannis, V., Chatzimisios, P., Vazquez-Gallego, F., Alonso-Zarate, J.: A survey on application layer protocols for the internet of things. Trans. IoT Cloud Comput. **3**(1), 11–17 (2015)
26. Seggelmann, R., Tuexen, M., Williams, M.: Transport layer security (TLS) and datagram transport layer security (DTLS) heartbeat extension. Request for Comments (RFC) **6520**, 1721–2070 (2012)
27. Sheffer, Y., Holz, R., Saint-Andre, P.: Summarizing known attacks on transport layer security (TLS) and datagram TLS (DTLS). Internet Engineering Task Force Request for Comments: 7457 (2015)
28. Siddiqui, T., Alazzawi, S.S.B.: Security of Internet of Things. Int. J. Appl. Sci. - Res. Rev. **5**(2:8), 1–4. https://doi.org/10.21767/2394-9988.100073. ISSN 2394-9988

Secure Data Transmission Techniques for Privacy Preserving Computation Offloading Between Fog Computing Nodes

Yash Ketan Patel(✉), Krunal Dipakbhai Patel(✉), and Payal Chaudhari(✉)

Pandit Deendayal Energy University, Gandhinagar, India
{yash.pce17,krunal.pce17,payal.chaudhari}@sot.pdpu.ac.in

Abstract. With the advent of technology to digitize the world, an increase in the storage & computational power to store and process the data requirements of the consumers brought Cloud Computing in existence. But computation on cloud is expensive and continuous expansion of this field leads to an arising security concern. This led to the creation of Fog Computing, a layer that transcends the barrier between cloud and edge devices, that operates on most of the end user's data at the lower level itself, minimizing the data sent to the cloud, hence reducing the computational cost and ensuring more security. There are still some prevalent issues in the basic foundations of the Fog Layer that pose a threat in its functionality. This paper proposes a unique architecture that solves some of the fundamental privacy and security issues of Fog Layer without affecting its baseline performance and efficiency. The architecture relies on establishing a layer of Data-Tables over the communicating nodes and a novel Encryption Procedure encompassed within the data transfer mechanism. The objective is to ensure a smooth flow of computation and data transfer while providing the necessary privacy and security standards which is quite essential nowadays in the paradigm of Fog Computing.

Keywords: Fog computing · Privacy · Security · Cloud computing and edge devices · Software architecture · Encryption · Cryptography

1 Introduction

An energy company "Envision" that drives its agendas from non-conventional renewable energy operates on around 20,000 turbines and approximately 3 million sensors accumulating about 20 TB of data at specified intervals of time. Moving such a high amount of data to a data center takes time and bandwidth, this is where Envision implemented fog computing to enhance the overall efficiency and make the system less expensive in transaction-based cost-intrinsic operations of cloud computing [1].

Fog computing is like a bud in a developing plant and hence it offers a million opportunities to operate and manipulate in areas which explore a completely different paradigm of possibilities that focuses on decreasing cost and increasing productivity than working only in cloud & edge computing.

M. Bhattacharya et al. (Eds.): ICICCT 2021, CCIS 1417, pp. 87–101, 2021.
https://doi.org/10.1007/978-3-030-88378-2_8

Cloud computing is predominantly a world of hassle-free infrastructure-less management, but the cost of operations based on the number of transactions is significantly higher when a project is involved in business ventures that involve constant entanglement of transactions between their application servers and clients.

Security and privacy issues are seen to be encountered by most of the systems in this ever-expanding digital arena and hence some provisions are to be made for strengthening the overall system technically and also ensuring people's acceptance into this budding technology. The excerpt of the idea for the proposed approach in this paper is to deal with different issues encountered while ensuring privacy and security of the network and its nodes, such as encrypting in a very low specification hardware and software compliance, maintaining the overall performance and efficiency just as it was, before integrating the proposed architecture and algorithms into original Fog foundations.

2 Literature Survey

A literature survey of existing techniques was carried out for secure offloading of computation among fog nodes. After analysis, the results attained points out the limitations presented in existent research work. This motivates us to formulate the research problem and design its solution.

Why Fog Computing with Cloud Computing: [8] The resources at the fog level are very scarce as compared to that at the cloud level. So, fog computing cannot completely replace the cloud computing. The fog network can be used in minimizing the computations done at the cloud. The cloud would only be used to store the data as a gargantuan data storage and processing facility whereas smaller operations can be grouped and regrouped by keeping in track the dependencies among them and process these bunch of smaller operations as a group of clusters at the fog level. If huge computations are to be done, which could not be carried out at the fog level, then only the data would be processed at the cloud.

How the Files are Transferred Between the Devices via Wi-Fi: With Wi-Fi direct (P2P) [5] the android devices can connect over a wireless network to exchange their data via Wi-Fi without any intermediate access point. With Wi-Fi direct the devices are capable of sharing their data over much larger distances than those supported by Bluetooth.

How the Server is Deployed and How it Manages the Nodes in the Fog Network: Fog server can be deployed on any network device which has the minimum distance from all nodes. However, minimal research has been conducted on where to deploy the fog server. Many factors need to considered before any further materialistic development on the topic of fog server deployment. The fog server deployment algorithm cited in the paper - [4] considers the data movement distance and computing resource in fog computing environment, and illustrates how it is deployed by simple example.

Light-Weight Encryption Algorithms: As the fog level has lower computational capability than the cloud layer, the encryption algorithms used in fog are required to have lower computational costs. [7] This cited paper portrays a hybrid proxy method of encryption that entangles lightweight symmetric & asymmetric encryption algorithms to create end-to-end secure communication related to transactions in fog level to IoT (Internet of Things) computing. [7] It claims to have high efficiency in the computational cost.

Various Contributions Related to the Paper's Objective: [9] This paper explains various security threats and challenges in fog computing and securing the data using ECC (Elliptic Curve Cryptography)). [10] Here, the Cipher text policy attribute-based encryption (CP-ABE) is used to establish secure communications among the participants to achieve confidentiality, authentication, verifiability, and access control, we combine CP-ABE and digital signature techniques. [11] This paper gives the basic understanding of fog computing and its various applications. [12] This shows various privacy and security challenges faced in cloud computing.

3 Problem Formulation

After a research of different areas concerning the privacy and security issues faced by different progenitor applications that were the stepping stones of what fog computing thrives on and keeps progressing into vast paradigm of customized and complex applications, some of the key issues that remain at the core of fog computing were sorted among many other prevalent issues to be resolved.

3.1 Privacy Issues

- Issue-1: A Data privacy breach arises when the data from a particular node is transferred to a neighbouring fog node for computation offloading and that neighbouring node disappears after receiving the data, without participating in computation. It means that during the process of completing the task, if the neighbouring node leaves the task with data of the particular node then a privacy breach is encountered. In fog computing, the processing is ensured to be performed at this level itself and hence the nodes connected in fog network release some computation and storage resources for facilitating in execution of tasks of its fellow nodes. If a neighbouring node (say node-1) has been assigned as a peer by the fog server node to share its resources for a data processing request announced by a particular node (say node-2), then the data sent to node-1, for processing is stored in its temporary storage. In this situation, if node-1 leaves the network during the execution of node-2's requested task then, node-1 carries with itself the private data of node-2. Hence it is a breach of privacy.
- Issue-2(External node stealing data): While transferring data from node D-1 to D-4(say), if the node D-4 is not an immediate neighbour of the sender node(D-1), then "routing" has to be done between inter-fog networks or intra-fog networks, to get to the destination node D-4. In that case, a hideous node may try to steal data from between the routing nodes i.e. if data from node D-1 goes through two adjoining networks and reaches D-4, then in between, another node, say D-15, may try to steal or interfere with the data.

3.2 Security Issues

- Issue-1: If any of the nodes try to communicate a suspicious file or a virus-equipped file in the data transfer then, such file should be traced back to the original sender, and sometimes such nodes route the virus through different nodes, to refrain getting caught. Such scenarios have many diverse and varying consequences to interconnected networks and hence can take down a cloud too. In such scenarios, there are two cases:
  - Case-1: Here, the sender node, which sent the virus is the direct sender and does not involve any routed paths to the target node. Hence tracking the original sender node is much easier than Case-2.
  - Case-2: Here, the sender node is multiple hops away from the target node and hence the network needs to backtrack the whole route.
- Issue-2: Any device or node can be tampered physically at the installed location. As all the nodes(devices) are connected to the fog network, the network can try to ensure the device's physical safety at the fog level.

4 Methodology

The problems mentioned in the above section seem to be pretty basic and simple in the context of small networks, but as soon as the networks merge or any inter-network or intra-network operations arise then such small flaws lead to destructive directions or uncertain paths where it shall not proceed in contrary to the predestined flow.

Hence to approach such cardinal issues, a systematic architecture has to be developed that forms a basis to solve all the listed problems simultaneously and in turn yield a smooth flow just as it was before integrating the proposed solution, without any lag/jitter/delay to be incurred due to the architecture other than the natural network issues.

The architecture to be developed is divided into phases of many operations applied at different levels and stages in the project in the form of dependent and independent modules (to maintain agile development flow), only to be linked together as a single functioning module or an operational prototype at the denouement of the project.

4.1 Phase I (Explaining the Basic Architecture)

1) Components: There are four main components comprising a network of connected nodes and Data-Tables: Nodes D-1 to D-8 are connected via WiFi-Direct to each other. That is, all nodes from D-1 to D-8 can communicate with each other independently for file/message transfer. One can visualize it similar to having a special link from each node to all other nodes.

- **Request Table (ReqT):** This table stores the records of all the requests made by different nodes. The request contains the information about: The "Operation" to be performed, the "Characteristics of the Data" to be transferred to other nodes for processing. the "Minimum Requirements" related to the current state and configuration of the device/node to perform the specified operation.

- **Resource Table (ResT):** This table holds the data of current statistics about the RAM, free storage space, and number of CPU cores along with CPU's configuration of all the nodes that are present in the network.
- **Allocated Relationship Table (ART):** After the data provided in the request (from "Request Table") of a particular node is divided into different sets, the information about which set of the data is to be sent from the Request-node to the another Allocated-nodes for processing is mentioned in this table.
- **Broadcast Table (BrT)**: This Table involves records containing information about any notices or alerts to be broadcast to all the other nodes. It is used in solving Security Issue-2 for broadcasting an alert if a node finds any changes in the Board/Port information of its assigned node.

2) Architecture Concept

- As shown in Fig. 1 D1 to D8 are Fog nodes and there are two tables namely, Resource Table & Request Table.
- Resource Table states the free resources currently available. It states the requests that have occurred in the fog network and the current status of the requests: For example, if node D-1 has a task of "Stitching 2 frames into 1 single frame". i.e. Pic1(1) & Pic2(1) frames are to be stitched to stitch-1 and similarly Pic1(2) & Pic2(2) frames are to be stitched to stitch-2 and node D-1 has such 100 frames i.e. Pic1(1) to Pic1(100) & Pic2(1) to Pic2(100), then at the end it has to have a total 100 stitched frames. So, if node D-1 itself processes 44 frames then it leaves out a request to process remaining 66 frames to other nodes which is depicted in the Fig. 1's "Request Table".
- After floating these requests in the network as shown by the Green Diamond <1> in Fig. 1, the Resource Availability Table is checked for nodes that satisfy the minimum resource requirements of the task and filters out the devices to whom the request is to be broadcasted as shown by Green Diamond <2> in Fig. 1. Note that the request is broadcasted to D-3, D-4 in this example. After broadcasting the request, the requested task is divided into independent sets/modules according to the processing power and available storage capacity of D-3 and D-4, illustrated by the Green Diamond <3> in Fig. 1.
- Here, D-4 has higher specification of free resources (RAM-8 GB, Storage: 80 GB, Cores: 8), so it is allocated 44 frames to process out of 66 frames and the request table is updated as shown in Fig. 2.
- Then the remaining 22 frames are allocated to D-3 for processing, which is next in order of higher to lower specifications of free resources.
- The Request table is updated and the information about which node is processing which set/module is stored in the Allocation Relationship Table for maintaining track of relations among nodes and their allocated sets of data.

3) Encryption Procedure

Generally, while using java.io.InputStream and java.io.OutputStream to transfer data bytes from one node to another, The data file is divided into packets of bytes, having packet-size in the multiples of 1024. For example, a total of 1024 bytes are sent from the client to the server and each byte consists of 8 bits. Now, 3 numbers are to be randomly

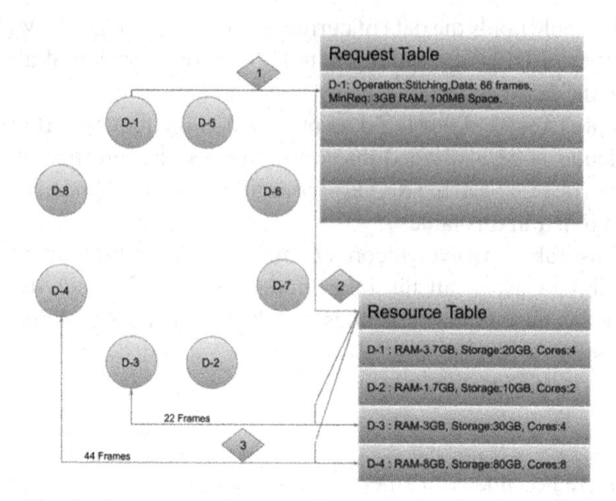

Fig. 1. General Architecture: Network of nodes, ReqT, & ResT

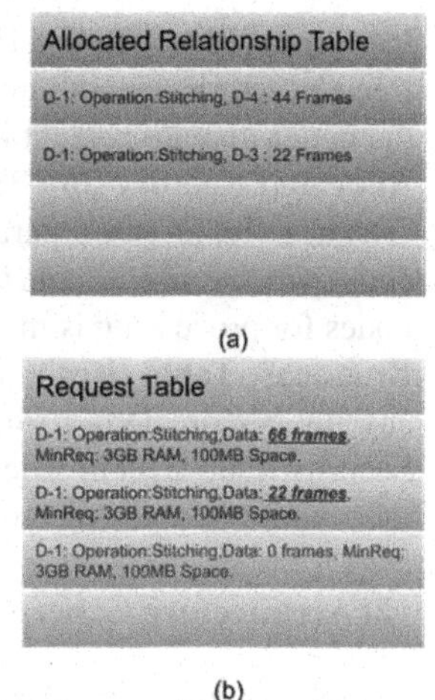

Fig. 2. Updated Status of (a) ART, (b) ReqT after allocating frames

selected from (1 to 1024) to form a key pair (here the key pair used in the example is shown in Fig. 3: 6,28,84) and from each of these selected bytes, a random bit number is flipped (here bit number 4), i.e. if bit value is 0, then it is set to 1 and vice versa. The algorithm generates key pairs periodically after 30 s (which can be customized and set according to user preference). Hence the key pair generated in each transaction between client and server will practically be different.

According to calculations: 178,490,368 trials will have to be made to find this particular set of 3 digits within 30 s of time, which is near to impossible. Because, even if the combination is detected after a lot of time, the key existing at that moment of time will have been completely changed and hence interpreting the intercepted communications will be very difficult.

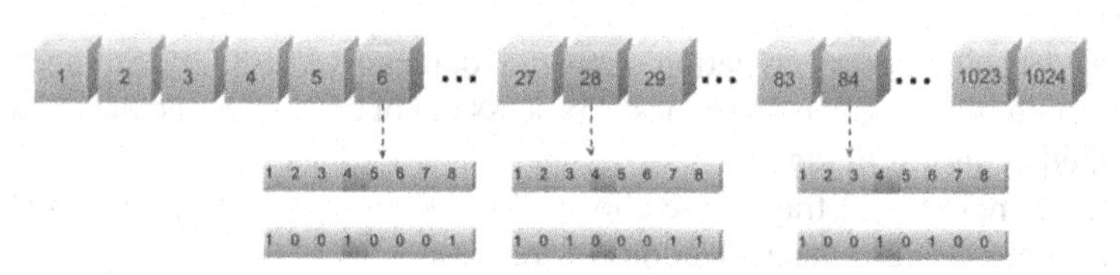

Fig. 3. Value of bit 4 of Byte Number 6,28,84 i.e. 1,0,1 will be flipped to 0,1,0 after encryption respectively

4.2 Phase II: Architecture Explanation

A) Privacy Issues:

1) Privacy Issue-1

After the architecture setup mentioned in Sect. 4.1 has been established, the main algorithm and technique of ensuring privacy is implemented.

In this technique, as soon as the task is allocated, the cryptographic algorithm is set. Wherein, the nodes participating in the fog network have an application that has two functional codes: sending encrypted data packet, decrypting received packet.

Suppose the sender is sending images to be stitched to the receiving node. Now, if the images are sent directly to the receiving node, then it can take the images and escape the operation and that would be a breach of privacy. To avoid such malpractices, the encryption algorithm proposed is applied.

- Whenever the sender sends the image/file/raw data etc., it passes through the encryption procedure mentioned in Sect. 4.1.
- After encryption, the data packets are sent to the receiver, and the encrypted content avoids eavesdropping of data while transferring it from one node to another.
- The receiver gets the encrypted file and the node has to check the global key-combination table, which has key-combination values like (6,28,84) provided by each node for a duration which can be customised by the node and which changes periodically. Note that entry of key combination in the table will only be visible to the node which has accepted the request for participating in the operation.
- Apart from this, the receiver cannot directly interpret the data received and store it somewhere else for misuse, because the algorithm works secretively and the receiver wouldn't even know when the operation gets completed. The file saved on the receiver's end is an encrypted one and is decrypted only inside the application whenever it has to be processed, hence any input/output stored in the storage disk on the receiver's end is always encrypted.
- Also, as the file saved on the receiver's end is an encrypted one, even if the receiver terminates the operation and gets disconnected from the network with the sender's file, the receiver won't be able to use it or extract something anyway as that node won't have the key combination.
- Along with this, after the data processing finishes and a response is ready, the response output is sent in the reverse way as the process mentioned above, it is only that in this response transfer, the acknowledgement message with the response is encrypted through the application and then sent to the previous sender, where the sender receives the package through the application after getting the decrypted data by the algorithm.
- Hence, the main concept of framing the architecture in this way is that no particular node can steal data from other nodes, and hence there should not be even a slightest probability or chance for them to try and steal data from other nodes for malpractices.

2) Privacy Issue-2

- If data from node D-1 goes through two adjoining networks and reaches D-4, then in between, another node, say D-15, may try to steal or interfere with the data. Then, such an attack is not possible by the architecture mentioned in Fig. 1. As, every request is recorded in the Request Table, and every node that has accepted the request is documented in the Allocated Relationship Table. So, if a node, say D-15, tries to take the data packets from in between router or nodes then, it is checked for its authenticity in the Tables.

- Note that, D-15 is a node that is trying to take data, hence it is trying to accept data packets and, nodes that have access to the data are documented in the Allocated Relationship Table from the very beginning. However, an entry of D-15 does not appear in the Table and hence is not allowed to touch any data packets in the transfer happening from D-1 to D-4. Also, even if D-15 intercepts the package, the data is still of no use to the node, as it would receive the encrypted packet and won't have the decryption key that is only accessible to the nodes that are actually listed in the Allocation Relationship Table as shown in Fig. 2.

B) Security Issues:

1) Security Issue-1

- The algorithm to trace back to the original node is shown in Fig. 4. In reference to Fig. 4, the colour coding is as follows: Green nodes: Safe Communication [D-1, D-3, D-4], Red nodes: Virus Carriers (Original) [D-6, D-7], Orange nodes: Directly affected nodes [D-2, D-8], Yellow nodes: Indirectly affected nodes & hence secondary virus carriers [D-5].
- These scenarios have many diverse and varying consequences in interconnected networks which can even take down a cloud network. There are two cases and each can be resolved by using the architecture mentioned in Fig. 4.
- Case-1: Here, the sender node, which sent the virus is the direct sender and does not involve any routed paths to the target node. Hence tracking the original sender node is easier than Case-2.

  - D-6 is a directly affecting virus node, hence its target node will communicate the alert, i.e. D-2 itself. So, in those cases, the trace-back from D-2 to D-6 can be done easily by checking the parent address of the Allocated entry of D-2(the affected node) in the Allocated Relationship Table. The entry in Allocated Relationship Table will obviously show D-6 to be D-2's parent as shown in 3rd row of Allocated Relationship Table in Fig. 4 and as D-6 does not have any incoming edges, it is deemed to be the virus commuter.

- Case-2: Here, the sender node is multiple hops away from the target node and hence the network needs to backtrack the whole route.

  - Now, node D-7 is an indirect virus node that aims to target the node D-5, not directly, but by enacting a decoy entity in the process and makes it look like some other node is the original sender and not itself. In the example shown above, the node D-7 first sends a request in Request Table, which is accepted by D-8 (Here D-8 is not the target node, still it will also get infected). Afterwards, D-7 is waiting for D-8 to initiate a request and get it accepted by D-5. Hence the virus's first entry point in the network started through D-8.

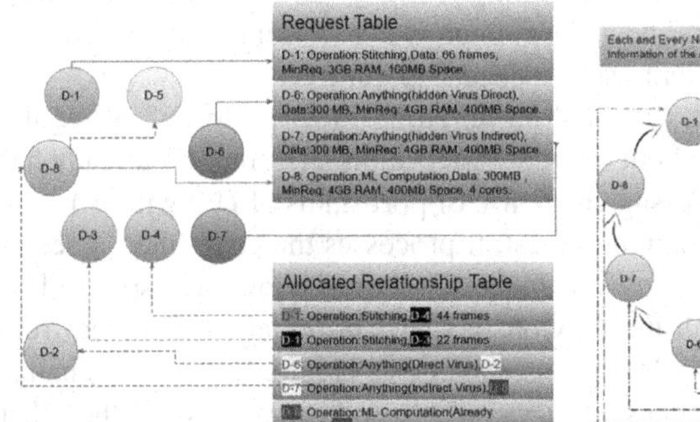

Fig. 4. Another Form of the Architecture used to resolve Security Issue-1 (Color figure online)

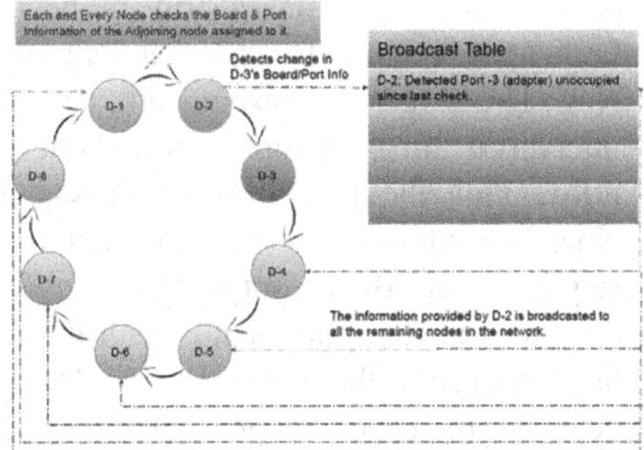

Fig. 5. Another Form of the Architecture used to resolve Security Issue-2

- Now, the main agenda here is, that the indirect virus source node tries to initiate a chain reaction which makes the virus reach more and more nodes and hence end by corrupting the whole network. To prevent such attacks a step-wise procedure has been proposed. Step-1: Whenever a node detects that it has got contaminated by rouge files, it sends a request to report an attack in the Reports & Queries table. Step-2: After which, the network tries to backtrack in the Requests Table to check that which other node accepted D-8's request and immediately terminate those transactions. Step-3: Next it tries to investigate to get the sender node, now let's assume that the network got to know about the contamination at node D-5. Step-4: Then the network checks the value D-5 for entries in the right-hand side of the Allocated Relationship Table, and traceback to the entry's parent node, along with checking the data packet that carries the virus. Step-5: Similarly, performing a recursive operation, the network reaches the root node, which had generated the virus, because at this node there would be no entries in the right-hand side of the Allocated Relationship Table such that the data-packet carrying the virus eventually traces back to the root node. Henceforth, the network can detain that node for disbanding its further operations.

4) Security Issue-2

- Solution to this issue is practically achievable by a well-defined architecture, just as the one mentioned in Fig. 5.
- Every node in the fog network will have the responsibility to check on another node, i.e. D-1 will look over D-2, D-2 will look over D-3, … D-8 will look over D-1. (Circular ring formation). Here one can probably wonder as to how can any node get to know if a node or device is physically tampered with.
- For that, every node has a computer coded algorithm with itself.

- Now, this code contains some instructions that are command line statements to extract the port's statistics which indicates which port is plugged in or which port is empty. Now, let's say, D-1 is an Arduino, D-2 is a Raspberry Pi, D-3 is a Smart Home, D-4 is Alexa.
- In the above case suppose that D-2 was functioning properly, but after a while someone tampered with the camera strip and now, the camera is not working or detached.
- In that case the port to which camera was attached will get free.
- Now, D-1 is given a responsibility to send an initiation message to D-2 at regular intervals of time stating "Inform about all your ports status". (Here, D-1 already has information about the previous message/response of port status of D-2 with it.)
- Now, as soon as D-2 gets the initiation request, it processes the code (mentioned in step 1) and sends the output to D-1 as a response. After receiving the response D-1 compares all the ports status with the previously saved information gathered from last response, if there is no change that means, the device is safe. If there is a change detected, then the change is broadcasted in the whole fog network to all the other nodes (such as smartphone (D-3), Alexa (D-4)). So that the user gets informed about any tampering in any device through other sources (such as Smartphone, Alexa). And then the user can look through complications in the D-2 device.
- Another case arises when the node D-2 is completely disconnected from its power source by any means. In that case, D-1 keeps sending a threshold number of initiation requests to D-2 and if there is no response or if the fog network has detected that it has left the network. Then, D-1 can broadcast the message that "D-2 is either physically disconnected or it has no network connectivity" to all other nodes, which can inform the user about the issue automatically.

5 Results

Customised authentication algorithm is implemented for a network comprising of a finite number of nodes using Firebase Database. Node Analytics such as timestamp capturing is also performed to have a pattern of time-slots when the node generally remains active. The methodologies designed for key generation on client node, key sharing between nodes and file transfer between nodes are implemented only for interaction of two nodes currently in these prototype results. For testing on a network of two or more nodes all the architectures have to be developed. Here for a feasible demonstration and development of a prototype, an Android application is developed.

Comparison with State-of-the-Art Techniques: Table 1 shows a tabular comparison of the Encryption and Decryption time of the proposed approach with the well-known encryption methods RSA [9] & AES [13], note that both use fixed-size bit keys whereas this approach follows a dynamic key pattern proportional to the file size. Clearly this approach outperforms RSA & AES in runtime, quality and execution.

App Version-1(Authentication Module): The module fetches the user's Wi-Fi MAC address and sends it to the database (Firebase).

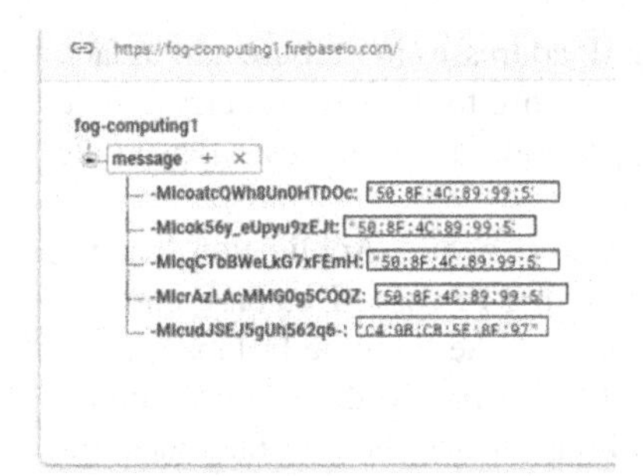

Fig. 6. Firebase database snapshot (Color figure online)

Table 1. Comparative results

Method) File Size – bits encrypted (key)	Encryption Time (in ms)	Decryption Time (in ms)
A) 143Kb File - 429 bit	6.48	30.81
A) 358Kb File - 1074 bit	50.32	71.61
A) 670Kb File - 2010 bit	98.04	116.13
A) 0.93Mb File - 2790 bit	142.17	155.81
B) 64 byte File- 256 bit	869	3876
B) 128 byte File - 256 bit	2870	6654
C) 500Kb File	90	-
C) 5Mb File	626	-

A) Proposed Approach (key size based on file size)
B) RSA – cited paper 256 bit key
C) AES – cited paper 256 bit key

- The code approach available on community forum only supported fetching MAC address till Android Marshmallow which is an older version of Android's Operating System.
- Hence a unique customised algorithm was created that supports all versions of Android's Operating System for availing the requirements needed to construct the Authentication Module and integrate it with the interface.
- The snapshot shown in Fig. 6 is captured from the Firebase Real-time Database, wherein every time the user opens the application, the entry of the device's Wi-Fi MAC Address gets into the database. The "Blue Boxes" drawn over the IP addresses shown in Fig. 6 have the same IP addresses which depicts that multiple entries of MAC addresses are made into the database whenever the user opens the application, which can be considered as a factor in evaluating the user's behavioral pattern in using the application.

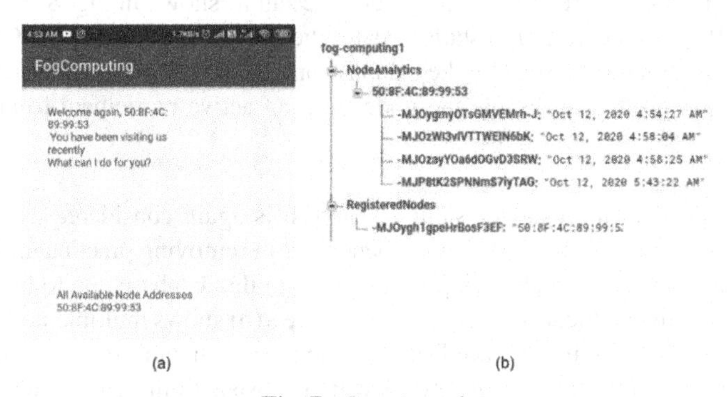

Fig. 7. Case 1 result

App Version-2(Node Analytics Module): This updated version involves maintaining the user's datetime log whenever he/she opens the application for their session tracking, checking availability and other purposes.

- This version also removes redundancy of storing IP addresses in firebase and in fact, it checks 3 different scenarios that can occur according to the node's current state and responds to the user via a message in the application just as the user logs in the application.
- Case-I (If the node is an existing member of a fog network or not): Application displays the message: "Welcome Again" and the corresponding Firebase Database entry as shown in Fig. 7(a) and Fig. 7(b) respectively. Note that, the Firebase Database entry shows only one record present in both Registered-nodes and Node-Analytics topics indicating it is already a part of the fog network and hence the interface message displays "Welcome Again".

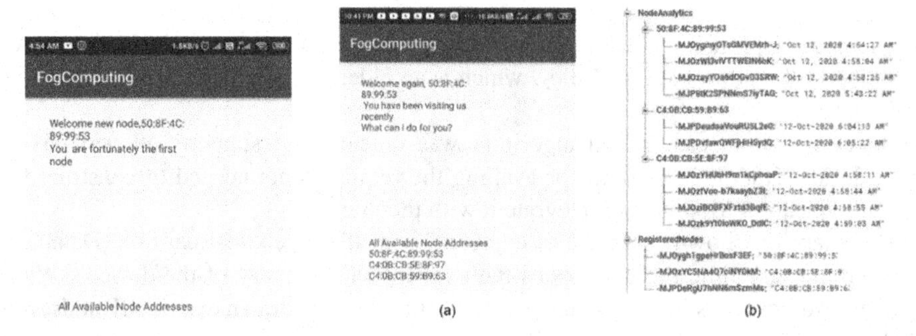

Fig. 8. Case 2 result

Fig. 9. Case 3 result

- Case-II (If a new node appears, then the entry is made only once into the database): If the node is the first node of the network, Application displays the message: "You are Fortunately the First node in the Network" and Available node addresses List is empty as currently there are no other nodes present as shown in Fig. 8.
- Case-III (If the user deletes/reinstalls/updates the application): As the MAC address of the device is already entered into the database once, it is going to be preserved forever until the user prefers to disable the node from an active participant to an in-active node.

Hence, whenever the user reinstalls the app, it is again considered as an existing node as shown in Fig. 9(a), which is very beneficial in removing practical and potential threats of data imbalance such as duplicate entries in the database due to human errors such as deleting the application accidentally. Figure 9(b) shows multiple node entries in Registered-Nodes Topic in Firebase Database, and for each entry in Registered-Nodes, there is a separate entry in Node-Analytics which consists of the Timestamp records of each node whenever its state changes to active.

App Version-3(Periodic Key Generation Module): Figure 10 shows the snapshot of the file-Transfer Application, where both the images are of clients and the White "Dotted" Boxes show the different values after Key Generation such as {151,14,369} and {593,665,660} in Fig. 10(a) and 10(b).

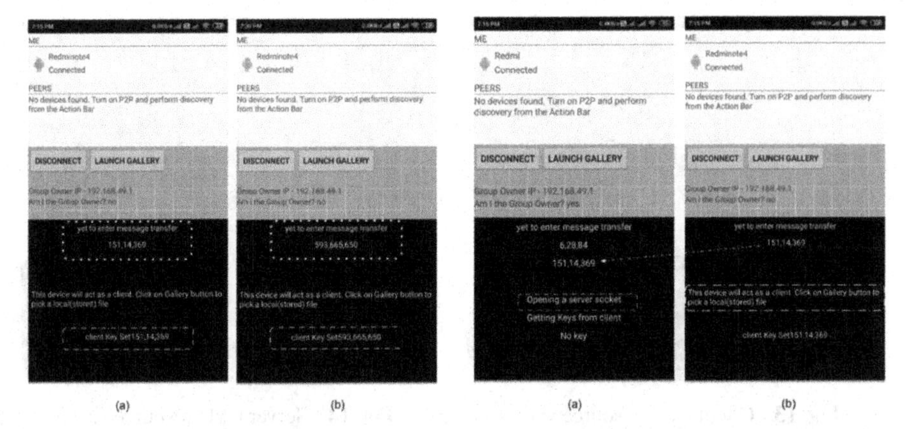

Fig. 10. Key generation result

Fig. 11. Key sharing result

App Version-3.1(Key Sharing Between Client and Server): Figure 11 depicts the Key being shared from the client to server. Here, the White "Dotted" Boxes represent that Fig. 11(a) is a Server, as ("Opening a Server Socket") is written there and Fig. 11(b) is a Client respectively. A White Arrow from Fig. 11(b) to Fig. 11(a) illustrates that the key shown in Client is also displayed in the Server.

Fig. 12. Encryption Results: (a, c) Encrypted, (b, d) Original

Post File Sharing Results: After sharing the Encrypted files from the Client to the Server, the results shown in Fig. 12 were obtained. The Encryption involved only flipping 1 bit from 3 random bytes of the entire file, i.e. if it is a 1 KB file, then only a total of 3 bits out of approximately 8,192 bits are flipped and the results depict complete distortion of the encrypted image in comparison to the original image. This Encrypted Image is received by the Server and stored in its local storage and when it needs to be processed it is opened and read in the application and then decrypted to operate as per further requirements, hence tackling all Privacy Issues that are mentioned in Sect. 3.1.

App Version-3.2(Fetching Resource Table Values): The Images depict partial implementation of the Resource Table wherein the "Dotted" Boxes show the number of CPU cores at different instances for the Client & the Server as shown in Fig. 13 and Fig. 14 respectively.

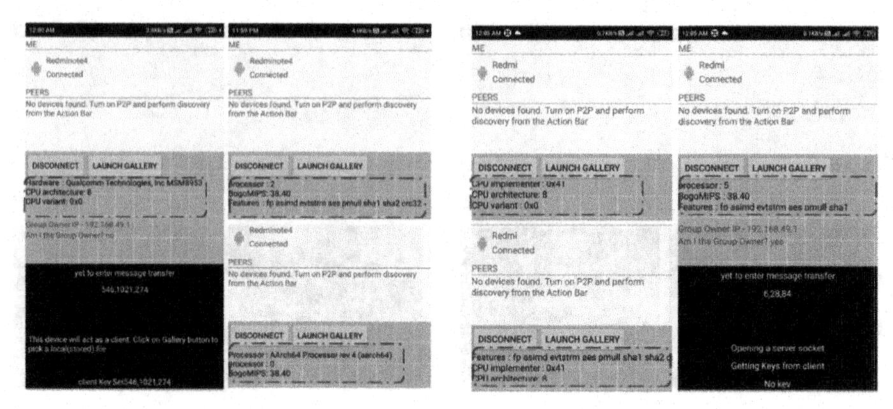

Fig. 13. Client CPU resource values **Fig. 14.** Server CPU resource values

6 Future Scope

Many modules discussed in methodology have advanced implications when integrated with other technologies such as Blockchain, IoT, Big Data, etc.

If the architecture mentioned in the proposed approach is integrated along with Blockchain Technology, the additional advantages of Blockchain's authentication concept will also improve upon the current user authentication policies mentioned in this project. Also, the network can also decrease the trust level of that suspicious node and hence even if it tries to enter the network again at some another time, it can be detected as a suspicious node and hence interaction with such a node would inherently be limited. If the Fog server node is not in vicinity of the devices (D1and D2), then an algorithm can recalculate the distance between nodes and give server rights to nodes with optimal and shortest distance of the node from D1 and D2 respectively.

The proposed architecture and encrypted algorithms can also be applied in a different form as a variation to the original structure in applications concerning the domains related to Construction Management, Transportation Management (Vehicular Telematics), Energy Providing Companies, Satellite Communication (Orbital Sciences), Space Sciences, Automation Industry, Wireless Power Technology. The encryption/decryption algorithm proposed can be evolved into more complex representations for group of devices to carry out efficient functioning and hence such an approach can naturally evolve the way in which cryptography advances into the future wherein the cryptographic modules do not increase the overall cost of operations.

7 Conclusion

Hardware Utilization as Never Before: The devices are continuously contributing their resources and computational capabilities to their fellow nodes in the fog networks maximum utilization gives maximum throughput and efficiency of the system than that achieved when using the device on regular basis.

Decreasing Expenses for Cloud Infrastructure Users: The Fog Computing nodes are established on the principle of maintaining cache, processing most of the data at the Edge Devices, sharing computation and storage resources over the network of Fog nodes, communicating the metadata (result of the processed data) and other relevant information to the Cloud.

Fog Layer Privacy and Security Solution: The proposed solution approach comprising the architectures for ResT, ReqT, ART, BrT and a novel algorithm for encryption/decryption on data, that directly encrypts on the data bits itself and utilizes lowest resources in encryption/decryption, tackles all the challenges mentioned in Sect. 3.

References

1. Mukherjee, M., Shu, L., Wang, D.: Survey of fog computing: Fundamental, network applications, and research challenges. IEEE Commun. Surveys Tutor. **20**(3), 1826–1857 (2018)
2. Bonomi, F., Milito, R., Zhu, J., Addepalli, S.: Fog computing and its role in the Internet of Things. In: Proceedings of the MCC Workshop Mobile Cloud Computing, pp. 13–16 (2012)
3. Moroney L.: The firebase realtime database. In: The Definitive Guide to Firebase. Apress, Berkeley (2017). https://doi.org/10.1007/978-1-4842-2943-9_3. ISBN: 978-1-4842-2942-2
4. Lee, J., Chung, S., Kim, W.: Fog server deployment considering network topology and flow state in local area networks. In: 2017 Ninth International Conference on Ubiquitous and Future Networks (ICUFN), Milan, pp. 652–657 (2017). https://doi.org/10.1109/ICUFN.2017.7993872
5. Funai, C., Tapparello, C., Heinzalman, W.: Enabling multi-hop ad hoc networks through WiFi Direct multi-group networking. In: 2017 International Conference on Computing, Networking and Communications (ICNC), January 2017
6. Lin, Y., Okur, S., Dig, D.: Study and refactoring of Android asynchronous programming. In: 2015 30th IEEE/ACM International Conference on Automated Software Engineering (ASE), November 2015
7. Khashan, O.A.: Hybrid Lightweight proxy re-encryption scheme for secure fog-to-things environment. IEEE Access **8**, 66878–66887 (2020)
8. Osanaiye, O., Chen, S., Yan, Z., Lu, R., Choo, K.R., Dlodlo, M.: From cloud to fog computing: a review and a conceptual live VM migration framework. IEEE Access **5**, 8284–8300 (2017)
9. Diro, A.A., Chilamkurti, N., Nam, Y.: Analysis of Light-weight Encryption Scheme for Fog-to-Things Communication. IEEE (2018)
10. Alrawais, A., Alhothaily, A., Hu, C., Xing, X., Cheng, X.: An Attribute-Based Encryption Scheme to Secure Fog Communications (2017)
11. Yi, S., Li, C., Li, Q.: A survey of Fog Computing: Concepts, Applications and Issues (2015)
12. Mukherjee, M., Matam, R., Shu, L., Maglaras, L., Ferrag, M.A., Choudhury, N., Kumar, V.: Security and Privacy in Fog Computing: Challenges (2017)
13. Vishwanath, A., Peruri, R., He, J.S.: Security in Fog Computing Through Encryption (2016)

Minimize Penalty Fees During Reconfiguration of a Set of Light-Tree Pairs in an All-Optical WDM Network

Amanvon Ferdinand Atta[1(✉)], Gilles Armel Keupondjo Satchou[1], Joël Christian Adépo[2], and Souleymane Oumtanaga[1]

[1] Institut National Polytechnique Felix Houphouët Boigny, Yamoussoukro, Côte d'Ivoire
{amanvon.atta,armel.keupondjo,souleymane.oumtagana}@inphb.ci
[2] Université Virtuelle de Côte d'Ivoire, Abidjan, Côte d'Ivoire
joel.adepo@uvci.edu.ci

Abstract. Multicast connections are efficient for applications such as telemedicine, distributed computing, and distance learning in an all-optical WDM network. To re-optimize the use of network resources such as wavelength channels in the all-optical WDM network, the network operator uses the reconfiguration operation, which affects certain multicast connections. Note that during reconfiguration, a pair of light-trees contributes to the identification of each multicast connection. A light-tree is a path used to transmit an optical flow of a multicast connection. However, the interruption of the flow during reconfiguration of a pair of light-trees may result in the operator having to pay a penalty fee. Thus, the problem studied here is to reconfigure a set of light-trees pairs while minimizing penalty fees. Existing methods focus on reconfiguration that aims to minimize flow interruptions. However, minimizing flow interruptions does not induce the minimum total penalty fees because the penalty fees due to flow interruptions are not always the same for all multicast connections. Therefore, we propose a method that considers the fact that the penalty fees are not the same for all multicast connections to minimize the penalty fees during reconfiguration. The experimental results confirmed the effectiveness of the proposed method.

Keywords: Multicast connection · Light-tree pairs reconfiguration · Flow interruption · Penalty Fees · WDM network

1 Introduction

Wavelength division multiplexing (WDM) is a mature technology that solves the electronic bottleneck problem [1]. Also, WDM provides a large bandwidth by allowing different optical flows to be sent simultaneously over a single optical fiber network [2]. A WDM optical network consists of a set of optical fibers connected by switching nodes [3]. If the transmission of optical flow within a WDM network does not undergo any optical-electrical conversions, although the connections may be controlled by electronics, then the WDM network is an all-optical WDM network [4].

M. Bhattacharya et al. (Eds.): ICICCT 2021, CCIS 1417, pp. 102–116, 2021.
https://doi.org/10.1007/978-3-030-88378-2_9

An all-optical WDM network can support unicast and multicast connections. A multicast connection allows the transmission of data flow from one source node to multiple destination nodes. Therefore, a network operator must use multicast connections to make available applications such as telemedicine, IP-based Television (IPTV), weather forecasts and distance learning. According to the annual report of the equipment manufacturer Cisco, the popularity of these applications is increasing and will continue for several years [5]. Therefore, each network operator must carry out certain operations with great care to meet customer needs. One of these operations is routing, which determines the path used by each multicast connection to transmit the optical flow while reducing the probability of blocking in the all-optical network. A common solution for this operation is called a light-tree [6]. A light-tree can be defined as a tree-shaped path [7] formed by a set of wavelength channels. The tree-shaped path allows the transfer of an optical flow (in an all-optical manner) from a source node to some destination nodes. Note that, each link of this tree-shaped path is called a wavelength channel. In addition, all wavelength channels of this tree-shaped path use the same wavelength.

Another important operation that has, retained our attention is reconfiguration. For each customer connection request, the network operator configures a light-tree. In other words, the network operator may end up with many light-trees established on the network at a given time owing to the high popularity of multicast applications. For efficiency, the network operator must continually change the configuration of these light-trees by replacing the initial light-trees with the new (or final) light-trees: it is said that the operator performs reconfiguration of a set of light-tree pairs. Note that each light-tree pair (initial light-tree, final light-tree) concerned one multicast connection. Reconfiguration is unavoidable when an event such as a large number of connection requests, occurs on the network [8]. Reconfiguration responds to a need to optimize the allocation of network resources. However, if reconfiguration is performed improperly, then it may cause some interruptions of the optical flow. In accordance with Service Level Agreements (SLAs), some customers whose connections are affected by flow interruptions may claim financial penalty fees [9]. As a result, if reconfiguration is carried out improperly, then the network operator may have an enormous total penalty fee for reimbursement to some of its customers.

Let M be a set of multicast connections such that each multicast connection is implemented by an initial (established) light-tree, which transports an optical flow. The set of initial (established) light-trees forms the initial configuration. At the end of the reconfiguration process, this set of initial (established) light-trees must be replaced by a set of final light-trees to be established. This set of final light-trees to be established forms the final configuration. The reconfiguration problem studied here is to find the sequence of configurations to migrate the optical flow from the initial configuration to the final configuration such that this sequence induces the minimum total penalty fees. This problem is tricky to solve because most of the time, a final light-tree (belonging to the final configuration) requires a wavelength channel that is already used by an initial light-tree belonging to the initial configuration. This situation can create a deadlock state during the reconfiguration process.

Previous works (such as [10, 11]) have focused on the reconfiguration that aims to minimize flow interruptions. However, the minimization of flow interruptions does not induce the minimization of the total penalty fees. Indeed, multicast connections that are affected by flow interruptions may be multicast connections with high penalty fees. Therefore, we propose a method that is more efficient solution of the reconfiguration problem studied here. The major contributions of this study are outlined as follows:

1. Use of a vertex-weighted digraph to model the dependencies, which are the origin of the deadlock state during the reconfiguration process.
2. The proposition of an algorithm that uses this graph to compute the minimum cost (i.e. penalty fees) feedback vertex set, which allows minimizing total penalty fees during the reconfiguration process.

The rest of this paper is structured as follows: Sect. 2 presents related previous methods in more detail. Sect. 3 presents our specification of the problem. Sect. 4 presents the methodology proposed to solve this problem. Sect. 5 introduces a set of numerical experiments to evaluate the performance of the proposed method. Sect. 6 summarizes our contributions and the performance of the proposed method.

2 Related Works

Reconfiguration works that concern multicast connections are divided into two groups: the group of reconfiguration works that affect only one multicast connection and the group of reconfiguration works that affect several multicast connections.

In the first group, the aim is to reconfigure a light-tree pair (an initial light-tree, a final light-tree) of a given multicast connection. In short, the problem is to migrate an optical flow from the initial light-tree to the final light-tree without flow interruption. In [10, 12], several methods based on a so-called branch approach were proposed to solve the problem.

In the second group of work, the problem is to reconfigure a set of light-tree pairs while minimizing the total number of flow interruptions. A solution to this problem is to use one of the methods belonging to the first group of works to reconfigure a set of light-tree pairs in parallel. This solution may cause a high total number of flow interruptions. In fact, the final light-tree of a connection x may require a wavelength channel that is used by the initial light-tree of another connection y and vice versa. The dependencies between connections create a deadlock state that must be resolved to minimize the total number of flow interruptions. Another solution is to use the Minimum Feedback Vertex Set Approach [13–15] (MFVSA) that is used to reconfigure a set of lightpath pairs. This approach involves temporarily deleting a set of lightpaths having minimum cardinality to overcome the deadlock state. However, a lightpath [16] concerns a single destination node as opposed to a light-tree, which concerns several destination nodes. Therefore, temporarily deleting a set of initial light-trees having the minimum cardinality does not involve the minimum total number of flow interruptions. In other words, MFVSA does not allow the reconfiguration of a set of light-tree pairs while minimization the total number of flow interruptions. To correct this, a method called Light-tree Set Reconfiguration Algorithm [11] (LSRA) has been proposed. Unlike MFVSA, the LSRA method temporarily deletes a set of initial light-trees, causing the minimum total number of flow interruptions.

Although LSRA allows the reconfiguration of a set of light-tree pairs while minimizing the total number of flow interruptions, it does not allow minimization of the total penalty fees. Indeed, LSRA temporarily deletes a set of initial light-trees, causing the minimum total number of flow interruptions. However, the penalty fees associated with the connections represented by these initial light-trees can be high. This can result in a high total penalty fee. In other words, a set of initial light-trees causing the minimum total number of flow interruptions is not necessarily a set of initial light-trees causing the minimum total penalty fees. Therefore, LSRA is not an efficient method for solving the reconfiguration problem.

3 Problem Specification

3.1 System Description

The WDM network is modeled as an undirected graph $G(V, E)$ where V is the set of optical nodes and E is the set of fiber links. Let $M = \{m_1, m_2, ..., m_n\}$ be a set of established multicast connections in the WDM network, where $n = |M|$. Each multicast connection $m_k \subseteq M$ is also denoted by $m_k =< s_k, D_k, PF_k >$ where s_k is the source of the multicast connection m_k, D_k is the set of destination nodes of m_k, and PF_k the penalty fees (in Monetary Unit (MU) such as Dollar or Euro) per time unit for each interruption of the flow to a destination node of the connection m_k in accordance with the Service Level Agreement (SLA) between the network operator and the customer concerned by the connection m_k. The optical flow of the multicast connection m_k was initially carried by the light-tree $T_k^0 = (V_k^0, E_k^0, \lambda_k^0)$. Note that $V_k^0 \subseteq V$, $s_k \in V_k^0$, $D_k \subseteq (V_k^0 \backslash \{s_k\})$, $E_k^0 \subseteq E$. λ_k^0 is the wavelength used to create one wavelength channel on each element of E_k^0. At a given time, the network operator decides to perform reconfiguration. Therefore, for each multicast connection m_k of M, he computes a final light-tree $T_k^f = (V_k^f, E_k^f, \lambda_k^f)$ that will be used to carry the optical flow of m_k at the end of the reconfiguration process. Note that $V_k^f \subseteq V$, $s_k \in V_k^f$, $D_k \subseteq (V_k^f \backslash \{s_k\})$ and $E_k^f \subseteq E$. λ_k^f is the wavelength used to create the wavelength channel on each element of E_k^f. It is assumed that no spare wavelengths are available during the reconfiguration process. This means that each connection cannot temporarily use a light-tree using a different wavelength. Therefore, deadlock states will result in flow interruptions, and penalty fees, which must be reduced.

3.2 Problem Formulation

The problem is formulated as follows:

- Input: A set of multicast connections $M = \{m_1, m_2, ..., m_n\}$, the initial configuration of light-trees $C_0 = \{T_1^0, ..., T_n^0\}$ for M and the final configuration of light-trees $C_f = \{T_1^f, ..., T_n^f\}$ for M, where T_k^0 and T_k^f are, respectively, the initial light-tree of multicast connection m_k and the final light-tree of m_k.
- Output: Sequence of configurations $C =< C_0, ..., C_f >$, where $C_i = \{T_1^i, ..., T_n^i\}$ is the configuration of the light-trees set used by the set of multicast connections M at step i of the reconfiguration process. Specifically, T_k^i is the light-tree used to carry the optical flow of the multicast connection m_k at step i of the reconfiguration process.

- Objective: Minimize total penalty fees causes by C.

The total penalty fees generated by the reconfiguration process is denoted by $PEN\_FIN(C)$. Let $PEN\_FIN(C_i)$ be the sum of penalty fees associated with connections that light-trees in the configuration C_i is a null-graph [17]. It is important to note that if a connection that light-tree in the configuration C_i is a null-graph, then C_i causes flow interruptions to the destination nodes of this connection. It is assumed that one unit of time is required to switch from one configuration to another configuration of the sequence of configurations C. $PEN\_FIN(C_i)$ is obtained using Eq. 1:

$$PEN\_FIN(C_i) = \sum_{m_k \in M} INT(C_i, m_k) * PF_k * |D_k| \quad (1)$$

Where

$$INT(C_i, m_k) = \begin{cases} 1, & \text{if the light-tree of } m_k =< s_k, D_k, PF_k > \text{ that belongs to } C_i \text{ is a null-graph} \\ 0, & \text{otherwise} \end{cases}$$

Equation 1 computes the penalty fees for one configuration. Hence, the total penalty fees $PEN\_FIN(C)$ induced by the sequence of configurations C is obtained using Eq. 2:

$$PEN\_FIN(C) = \sum_{C_i \in C} PEN\_FIN(C_i) \quad (2)$$

4 Methodology

Section 2 shows that the previous method called LSRA allows the reconfiguration of a set of light-tree pairs while minimizing the total number of flow interruptions. However, LSRA is not efficient to reconfigure a set of light-tree pairs while minimizing the total penalty fees. Indeed, to overcome the deadlock state in the reconfiguration process, LSRA temporarily deletes a set of initial light-trees such that the connection of each of these initial light-trees belongs to a feedback vertex set having a minimum total number of flow interruptions that is not necessarily a feedback vertex set having minimum total penalty fees. Therefore, here, we propose a method that temporarily deletes a set of initial light-trees such that the connection of each of these initial light-trees belongs to a feedback vertex set having minimum total penalty fees. This feedback vertex set is computed from a graph that models all the dependencies between connections concerned by the reconfiguration. Hence, we first, present the tool used to model all dependencies and then explain our method in more detail.

4.1 Dependencies Modelling

The reconfiguration problem studied here (see Sect. 3) is tricky to solve because dependencies may exist between connections. In the following, a vertex-weighted graph is used to model all dependencies. This dependency graph is used by our method to compute a feedback vertex set having the minimum penalty fees. The dependency graph, which is a vertex-weighted graph is built according to the following process:

- Each connection concerned by the reconfiguration is a vertex of the dependency graph
- If the final light-tree of the connection x requires one or more wavelength channels already used by the initial light-tree of connection y, then a directed arc must be created from vertex x to vertex y.
- A weight is assigned to each vertex according to a weighting function. Formally, the dependency graph, which is weighted by the vertices, is denoted by $G_f(V_{G_f}, E_{G_f}, f)$, where V_{G_f} is the set of vertices, E_{G_f} is the set of arcs and f is the weighting function that is defined as follows:

$$f : V_{G_f} \to N$$

$$x \mapsto f(x)$$

Let $m_k =< s_k, D_k, PF_k >$ be the connection corresponding to vertex (or node) x of the graph $G_f(V_{G_f}, E_{G_f}, f)$. Eq. 3 is used to compute the weight of vertex x:

$$f(x) = f(m_k) = PF_k * |D_k| \tag{3}$$

It is important to note that, $f(x)$ represents the penalty fees caused by the deletion of the light-tree of the connection represented by vertex x at a reconfiguration step.

In the following, the process used to build the dependency graph is illustrated. Figure 1 shows an example of the reconfiguration problem studied here. Let $m_1 =< s_1, \{d_1, d_2\}, 10 >$, $m_2 =< s_2, \{d_3, d_4, d_5\}, 3 >$ and $m_3 =< s_3, \{d_7, d_8\}, 5 >$ be three multicast connections established on an all-optical WDM network. We assume that only one wavelength channel can be established on each optical fiber. Figure 1a contains the initial light-tree (represented by the set of wavelength channels in solid grey) of m_1, the initial light-tree (represented by the set of wavelength channels in dashed grey) of m_2 and the initial light-tree (represented by the set of wavelength channels in dotted grey) of m_3. Also, Fig. 1b contains the final light-tree (represented by the set of wavelength channels in solid grey) of m_1, the final light-tree (represented by the set of wavelength channels in dashed grey) of m_2 and the final light-tree (represented by the set of wavelength channels in dotted grey) of m_3.

We assume that all wavelength channels of both (current and final) configurations use the same wavelength. Figures 1a and 1b allow to detect the following dependencies:

- The connection m_1 depends on the connection m_2. Indeed, the final light-tree of m_1 requires the wavelength channel $b \to c$ that is already used by the initial light-tree of m_2.

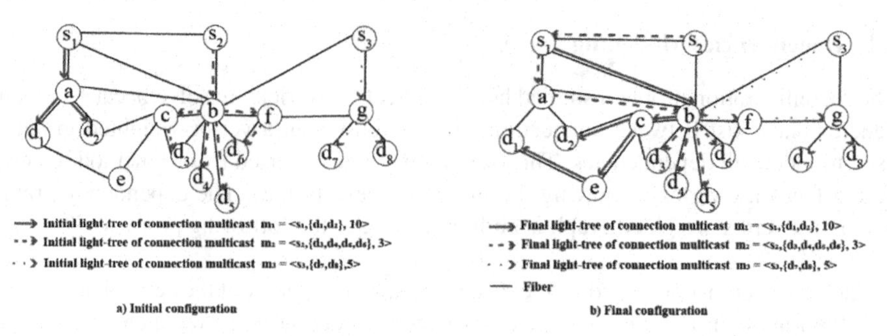

Fig. 1. Instance of reconfiguration problem

- The connection m_2 depends on the connection m_1. Indeed, the final light-tree of m_1 requires the wavelength channel $s_1 \rightarrow a$ that is already used by the initial light-tree of m_1.
- The connection m_3 depends on the connection m_2. Indeed, the final light-tree of m_3 requires the wavelength channel $b \rightarrow f$ that is already used by the initial light-tree of m_2.

From the above dependencies and the building process of the dependency graph, it is easy to deduce the dependency graph (i.e., a vertex-weighted graph), as illustrated in Fig. 2.

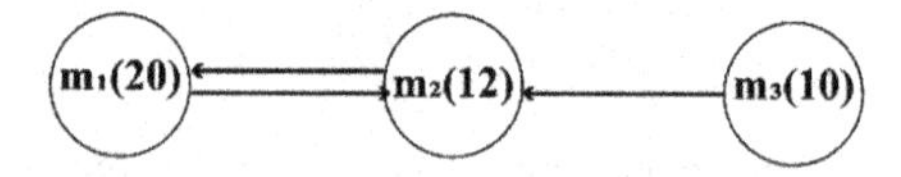

Fig. 2. Dependency graph corresponding to the problem instance

4.2 Proposed Method

In this section, we introduce an algorithm to minimize the penalty fees during the reconfiguration process. In other words, our algorithm denoted by LPARLS, uses the vertex-weighted dependency graph to model all dependencies and based on this graph, LPARLS deletes light-trees of connections that cause minimum penalty fees in order to overcome the deadlock state in the reconfiguration process.

Algorithm 1 Low-Penalty Algorithm for Reconfiguration of Light-tree Set LPARLS

Input: C_0, C_f // C_0 : Initial configuration; C_f : Final Configuration

$M = \{m_1, m_2, ..., m_n\}$ // Set of multicast connections

Output: $C =< C_0, ..., C_f >$ // Sequence (or list) of configurations

1: Initialize the current configuration Cur_C with the value of C_0; then C_0 is added to C as the first element of C.

2: Build the dependency (vertex-weighted) graph $G_f(V_{Gf}, E_{Gf}, f)$ according to C_0, C_f and M.

3: Compute the Minimum Cost Feedback Vertex Set $MCFVS$ from $G_f(V_{Gf}, E_{Gf}, f)$.

4: Compute a new value of Cur_C by the deletion of the initial light-trees of the elements of $MCFVS$; then add this new value of Cur_C to C.

5: **Repeat**

6: Compute a new value of Cur_C by establishing the final light-trees of the set of free connections $FC \subseteq M$ (i.e., nodes in V_{Gf}) and deleting the initial light-trees of each element c of FC if $c \in V_{Gf} \setminus MCFVS$; then add this new value of Cur_C to C.

7: **Until** Cur_C is equal to C_f

8: **Return** C

First, the first value belonging to the sequence of configurations C is set to C_0. Then, the dependency (vertex-weighted) graph is built (refer to line 2) as described in Sect. 4.1. This dependency graph shows a deadlock state during the reconfiguration process. The deadlock state is represented by at least one cycle in the dependency graph. To overcome this deadlock while causing minimum penalty fees, the minimum cost feedback vertex set (from the dependency graph) is computed (refer to line 3) using Demetrescu's algorithm [18]. The cost of a feedback vertex set is the sum of the weights of the elements of the feedback vertex set. Let us remind you that a feedback vertex set of a graph is a set of nodes such that the deletion of these nodes breaks all cycles in the graph. Hence, the initial light-trees that the connections are represented by the nodes belonging to $MCFVS$ are deleted from the current configuration in order to update the current configuration. This (new value of) current configuration is appended to C (refer to line 4). Finally, the light-tree pair (initial light-tree, final light-tree) of each free connection is reconfigured (i.e., final light-tree is established then eventually initial light-tree is deleted) to update the current configuration and the updated current configuration is added to C (refer to line 6). This last instruction is repeated until the updated current configuration is equal to the final configuration.

We explain LPARLS using the problem instance illustrated in Fig. 1. For this problem, Fig. 2 shows the dependency graph built by LPARLS.

From Fig. 2, it can be seen that the reconfiguration process is in a deadlock state because the dependency graph contains the cycle formed by connections $m_1 =<$

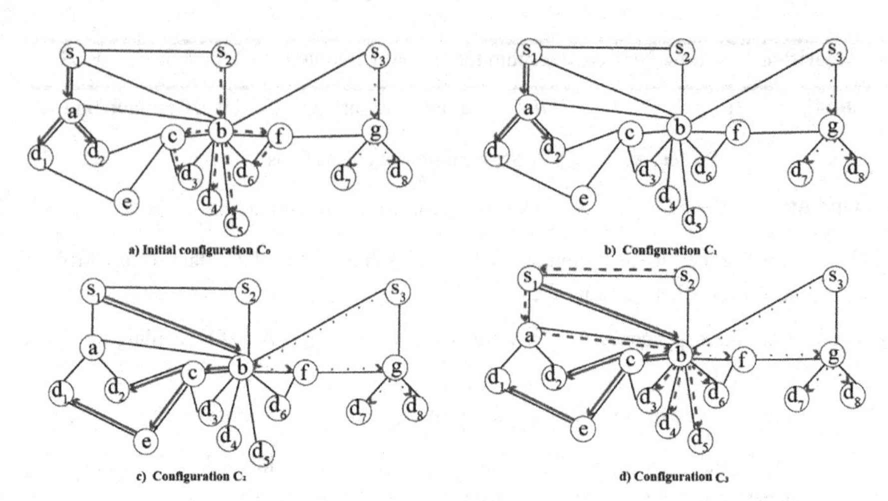

Fig. 3. Illustration of sequence of configurations returned by LPARLS

$s_1, \{d_1, d_2\}, 10 >$ and $m_2 =< s_2, \{d_3, d_4, d_5\}, 3 >$. The deletion of the light-tree of the connection m_1 causes lower flow interruptions (i.e., towards two destination nodes) than the deletion of the light-tree of the connection m_2 (i.e., towards four destination nodes). However, according to the weights of nodes in the dependency graph, the deletion of the light-tree of the connection m_2 causes lower penalty fees than the deletion of the light-tree of the connection m_1. Therefore, the minimum cost feedback vertex set computed by LPARLS is $\{m_2\}$. Then the initial light-tree of m_2 is deleted from the initial configuration to obtain the current configuration C_1 (see Fig. 3b). Next, LPARLS computes the set of free connections, which is $\{m_1, m_3\}$. According to line 6 of Algorithm 1, the final light-tree of each connection belonging to $\{m_1, m_3\}$ is established and its initial light-tree is deleted to obtain a new value of the current configuration (see Fig. 3c). This current configuration is not equal to the final configuration because the final light-tree of m_2 is not established. Hence, the new set of free connections is $\{m_2\}$. Then the new value of the current configuration (see Fig. 3c) was obtained by establishing the final light-tree of m_2. This current configuration is equal to the final configuration, as shown in Fig. 1b: LPARLS stops.

5 Experimental Study

Experiments were conducted to confirm the effectiveness of the proposed method (i.e., LPARLS). In other words, the purpose of these experiments is to confirm that LPARLS generates lower total penalty fees than the LSRA method. The Python language was used to conduct these experiments on a laptop computer equipped with an Intel Core i5-9300H processor at 2.40 GHz, 8 GB RAM, and a Windows 10 operating system. The metric evaluated in the experiments, settings, and results of these experiments are presented in the following subsections.

5.1 Performance Criterion

The main criterion used to confirm the effectiveness of our method is the total penalty fees, which is expressed by Eq. 2 in the problem formulation subsection (see Sect. 3.2). A secondary criterion is used in this study. This criterion is the total number of flow interruptions, which formula is given in [11]. The total number of flow interruptions is used to show if LPARLS generates a high number of flow interruptions.

5.2 Experimental Setup

During the experiments, a random graph generation process similar to that used in [11] was applied. Thus, the well-known Waxman approach [19] is applied to generate a randomly graph that represents the all-optical WDM network composed by 220 nodes. In this approach, the vertices are placed randomly in a rectangular coordinate grid by generating uniformly distributed values for their and coordinates. An edge between two nodes u and v is added using the probability function:

$$F = \lambda * exp(-\frac{d(u, v)}{\gamma * \delta}) \tag{4}$$

where $d(u, v)$ is the Euclidean distance between the nodes that form the endpoints of the edge (u, v), δ is the maximum Euclidean distance between any two nodes, $0 \leq \lambda$, $\gamma \leq 1$. Higher values of λ produce graphs with higher edge densities, while small values of γ increase the density of short edges relative to higher ones. In the experiments, λ and γ were set respectively to 0.7 and 0.9.

The all-optical WDM network represented by the randomly generated graph was used to generate instances of the problem of reconfiguration of a set of light-tree pairs. It is assumed here that the number of multicast connections (noted $|M|$) involved in the reconfiguration is equal to 24. The range of the number of destination nodes $\Delta D_1 =$ [2;10], $\Delta D_2 =$ [11;20], $\Delta D_3 =$ [21;30]). Another experience parameter is the Penalty Fees PF associated with the flow interruption to a destination node of a given connection. It is assumed that the higher the PF value, the higher the priority level of the connection associated with PF from the network operator's point of view. We set three possible values for PF, which are 10 MU (Monetary Unit), 25 MU, and 50 MU. In other words, if the penalty fees associated with the interruption of flow to a destination node of a given connection is 10 MU, then that connection is classified as a low-priority connection. If the penalty is 25 MU, the connection is classified as a medium-priority connection. If the penalty is 50 MU. Then the connection is classified as a high-priority connection. There are three possible distributions of the different types of connections in the set of connections:

- Distribution with high-priority minority connections: ten (10) connections are low-priority connections, ten (10) connections are medium-priority connections, and four (04) connections are high-priority connections.
- Distribution with an equal number of connection types: eight (08) connections are low-priority connections, eight (08) connections are medium-priority connections, and eight (08) connections are high-priority connections.

- Distribution with high-priority majority connections: five (05) connections are low-priority connections, five (05) connections are medium-priority connections, and fourteen (14) connections are high-priority connections.

For each type of distribution, an experiment was conducted for each range value of the number of destination nodes ΔD_i(with $i \in [1; 3]$) by generating 100 instances of the reconfiguration problem. Each instance of the problem is generated as follows:

- Twenty-four (24) multicast connections are generated: 24 source nodes are uniformly selected from [1; 220] and for each multicast connection, the number of destinations is randomly selected from ΔD_i.
- The type of distribution considered (i.e., with high-priority minority connections or with an equal number of connection types or with high-priority majority connections) is created by randomly assigning each of the 24 connections to a connection type (low, medium or high-priority).
- The initial configuration is created: For each multicast connection, the initial light-tree is obtained using the Dijkstra algorithm [20] and a wavelength.
- The final configuration is created: For each multicast connection, the light-tree is obtained using the Prim algorithm [21] and a wavelength such that each multicast connection depends on two other multicast connections.
- These two configurations (i.e., initial and final) are taken as inputs by LSRA and LPARLS (i.e., our method). At the end of each experiment, the average of the values of the performance criterion for each of these two methods was retained.

5.3 Results Analysis

Let us remind you that, we have established twenty-four (24) connections and based on three levels of priority (low priority, priority, medium and high priority), we have distributed them according to three types of distribution (with high priority minority connections, with an equal number of types of connections and high priority majority connections). For each type of distribution, we successively considered $\Delta D_1 = [2;10]$, $\Delta D_2 = [11;20]$, and $\Delta D_3 = [21:30]$ as the range of the number of destination nodes of any multicast connection. To obtain representative results, one hundred (100) instances of reconfiguration problems were generated and then the average of the penalty fees were computed.

Figures 4a, 4b and 4c, show, for each type of distribution considered, the penalty fees according to the different values when the range of the number of destination nodes denoted by ΔD takes the values $\Delta D_1 = [2;10]$, $\Delta D_2 = [11;20]$ and $\Delta D_3 = [21:30]$ respectively. From Figs. 4a, 4b and 4c, the following observations were made:

- Regardless of the type of distribution considered, LPARLS produces fewer penalty fees than LSRA for each range of the number of destination nodes. The reason is that LSRA temporarily deletes initial light-trees from a feedback vertex set having a minimum total number of flow interruptions while LPARLS temporarily deletes initial light-trees from a feedback vertex set having minimum total penalty fees. Indeed, several nodes (or vertices) belonging to the feedback vertex set having a minimum total number of flow interruptions can be high-priority connections. Thus, the total penalty fees generated by this feedback vertex set is greater than or equal to the total penalty fees generates by the feedback vertex set having minimum total penalty fees. In other words, a feedback vertex set having a minimum total number of flow interruptions is not necessarily a feedback vertex set having minimum penalty fees.

- For each distribution type, the penalty fees increase with the range of the number of destination nodes considered. In other words, the greater the number of nodes destination plus penalty fees are important. This is consistent with Eq. 2.
- Penalty fees are higher in the case of distribution with high-priority majority connections. This observation is explained by the fact that in the case of distribution with high-priority majority connections, there is a higher number of high-priority connections than low and medium-priority connections. Thus, the feedback vertex set having the lowest penalty fees is susceptible to contain high-priority connections that have the highest penalty fees.

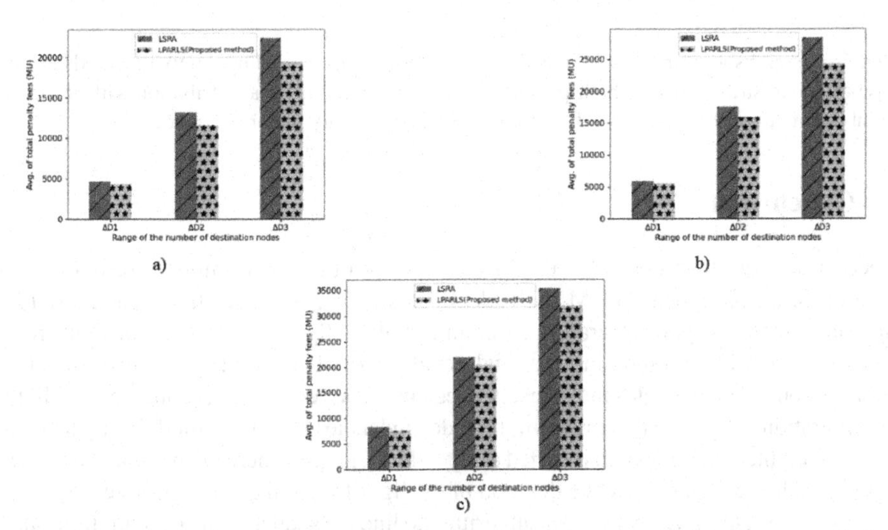

Fig. 4. Comparison of the total penalty fees caused by the two methods in the case of: a) a distribution with high-priority minority connections; b) a distribution with an equal number of connection types; c) a distribution with high-priority minority connections.

Regardless of the type of distribution considered (see Figs. 5a, b and c), LPARLS produces more flow interruptions than LSRA. The reason is that LSRA temporarily deletes current light-trees from a feedback vertex set having minimum total number of flow interruptions while LPARLS temporarily deletes current light-trees from a feedback vertex set having minimum total penalty fees.

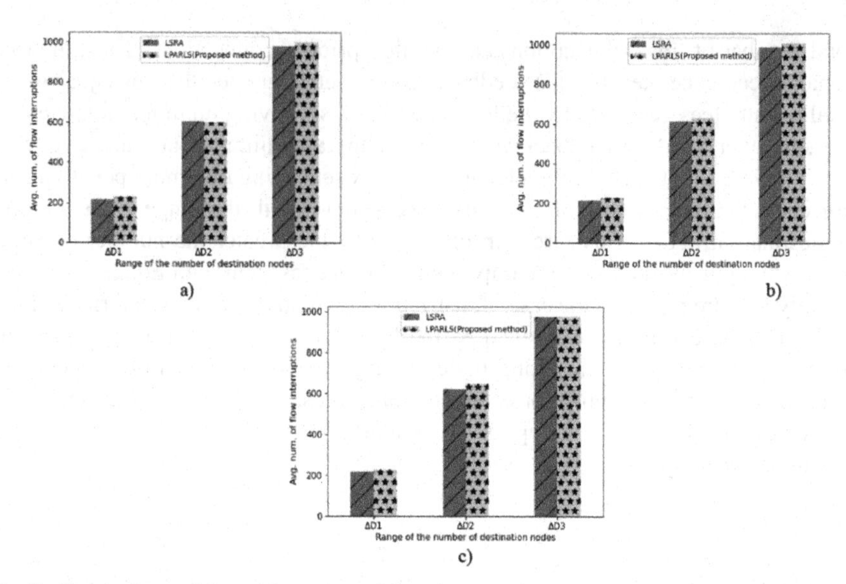

Fig. 5. Comparison of the total number of flow interruptions caused by the two methods in the case of: a) a distribution with high-priority minority connections; b) a distribution with an equal number of connection types; c) a distribution with high-priority minority connections.

6 Conclusion

Reconfiguration is an operation used by a network operator to improve resource utilization in an all-optical WDM network. In this paper, we study the reconfiguration problem that consists of migrating from an initial configuration of a set of light-trees to a new one while minimizing the total penalty fees. This problem is tricky to solve because some dependencies may exist between multicast connections concerned by both configurations. To solve this problem, first, dependencies have been modeled by using a vertex-weighted graph, which is called a dependency graph, where the weight of a vertex is the penalty fees caused by the deletion of the light-tree of the connection represented by this vertex. Next, we proposed an algorithm to find the sequence of configurations that minimizes the total penalty fees. Finally, the effectiveness of our proposed method was experimentally confirmed. Indeed, the proposed method (i.e., LPARLS) causes lower penalty fees than the method called LSRA. Therefore, LPARLS is an efficient method if the network operator aims to reimburse low penalty fees (caused by flow interruptions during reconfiguration) to some of its customers.

Future work may study the trade-off between the total penalty fees and the total number of flow interruptions during the reconfiguration of a set of light-trees pairs.

References

1. Somani, A.K.: Survivability and Traffic Grooming in WDM Optical Networks. Cambridge University Press, Cambridge (2005)

2. Chadha, D.: Optical WDM Networks: From Static to Elastic Networks. IEEE Press/Wiley, Hoboken, New Jersey (2019)
3. Dutta, M.K., Chaubey, V.K.: Performance analysis of all-optical WDM network with wavelength converter using Erlang C traffic model. In: Das, V.V., et al. (eds.) BAIP 2010. CCIS, vol. 70, pp. 238–244. Springer, Heidelberg (2010). https://doi.org/10.1007/978-3-642-12214-9_39
4. Kaminow, I.P., et al.: A wideband all-optical WDM network. IEEE J. Select. Areas Commun. **14**(5), 780–799 (1996). https://doi.org/10.1109/49.510903
5. Cisco: Cisco annual internet report (2018–2023). https://www.cisco.com/c/en/us/solutions/collateral/executive-perspectives/annual-internetreport/white-paper-c11-741490.html
6. Barat, S., Keshri, B.N., De, T.: A cost function based multi-objective multicast communication over WDM optical fiber mesh network. In: Biswas, U., Banerjee, A., Pal, S., Biswas, A., Sarkar, D., Haldar, S. (eds) Advances in Computer, Communication and Control. Lecture Notes in Networks and Systems, **41**, 75–85. Springer, Singapore (2019). https://doi.org/10.1007/978-981-13-3122-0_8
7. Lin, H-C., Wang, S.-W.: Splitter placement in all-optical WDM networks. In: GLOBECOM 2005. IEEE Global Telecommunications Conference, 2005, November 2005, Vol. 1, p. 5 (2005). https://doi.org/10.1109/GLOCOM.2005.1577639
8. Li, H., Wu, J.: Survey of WDM network reconfiguration: topology migrations and their impact on service disruptions. Telecommun. Syst. **60**(3), 349–366 (2015). https://doi.org/10.1007/s11235-015-0050-5
9. Belhareth, S., Coudert, D., Mazauric, D., Nisse, N., Tahiri, I.: Reconfiguration with physical constraints in WDM networks. In: 2012 IEEE International Conference on Communications (ICC), Ottawa, ON, Canada, pp. 6257–6261 (2012). https://doi.org/10.1109/ICC.2012.6364833
10. Adépo, J.C., Aka, B., Babri, M.: Tree reconfiguration with network resources constraint. Int. J. Comput. Sci. Telecommun. **7**(1), 1–4 (2016)
11. Atta, A.F., Adépo, J.C., Cousin, B., Oumtanaga, S.: Minimize Flow Interruptions during Reconfiguration of a set of Light-trees in All-optical WDM Network, **20**(7), 77–85 (2020)
12. Cousin, B., Adépo, J.C., Babri, M., Oumtanaga, S.: Tree reconfiguration without lightpath interruption in wavelength division multiplexing optical networks with limited resources. Int. J. Comput. Sci. Issues (IJCSI) **11**(2), 7 (2014)
13. Jose, N., Somani, A.K.: Connection rerouting/network reconfiguration. In: Fourth International Workshop on Design of Reliable Communication Networks 2003 (DRCN 2003). Proceedings Banff, Alberta, Canada, pp. 23–30 (2003). https://doi.org/10.1109/DRCN.2003.1275334
14. Solano, F.: Slick lightpath reconfiguration using spare resources. J. Opt. Commun. Netw. **5**(9), 1021 (2013). https://doi.org/10.1364/JOCN.5.001021
15. Kadohata, A., Hirano, A., Inuzuka, F., Watanabe, A., Ishida, O.: Wavelength path reconfiguration design in transparent optical WDM networks. J. Opt. Commun. Netw. **5**(7), 751 (2013). https://doi.org/10.1364/JOCN.5.000751
16. Chlamtac, I., Farago, A., Zhang, T.: Lightpath (wavelength) routing in large WDM networks. IEEE J. Sel. Areas Commun. **14**(5), 909–913 (1996). https://doi.org/10.1109/49.510914
17. Harary, F., Read, R.C.: Is the null-graph a pointless concept? In: Bari, R.A., Harary, F. (eds.) Graphs and Combinatorics: Proceedings of the Capital Conference on Graph Theory and Combinatorics at the George Washington University June 18–22, 1973, pp. 37–44. Springer, Berlin, Heidelberg (1974). https://doi.org/10.1007/BFb0066433
18. Demetrescu, C., Finocchi, I.: Combinatorial algorithms for feedback problems in directed graphs. Inf. Process. Lett. **86**(3), 129–136 (2003). https://doi.org/10.1016/S0020-0190(02)00491-X

19. Jia, X.-H., Ding-Zhu, D., Xiao-Dong, H., Lee, M.-K., Jun, G.: Optimization of wavelength assignment for QoS multicast in WDM networks. IEEE Trans. Commun. **49**(2), 341–350 (2001). https://doi.org/10.1109/26.905896
20. Dijkstra, E.W.: A note on two problems in connexion with graphs. Numer. Math. **1**(1), 269–271 (1959). https://doi.org/10.1007/BF01386390
21. Prim, R.C.: Shortest connection networks and some generalizations. Bell Syst. Tech. J. **36**(6), 1389–1401 (1957). https://doi.org/10.1002/j.1538-7305.1957.tb01515.x

Computational Intelligence Techniques

Web-Based Real-Time Gesture Recognition with Voice

Ghadekar Premanand Pralhad, S. Abhishek, Tejas Kachare(✉), Om Deshpande, Rushikesh Chounde, and Prachi Tapadiya

Department of Information Technology, Vishwakarma Institute of Technology, Pune, Maharashtra, India
{premanand.ghadekar,s.abhishek18,tejas.kachare18,om.deshpande18, rushikesh.chounde18,prachi.tapadiya18}@vit.edu

Abstract. WebPredict.ai is a model web developed for converting image into text, followed by the conversion of the predicted text to speech. Images used in the proposed application are taken in the form of hand gestural inputs with the help of webcam. This application also helps to interact with google home or smart home of google. As the gestures captured by the webcam can be given for functionality which works for google home and the application can speak out the gesture inputs with different OS voices. These different OS voices are supported by user's browser and Device. This application was developed on HTML, JavaScript and with the help of Tensorflow.js. Tensorflow.js is an open-source library which is used to train, define, and run machine learning models entirely in the browser, with the help of JavaScript and some high-level API's. The runtime environment for the proposed model is provided by Node.js which helps in the development of web application and is written in JavaScript. In the proposed application, ML algorithm is running in the web browser, hence there is no need to install any driver or library. Users just need to open the webpage and the program is ready to run. Users can also open the webpage from a mobile device.

Keywords: CNN · Mobilenet · Nodejs · WebGL · Google nest · Gesture recognition

1 Introduction

Humans' communication with computers has always been a challenge. People have spent quite half the past century experimenting with various ways to interact with computers. Mouse and Keyboard have been standard input for computers. Most of our interaction with computers involves fingers and eyes touchscreen had come into practice which gave some major improvements and made use of multi-finger usage which made interaction easier. Hand gestures can add another dimension to communication. Voice recognition technology over the years has shown great progress, now with the help of Machine Learning, we are able to recognize facial expressions. Gesture Recognition has been powered by machine learning giving the best accuracy. Reading of gestures involves the recognition of images or recordings captured by the camera. Gesture, when identified,

M. Bhattacharya et al. (Eds.): ICICCT 2021, CCIS 1417, pp. 119–131, 2021.
https://doi.org/10.1007/978-3-030-88378-2_10

is translated to a selected command within the controlled application. Standard image processing techniques don't yield great results, gesture recognition must be backed by machine learning to unfold its full potential.

Gesture can be referred to as any non-verbal communication intended to communicate a specific message to another person. Gestures play a significant role in expressing the actions and emotions of humans. Movements of different body parts like hand, arms, legs, fingers, and head form gestures. Thus, Gesture recognition detects or originates from the face and hand or any physical movement, large or small. Motion sensor interprets any of the face, hand, or physical movement. With the tremendous advancement in technology, several techniques are introduced every other day. One such approach is gesture recognition. Gesture recognition is considered very crucial for human-computer interaction. It is a subject of matter in computer science that aims to decode human gestures or postures with mathematical algorithms.

In traditional methods, using input devices such as a mouse or a keyboard, people used to give input to the systems and digital gadgets. But this conventional method of giving the input data to systems is no longer cutting-edge and progressive because of the limitations of input devices. One solution to this drawback is the usage of gestures to provide input data or commands to digital devices. Users can interact with these digital gadgets using gestures. Thus, it constructs a better path between humans and machines. Hence in the paper, we have presented an approach with which digital gadgets can perform some tasks or operations by detecting gestures.

The proposed model which is a web-based application predicts hand gestures and converts the predicted gesture into verbal or voice commands. These commands activate the smart home of google or google nest. The Google Smart Home or Google Nest is a platform that lets users control connected devices through the Google Home app and Google Assistant, which is available on more than 1 billion devices. Google assistants make life easier by helping users do things like access media and manage their tasks. The proposed model also allows users to give gestures and access and manage media and functions. This application is a deep learning model integrated and deployed on the web using the library Tensorflow.js. The use of Tensorflow.js in the proposed model involves one of the significant factors called shareability. Tenserflow.js can be used to import the pre-trained model mobilenet. TensorFlow.js can run the Machine learning model in standard browsers like Firefox, Chrome, Edge, and Safari without any additional installations. As web applications can get shared easily on the Internet, it lowers the barrier for machine learning. It leads to an increase in the number of users using machine learning models for educational and development purposes. The wide use of such applications also motivates the developers to contribute to this field. From a developer's perspective, webs interactive nature, which interacts with device hardware such as the web camera, microphone, and accelerometer in the browser, allows easy integration between ML models and sensor data. The proposed model uses the interactive nature of the web to integrate it with the machine learning model for gesture prediction and the speaker feature to speak out the predicted text.

2 Literature Review

As a prototype, aByeongkeun et al. using Convolutional Neural Networks [1], proposed a run time sign language finger. American Sign Language is a specific and vital part of sign language recognition. The given system had depth sensors for capturing some additional information and to improve accuracy. It is a system for converting sign language to text or speech by capturing the given gestures continuously and processing them without using hand-held gloves and sensors. In this method, for recognition, only a few images are captured. As a prototype, a communication gap got created between ordinary people and mute people. communication gap got created between ordinary people and mute people.

The given system recognizes American Sign Language, which consists of 30 words in its vocabulary. Hidden Markov model (HMM) classified an appearance-based representation and hand tracking system and it was constructed. The RWTH-BOSTON-50 database has an error rate of 10.91%.

Another work related to this field was creating a sign language recognition system using pattern matching [2]. The main aim of this proposed work was to develop a plan that will work on sign language recognition. Many researchers have already introduced various sign language recognition systems and have implemented them using different techniques and methods. This proposed system is specialized in an approach which is to place the SLR system, which can work on Signs also as Text (which are going to be understandable by deaf and dumb persons and by average persons as well). The main task is going to be performed in two ways by the system. It will take input by the user in the form of text, which will then match the sign and vice-versa.

Sruthi Upendran [3] and et.al introduced "American Sign Language Interpreter for Dumb and Deaf People". Out of 24 static alphabets, 20 alphabets could be implemented in the procedure. The letters A, M, N, and S couldn't get recognized due to occlusion problems. The number of images used was limited.

TensorFlow.js users have developed applications that allow users with limited motor ability to control a web browser with their face [4] and perform real-time facial recognition and pose-detection [5]. Each of these examples is supported by a pre-trained convolutional neural network model called MobileNet. With the help of the Mediapipe framework, real-time gesture recognition was implemented, which could predict gestures such as "thumps-up", "five", and "spiderman" [6].

There have been methods where gestures could also be detected using unique bands. Myo band is used to detect foot gestures using acceleration signals or EMG. It is predominantly used for people with physical disabilities. The Myo controller within the sort of a band attached to the leg below the calf is liable for detecting a change in muscle tension. Myo sends some sort of data to the application. The application uses that data to control the virtual keyboard [7].

Also, Myo band is used in gaming platforms. It enabled using myoelectric signal to aid rehabilitation. The Myo band helps to find a solution by being interactive and gives the user a joyous experience. 'Myo armband' equipped with eight Electromyography (EMG) electrodes which are employed to retrieve data from the forearm muscle of the user [8].

"Real-time gesture recognition using deterministic boosting" [9] paper describes a system for automatic real-time control of a windowed OS entirely supported one-handed gesture recognition. Using a computer-mounted video camera, the system can securely interpret a 46-element gesture set on a 600 MHz notebook PC. The most recent sign-language recognition method is dependent on the representation of temporal constraints via Markov models [5, 6, 17, 21] to realize high-accuracy operation. The difficulty with systems supported by Markov models of temporal gesture behavior is that they restrict the range of gestures that will be accurately recognized. A gesture is reported to the OS if detected in three successive frames. Therefore, the effective frame rate is five fps, giving a test set of 3000 input images. The number of false positives was 4, like a 99.87% success rate. This paper has provided a new perspective to hand-gesture recognition, which attains exceptionally high recognition rates on long image sequences. This work was supported by the University of Oxford.

With the increase in the development of the Internet of things (IoT) technology, the interaction between people and things has become increasingly frequent. Simple gestures are used instead of complex operations for interacting with the machine; the smart data fusion features information then has gradually become an inquiry hotspot. The depth image of the Kinect sensor does not contain color information considering it and is susceptible to depth thresholds; in this paper, he authors has proposed a gesture segmentation method was supporting the color information fusion and depth information; to confirm the complete knowledge of the segmentation image, a gesture feature extraction method helped Hu invariant moment, and HOG feature fusion is proposed; and by the recognition of the optimal weight parameters, the worldwide and native features are effectively fused. Finally, the SVM classifier is employed to classify and detect the input gestures. The experimental results prove that the proposed system features a better gesture recognition rate and better robustness than the traditional method [10].

3 Methodology

This section explains the implementation of the proposed model (Fig. 1).

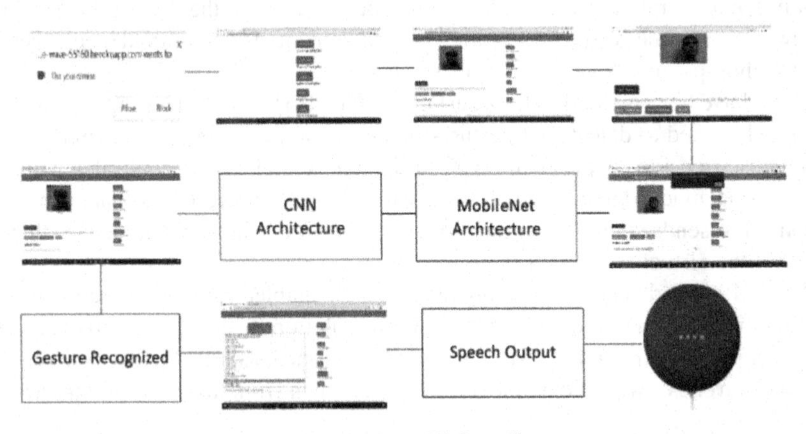

Fig. 1. Proposed model flow diagram.

The First step in the Flow diagram starts with integrating the webcam with the website. The training process will be the next step, with hand gestures. After the training process, a pop message will be alerted. The training processes hand gestures image go through mobilenet, and CNN architecture and later gestures can be recognized and displayed on the screen. After the recognition can choose the OS voice, and later, Speech output can be heard, and 6ywhich will activate Google Nest.

3.1 Asynchronous Execution

In a script called by a JavaScript function, the interpreter adds it to the call stack and then starts the function. When the present function is finished, the interpreter takes it off the stack and resumes execution where it left off within the last code. JavaScript is a single-threaded language at any given point in that time JavaScript is running at the most lines of JavaScript code, shared with tasks like page layout and event handling. Long-running JavaScript functions can cause page slowdowns or delays in handling events. JavaScript users rely on event callbacks and promises. A promise is an object which represents the ultimate completion or failure of a particular operation.

3.2 MobileNet

MobileNet is a CNN architecture model used for Image Classification and Mobile Vision. There are so many models available. But MobileNet model stands out differently from other models because it uses very little computation power to run and can apply transfer learning efficiently. So, it is best suited for embedded systems, Mobile devices, and computers without GPU or low computational power without compromising the results' accuracy. It is also a perfect fit for web browsers as browsers have limitations over computation power, graphic processing, and storage.

Mobilenet models are based on a streamlined architecture that uses depthwise separable convolutions and pointwise convolutions to build lightweight deep neural networks. In Depthwise Convolution a single convolutional filter is applied to each input channel, in contrast, Pointwise Convolution is a type of convolution which uses a 1×1 kernel that iterates through every single point. For MobileNet, a single filter to each input channel is applied by a depthwise convolution layer. Then pointwise convolution applies a 1×1 convolution and combines the output of the depthwise convolution with its own output. After that, the standard convolution filters and combines these inputs into a new set of outputs in one step. At that time, depthwise separable convolution splits this into two layers, one layer for filtering and another separate layer for combining. Computation power and model size can be reduced drastically using this factorization technique (Fig. 2).

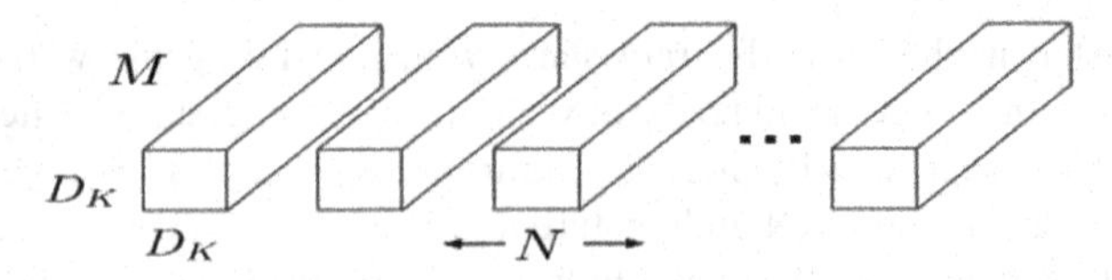

(a) Standard Convolution Filters

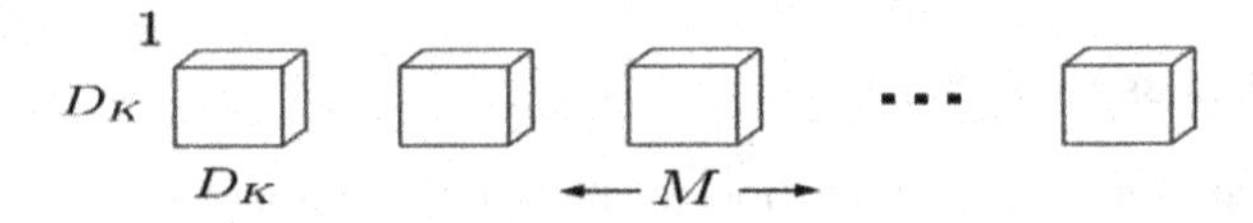

(b) Depthwise Convolutional Filters

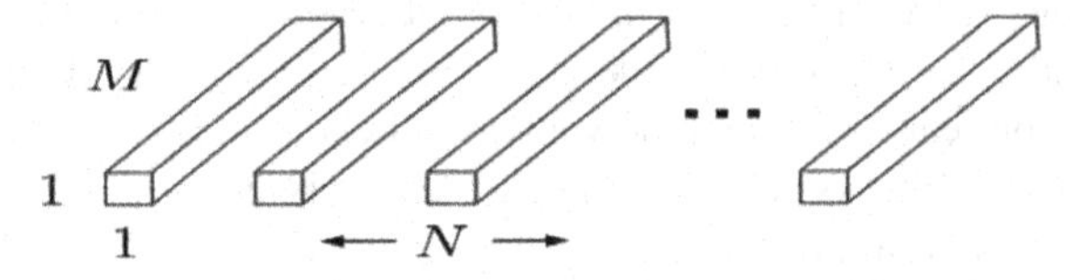

(c) 1×1 Convolutional Filters called Pointwise Convolution in the context of Depthwise Separable Convolution

Fig. 2. MobileNet architecture.

3.3 Activation Functions

Activation functions are mathematical equations used in Deep learning models that determine the output of a neural network. Activation Functions are attached to each neuron in the network, and it determines whether to activate that neuron or not. This decision is based on whether each neuron's input is relevant to the model's prediction.

Following is the list of Activation functions used in the proposed model:

1. Softmax Activation Function

Softmax Function is prominently used for classification problems where you must deal with multi-class problems. It is used for the normalization of the output of the neural network between zero and one. It is used to represent the certain "probability" in the output of the network. As we are handling with probabilities here, all the scores returned by the softmax function will add up to 1. Thus, the predicted is the item in the list whose score is the highest.

- Formula of Softmax Activation Function:

$$\sigma(\vec{z})_i = \frac{e^{z_i}}{\sum_{j=1}^{K} e^{z_j}} \tag{1}$$

2. RELU Activation Function (Rectified Linear Unit)

Rectified Linear Activation Function or RELU layer removes all negative values from the filtered image and replace it with zero's, and the positive values are kept as it is. RELU is used to avoid the values from summing up to zero. It is the most used activation function in neural network models as it reduces the computational complexity, easier to train, and achieves better performance (Fig. 3).

$$f(x) = \begin{cases} 0 \ if \ x < 0 \\ x \ if \ x \geq 0 \end{cases} \quad (2)$$

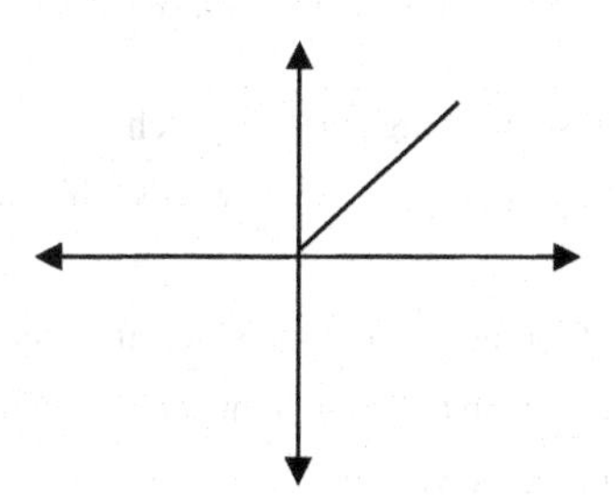

Fig. 3. Graph of Relu.

3.4 Text to Speech

Web Speech API is one necessary tool used to synthesis speech while converting text to speech and to recognize speech while converting speech to text. It is used for adding the feature of "Text to Speech" in a webpage using JavaScript. SpeechSynthesis API is easily available in all modern browsers — Chrome, Firefox, Microsoft Edge & Safari.

1. SpeechSynthesis

This interface controls the synthesis or creation of speech using the text provided. It is used to start, stop, pause, and resume the speech and get the voices supported by the device (Table 1).

Table 1. Methods available in "SpeechSynthesis" Interface

Speak ()	To start the speech
Pause ()	To pause the current speech
Resume ()	To resume the discontinued speech
Cancel ()	To cancel all the pending statements or speech generated, which are not yet played
getVoices ()	To get a list of all voices supported by the device

2. Speech Synthesis Utterance

This is used to create the speech or utterance using the text provided. Users can set the volume of the voice, the pitch of the voice, rate of speech using this feature. Users can also set the language of speech according to his/her requirement. After the object of this interface is created, then it is provided to the SpeechSynthesis object speak () method to get the voice output (Table 2).

Table 2. Properties of SpeechSynthesisUtterance Interface

Lang	To get and set the language of speech
Rate	To get and set the speed of the words which will be spoken
Rate	
Pitch	To get and set the pitch of the voice at which words will be spoken
Text	To get and set the text which must be spoken
Volume	To get and set the volume
Voice	To get or set the type of voice to be used
Onstart	Fired when speech has begun to be spoken
Onend	Fired when speech has finished
Onboundary	Fired when speech reaches a word or sentence boundary

3. window.speechSynthesis

JavaScript (JS) window has one property that is used to perceive the reference of the speech synthesis controller interface, which we will call the speak () method.

3.5 Algorithm

Declare mobilenet and other model variables and Initialize the webcam. Define a constant for webcam object stored in webcam.js. Initialize this constant by linking it to our webpage using the variable 'wc'. Need to create another Javascript file to save the dataset.

Define the function in which it sets up the webcam by calling the webcam. Setup and integrates our web application with the webcam. For the training, purpose load the mobilenet model followed by the creation of new tf.sequential model. Define a function called "loadMobilenet" which will load the pre-trained mobilenet model using instruction tfloaLayersModel. load the JSON model and use tf.loaLayersModel to load it into an object. The new model and its constructor can take inputs and outputs, which will be set to take the mobilenet inputs, namely the top of the mobilenet, and then get one of the output layers from the preloaded mobilenet i.e. Conv PW 13 relu as output. And return tf.model of the same. Now get one among the output layers from the preloaded mobilenet.

Need to define a new tf model class to make a new model, and its constructor can take inputs and outputs, which will set to take the mobilenet inputs, namely the top of the mobilenet and then Conv PW 13 relu as output. So that everything beneath that layer is going to be ignored once we connect a replacement set of layers to the present model. A function called to train in which need to create a new tf.sequential model. The first layer of this model would be the flattened output from the mobilenet model. The next layer will contain the activation function as "relu" and units 100, and the last layer with the activation function as "softmax" and units according to the number of gestures.

Need to create a class in the dataset file, and for the defined class, need to add two methods. The first method in which it takes a gesture with its label from the mobilenet model. Initially keep the xs as null. So set the xs to be the tf.keep for the gesture, and push the label into the label's array. For all subsequent samples, append the new example to the old. Do this by creating its temporary variable for the old set of xs called oldX, then tf.keep, and then abolish the oldX. Push the label to the array. The second method sets the y's to null, encodes labels passing it several labels. It will then one-hot encode and put the results into the dataset.ys. The inputs of this will be the results of the output of the truncated mobile net. for model.fit need to provide it the dataset x's and dataset y's training for ten epochs (Figs. 4 and 5).

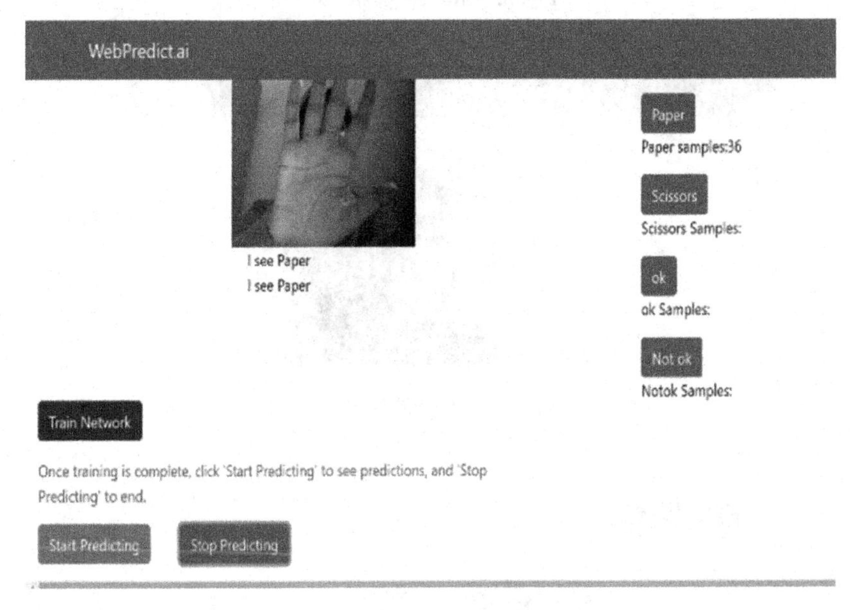

Fig. 4. Demo output.

Define a function "startPredicting" which calls the predict function need to initialize a variable for setting as the prediction starts, can append into the set. Define a function "stopPredicting" which stops the prediction process. After prediction stops the predicted text appending into the set will stop and need to convert the set into a string and print it in the UI as a paragraph in HTML i.e., < p > like Fig. 6.

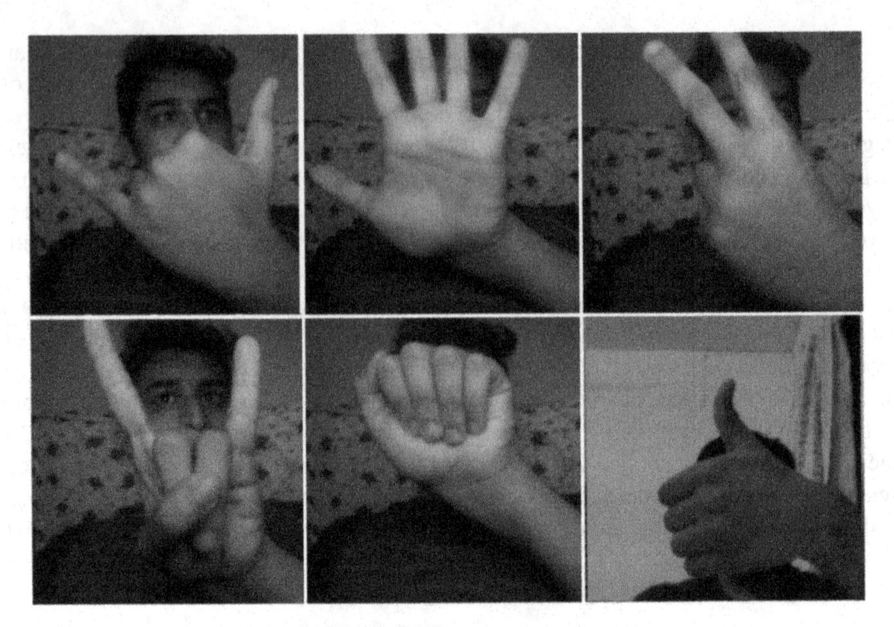

Fig. 5. Different gestures.

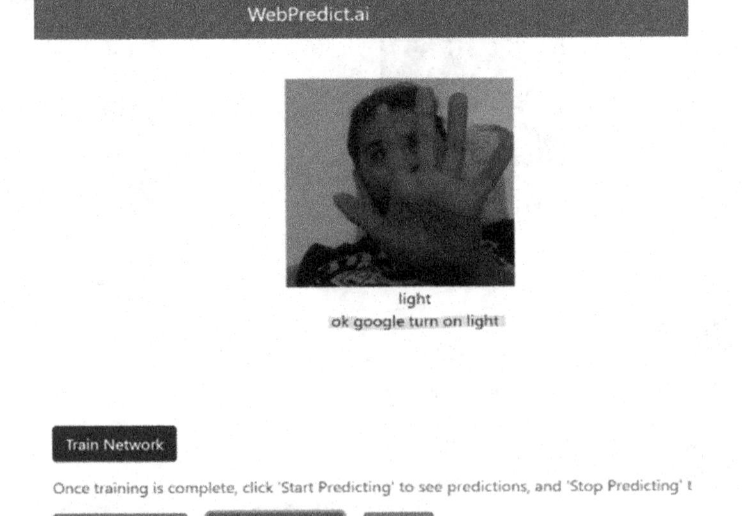

Fig. 6. String formation.

The DOM features of JavaScript need to check whether the text is predicted from the gestures that have formed and the string has occurred on the main web page.

For the string, we will addEventListener in which we call the class SpeechSynthesisUtterance. Can set functions for the voice volume, pitch, rate, and text for the string which will be spoken. SpeechSynthesis will help us for the word to speak () with different OS voices like Fig. 7 is an example for Microsoft OS device, and browsergoogle chrome. After the voice command is successfully spoken, will activate Google Nest or Google smart home.

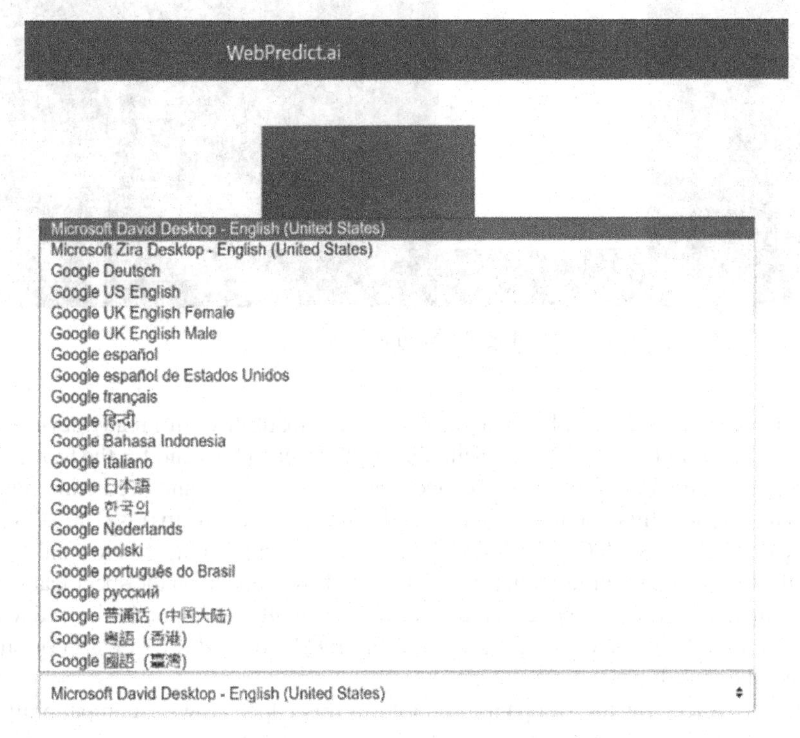

Fig. 7. OS voice google chrome.

4 Conclusion

The proposed model successfully recognizes the gestures captured by the webcam on the webpage and converts them into text, which can be seen on the webpage. The gestures captured by the webcam and recognize with an accuracy of 98.14 ± .025. It further converts that text into speech using a voice module with different OS voices. The output which is obtained in the form of voice is given as an input to the google home. Then the Google Home follows the command and performs the task assigned to it. The functions such as turn on and turn off light which is a functionality of google assistant, which commands another device that is connected with through Bluetooth (Fig. 8).

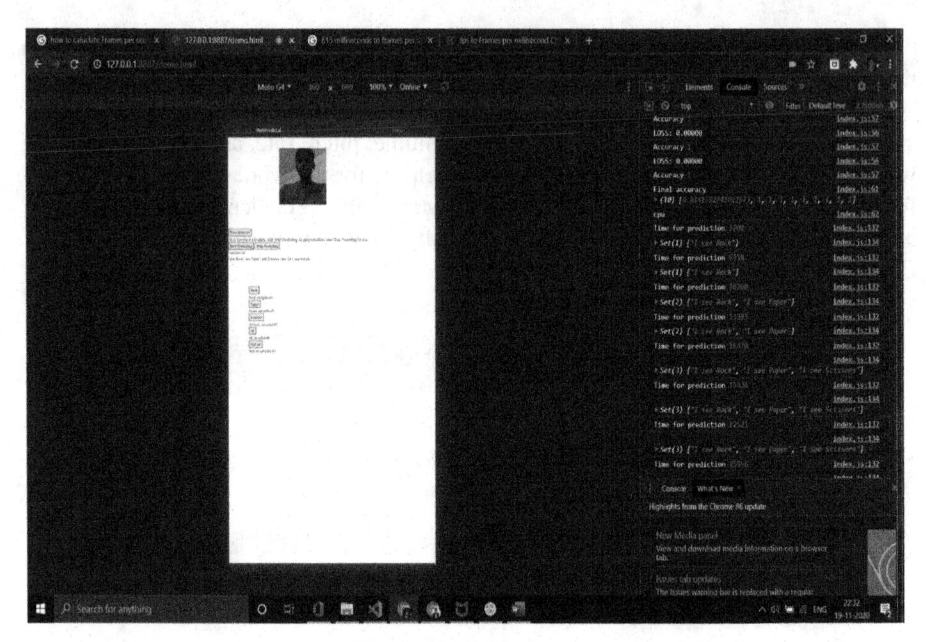

Fig. 8. Accuracy array.

Using tensoflow.js web platform maintaining compatibility with many devices and execution environment. WebGL backend the most powerful backend for the browser. In the proposed model the frames per second (fps) is 0.00166016 and prediction time as 615ms at the same time vanilla CPU backend frame per second (fps) is 0.0002702703 and prediction time is 3700ms. WebGL is the better performer compared to vanilla CPU cause the time taken for prediction is less. The speed will depend on the specifications of your computer or mobile device as the application is built to work on both the devices. This algorithm should be working with all the different kinds of skin color as the gestures are trained on run-time.

The OS voices for the webpage proposed in this paper browser compatibility for different browsers are (Table 3):

Table 3. OS voices.

Browser	Number of voices
Google chrome	33
Firefox	49
Safari	7
Opera	21
Edge	14
Internet explorer	NA

5 Future Scope

Future work will focus on improving the performance of the proposed system and on-device compatibility. In the future, we can extend or expand the functionality of the model. Users may use the proposed model to switch TV channels without a remote, control the audio of the TV, play games, and to surf the internet. It can also be used to handle all the electronic devices of our homes via hand gestures. It can be used in an automated sign language translator where we can convert the signs into text. In the future, we can also train the model to unlock the user's mobile and laptop.

References

1. Kang, B., Tripathi, S.: Real-Time Sign Language Fingerspelling Recognition Using Convolutional Neural Networks from Depth Map, June 2016. https://ieeexplore.ieee.org/document/7486481
2. Keni, M., Meher, S.: Sign Language Recognition System, December 2019. https://www.ijser.org/paper/Sign-Language-Recognition-System.html
3. Upendran, S., Thamizharasi, A.: American sign language interpreter system for deaf and dumb individuals, July 2014. https://www.researchgate.net/publication/286240761_American_Sign_Language_interpreter_system_for_deaf_and_dumb_individuals
4. Ramos, O.: Handsfree (2018). https://handsfree.js.org/
5. Friedhoff, J., Alvarado, I.: Move Mirror: An AI Experiment with Pose Estimation in the Browser usingTensorFlow.js. TensorFlow Medium, July 2018. https://medium.com/tensorflow/move-mirror-an-ai-experiment-with-pose-estimation-in-the-browser-using-tensorflow-js-2f7b769f9b23
6. On-Device: Real-Time Hand Tracking with Media Pipe Valentin Bazarevsky and Fan Zhang, August 2019. https://ai.googleblog.com/2019/08/on-device-real-time-hand-tracking-with.html
7. Dobosz, K., Trzcionkowski, M.: Text input with foot gestures using the Myo Armband. In: Miesenberger, K., Manduchi, R., Covarrubias Rodriguez, M., Peňáz, P. (eds.) Computers Helping People with Special Needs. ICCHP 2020. Lecture Notes in Computer Science, **12377**, 335–342. Springer, Cham (2020). https://doi.org/10.1007/978-3-030-58805-2_39
8. Bhattacharyya, A., Muzumder, O., September 2018. https://ieeexplore.ieee.org/abstract/document/8554686
9. Lockton, R., Fitzgibbon, A.W.: Real-Time Gesture Recognition Using Deterministic Boosting (2002). https://pdfs.semanticscholar.org/73f0/a4b107e6f30c2957bf454e4115f72190d50e.pdf
10. Tan, C., Sun, Y., Li, G., Jlang, G., Chen, D., Liu, H.: Research on gesture recognition of smart data fusion features in the IoT. Neural. Comput. App. **32**, 16917–16929 (2020). https://doi.org/10.1007/s00521-019-04023-0

A MapReduce Approach to Automatic Key File Updates for SPT (Squeeze, Pack and Transfer) Algorithm

Shiv Preet[1(✉)] and Amandeep Bagga[2]

[1] Department of Information Technology, iNurture Education Solutions Ltd., Banglore, Karnatka, India

[2] Department of Computer Science and Engineering, Lovely Professional University, Phagwara, Punjab, India

Abstract. SPT algorithm helps in reducing network congestion by using loseless compression methodology. It helps in improving network usage by efficiently utilizing the available network resources and reducing size of dataset by using inbuilt compression algorithm. This paper epitomizes the use of MapReduce for finding out symmetrical binary keys for SPT (Squeeze Pack and Transfer) algorithm from various data files available on public cloud domains. This paper describes the use of MapReduce functionality which can help in finding out best contender keys for creating generic binary key file for the SPT algorithm which is used for binary compression. This paper discusses the manipulation of MapReduce for creating an automated system which can extract regular binary patterns from data files available on public cloud domain and inculcating them in generic key file which can be used for compressing and decompressing files with the help of the SPT algorithm. It can not only eliminate the manual generation of the keys for SPT algorithm but also it helps in reducing the time required for creation and updates of the key files.

Keywords: HADOOP · MapReduce · Mappers · Reducers · Binary keys · Shuffler · Map · Big data · Artificial intelligence · Knowledge nugget · SPT algorithm

1 Introduction

Big data is booming with the expansion of cloud storage and diverse multimedia. People in general are storing their personal data like images, videos and other such files over the cloud [1]. Google Drive and Drop Box are two prominent examples which are being used massively for cloud data storage. Archive.org has lot of data files for their users. These data files are in the form of multimedia, text files, PDF, Audio files etc. People across the world access these data files to quench their infotainment as well as intellectual thirst [2]. Hospitality industry, Expert diagnosis systems etc. rely on big data to extract information from it and to improve their functionality automatically by discovering new knowledge nuggets and upgrading existing knowledge base. YouTube,

M. Bhattacharya et al. (Eds.): ICICCT 2021, CCIS 1417, pp. 132–140, 2021.
https://doi.org/10.1007/978-3-030-88378-2_11

Netflix, Amazon Prime has all their data available on public or private cloud servers. These companies use their proprietary algorithms which automatically track usage of their clients and recommend appropriate videos as per their likings. These algorithms extensively use artificial intelligence to study likings and disliking of the users and subsequently recommend them suitable video titles on the basis of their likings. All this happens without the intervention of human beings.

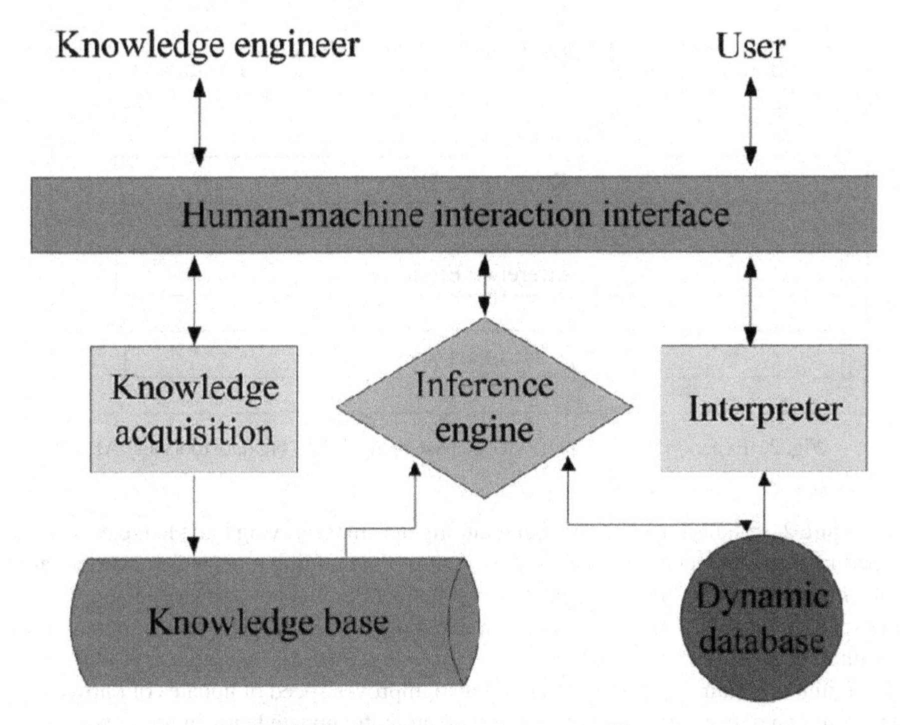

Fig. 1. Structure of expert system [3]

Figure 1 explains basic structure of expert system. Figure 1 depicts that in an expert system, knowledge base is the core on which every other aspect of the expert system relies. It is the knowledge base which provides information to the inference engine which in turn provides result to the user through user interface [4]. Knowledge can be fed in the knowledge base using human interaction or some other means. In the conventional expert systems, it was initially human interaction which was used to feed knowledge base [5]. Now a day, knowledge base in expert system can collection data from various automated aids by using algorithms created in artificial intelligence. Knowledge base can be updated on its own with minimal or no user interaction with the help of artificial intelligence. Machine learning helps an expert system in extracting new information from existing data or from new data inputs without any human intervention. New knowledge nuggets can be generated on the basis of updated rule base and obsolete knowledge nuggets can be discarded automatically [6]. It also means that these updates can be done on a faster pace using automated machine learning tools.

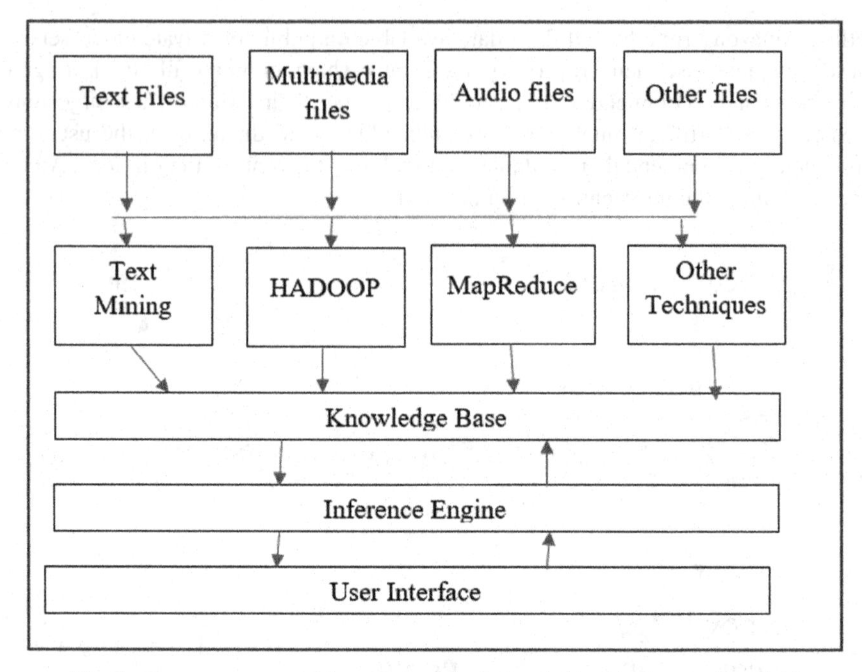

Fig. 2. Expert system with knowledge base being fed by big data tools and AI

Figure 2 explains the use of artificial intelligence in improving knowledge base of an expert system. Artificial intelligence uses various data mining tools which extract valuable information from the available data and helps in producing knowledge nuggets for an expert system [7]. These nuggets are updated in the knowledge base of expert system for improving its performance. Advantage of using artificial intelligence is that it not only eliminates human intervention but it also improves speed of updates of knowledge base. Human data entry is also error prone. Automatic update helps in removing errors while entering data in an expert system. It can check all the present information in the knowledge base and remove any kind of obsolete information on the fly [8].

2 MapReduce

MapReduce is an algorithm which distribute task over the network and combine the result to give final output to the user [9]. It has two functions namely MAP and REDUCE. MAP breaks down huge set of data into smaller subsets. Each subset is converted into records or tuples which have values associated with keys (Key and Value Pairs). Once the job of MAP function is done, output of MAP is passed to REDUCE function as its input via shuffler [10]. Now REDUCE function combines those inputs (output of MAP function) into bigger tuples. Shuffler acts as a catalyst in transferring output from mapper to Reducers. Mappers are those functions which process data subsets which have been given as input to MAP function [11]. Mappers can also be considered as the preliminary elements for MAP function. In the same way Reducers are those functions which act on the input (output of Mappers) given to reduce function. Should be numbered. Lower level headings remain unnumbered; they are formatted as run-in headings.

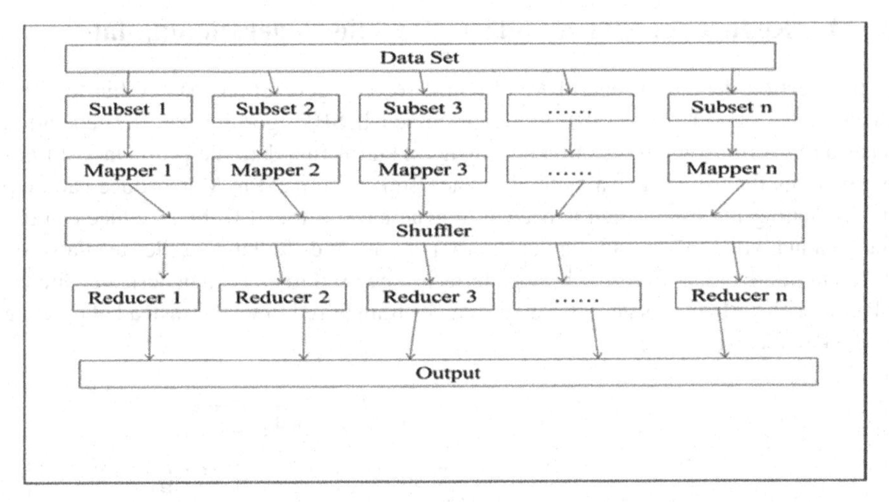

Fig. 3. Generic MapReduce algorithm.

Figure 3 shows basic working of MapReduce algorithm. Initially a huge dataset S is converted into smaller subsets.

$$S \Rightarrow \{s1,\ s2, s3, \ldots\ si, \ldots, sn\}$$

Each subset is given to MAP function as an input. MAP function assigns each dataset to a specified mapper from the set of mappers. Here mappers are preliminary elements in the map function.

$$M \Rightarrow \{m|m \text{ is a mapper in MAP function}\}$$

Each mapper processes dataset which has been given to it as an input. A temporary output is created from each mapper.

$$TO \Rightarrow \{to\ |to \text{ is a temporary output from mapper}\}$$

Each temporary output is transferred to shuffler which further transfers them to reducers as their input. Output of the mappers acts as input for the reducers. Shuffler acts as a temporary buffer for transferring the output from mappers to reducers. Each reducer processes its input and produce final output O.

$$0O \Rightarrow \{\ o|o \text{ is Final Output from Reducer}\}$$

This output is rejoined to create a single result file R which is the final outcome of MapReduce function.

3 MapReduce for SPT Standard Key File Generation/update

In this section detailed usage of MapReduce for key generation for key file for SPT (Squeeze Pack and Transfer) algorithm is discussed. In SPT algorithm, key file generation takes a lot of time as it has to find out regular patterns in a data file [12]. Once it finds out patterns then it assigns a symbol to each and every pattern. MapReduce can help in eliminating the time which is needed to generate key file [13]. MapReduce can use big data analytics to find out regular binary patterns in different data files available on different pubic cloud servers. These patterns can be used to create a generic key file for SPT algorithm [14]. This generic key file can be transferred to sender instead of creating a new key file every time.

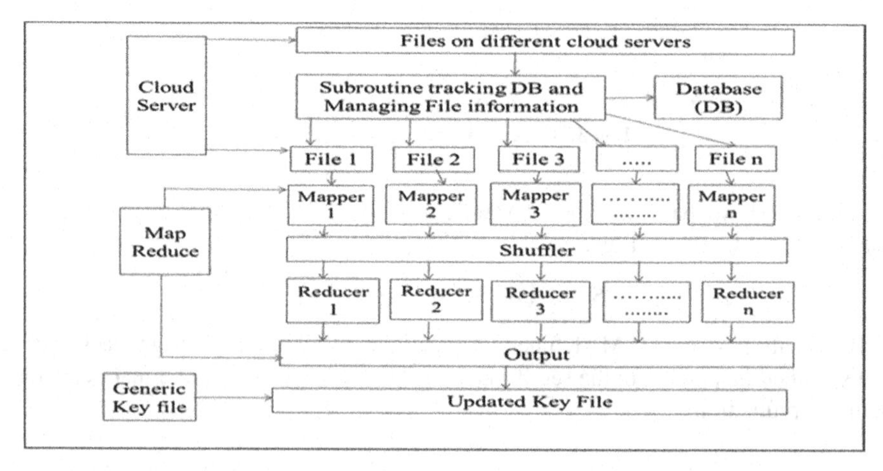

Fig.4. MAPREDUSE for creation / update of generic key for SPT algorithm.

Figure 4 shows detailed usage of MapReduce algorithm for generating generic key file for SPT (Squeeze Pack and Transfer) algorithm. It demonstrates three major steps in the generation of key file for SPT algorithm. First Step is the interaction with the cloud servers. Second step is the use of MapReduce functionality for generating key file. Third and file step is the generation of key file. Initially a generic algorithm checks for various data files available on public cloud servers. This generic algorithm can be a variant of original MapReduce algorithm where a subroutine will scrutinize all the popular data files on public cloud servers [15]. Information about all these files is stored in the database **DB**. This comprehensive database DB also has information of all earlier inspected data files. Database DB will have three fields, **File Name** (in string), **File Location** (in string) and **File Size** (in bytes). Database DB can be based on file databases and it can store data in txt or csv formats. But it is better if this database is stored on client server architecture [16]. SQL Server, ORACLE or some other renowned relational database management systems are better candidates for this database [17].

Now MapReduce function **MR** will use its **MAP** subroutine and map popular data files available on public cloud servers (These files will be assigned to MapReduce function by the subroutine which is monitoring database DB) to different Mappers [18]. It can assign each file to a unique mapper or it can decide to create subsets of the file and assign each subset to a different mapper [19]. This decision will be taken on the basis of size of file and number of Mappers available. Now each mapper will process the file and produce a temporary output of binary keys along with their frequency count of each binary key [20]. These temporary outputs will be sent to **shuffler** which sends it to **reducers** as their inputs [21]. Shuffler in a way acts as a buffer between mappers and reducers. It temporarily stores the output from mappers and then transfers it to reducers [22]. Task of the reducers will be to find total frequency count of each binary key. These Reducers will produce binary key file for SPT algorithm. Binary keys having frequency count less than a threshold will be discarded. This threshold value can be calculated using Max() and Min() functions as shown in Fig. 5 [23].

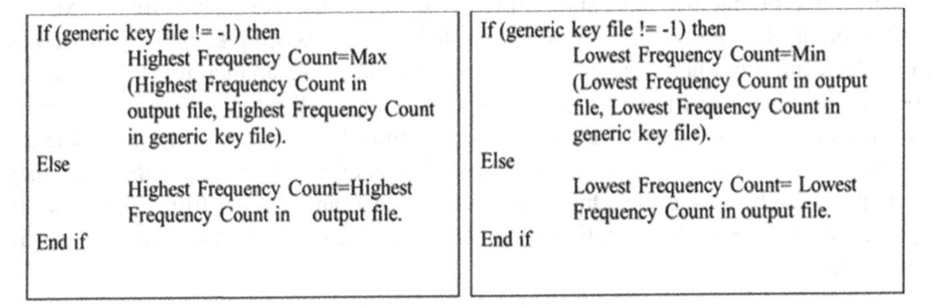

Fig.5. MAPREDUSE for creation/update of generic key for SPT algorithm.

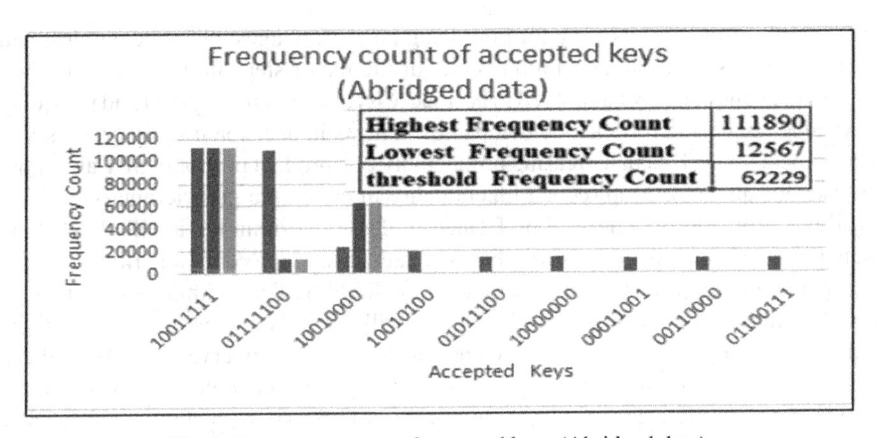

Fig.6. Frequency count of accepted keys (Abridged data).

Figure 5 demonstrates that how Highest Frequency Count and Lowest Frequency Counts are selected. Figure 6 shows the graphic version of abridged data for the accepted keys along with their count. It has shown both Highest Frequency Count and Lowest

Frequency Count along with Threshold Frequency Count which is the average of Highest Frequency Count and Lowest Frequency Count. Highest frequency count is selected using max () function on highest frequency count in the output file and highest frequency count in the existing generic binary key file if generic key file exists otherwise only highest frequency count of output file is considered.

Similarly, lowest frequency count is selected using min () function on lowest frequency count in the output file and lowest frequency count in the existing generic binary key file if generic key file exists otherwise only lowest frequency count of output file is considered.

Finally, Threshold frequency count is calculated using mean of highest frequency count and lowest frequency count.

Threshold frequency count = (Highest Frequency Count + Lowest Frequency Count)/2.

Binary keys having frequency count greater than or equal to threshold frequency count will be sorted on their frequency count in descending order. These binary keys will be sent as output to another subroutine. It will check these keys with already existing keys in binary key file. If keys in the output file and pre-existing keys (in Generic key file) match and if frequency count of those matching keys are smaller than the frequency count in the generic key file then these pre-existing keys will get new rank in the generic keys file on the basis of new frequency count. Similarly new keys from output file are updated in the generic key file and their position is also ranked as per the frequency count provided by output file. Once this process is over and all keys from output file are updated in the Generic key file, new symbols are provided to the freshly created keys in generic file.

4 Conclusion

SPT algorithm can pave the way for improving network congestion in the wireless and mobile networks. Key file generation is one of the major steps in the SPT algorithm. MapReduce functionality has proved to be a success for analyzing big data and providing knowledge nuggets to fortune 500 companies. It helps in decision making processes of managements as well as in increasing productivity of a product portfolio. SPT algorithm can take advantage of MapReduce functionality in creating a generic binary key file. It will not only help in optimized performance of the algorithm but it will also help in finding most common pattern among different data files. MapReduce algorithm can also help in automatic updates of generic binary key file of SPT algorithm without manual help. MapReduce Algorithm if used properly will not only increase the productivity of the SPT algorithm but also its efficiency without human intervention. As artificial intelligence is becoming an integral part of Information Technology industry, in the coming time MapReduce can play a pivot role in improving capabilities of the SPT algorithm.

References

1. Preet, S., et al.: An overview of the Internet of Things and its research issues. IJCTA (2016)

2. Preet, S., Luhach, A.K., Luhach, R.: Comparison of various routing and compression algorithms: a comparative study of various algorithms in wireless networking. In: Unal, A., Nayak, M., Mishra, D.K., Singh, D., Joshi, A. (eds.) SmartCom 2016. CCIS, vol. 628, pp. 128–134. Springer, Singapore (2016). https://doi.org/10.1007/978-981-10-3433-6_16
3. Sun, Q., Zhang, M., Mujumdar, A.S.: Recent developments of artificial intelligence in drying of fresh food: a review. Critical Rev. Food Sci. Nutr. **59**, 2258–2274 (2018)
4. Preet, S., Bagga, A.: Squeeze pack and transfer algorithm: a new over the top compression application for seamless data transfer over wireless network. IJITEE **08**, 175–179 (2019)
5. Williams, A., Mitsoulis-Ntompos, P., Chatziantoniou, D.: Tagged MapReduce: efficiently computing multi-analytics using MapReduce. Springer **68**, 240–251 (2011)
6. Ding, L., Liu, S., Liu, Y., Liu, A., Song, B.: Efficient processing of multi-way joins using MapReduce. Springer **5**, 1184–1195 (2015)
7. Li, Y., Chen, Z., Wang, Y., Jiao, L.: Quantum-behaved particle swarm optimization using MapReduce. In: Gong, M., Pan, L., Song, T., Zhang, G. (eds.) Bio-inspired Computing – Theories and Applications: 11th International Conference, BIC-TA 2016, Xi'an, China, October 28-30, 2016, Revised Selected Papers, Part II, pp. 173–178. Springer, Singapore (2016). https://doi.org/10.1007/978-981-10-3614-9_22
8. Arputhamary, B.: Skew handling technique for scheduling huge data mapper with high end reducers in MapReduce programming model. In: Jain, L.C., Peng, S.-L., Alhadidi, B., Pal, S. (eds.) ICICCT 2019. LAIS, vol. 9, pp. 331–339. Springer, Cham (2020). https://doi.org/10.1007/978-3-030-38501-9_33
9. Dhanani, J., Mehta, R., Rana, D., Tidke, B.: Back-propagated neural network on MapReduce frameworks: a survey. In: Tiwari, S., Trivedi, M.C., Mishra, K.K., Misra, A.K., Kumar, K.K. (eds.) Smart Innovations in Communication and Computational Sciences. AISC, vol. 851, pp. 381–391. Springer, Singapore (2019). https://doi.org/10.1007/978-981-13-2414-7_35
10. Zhang, X., Shui, J., Zhibin, J.: A scheduling method based on deadlines in MapReduce. In: Wang, X., Wang, F., Zhong, S. (eds.) Electrical, Information Engineering and Mechatronics 2011: Proceedings of the 2011 International Conference on Electrical, Information Engineering and Mechatronics (EIEM 2011), pp. 1585–1592. Springer, London (2012). https://doi.org/10.1007/978-1-4471-2467-2_189
11. Kalach, G.G., Romanov, A.M., Tripolskiy, P.E.: Loosely coupled navigation system based on expert system using fuzzy logic. IEEE **08**, 167–169 (2016)
12. Magro, D., Paulino, H.: In-cache MapReduce: leverage tiling to boost temporal locality-sensitive Mapreduce computations. IEEE **1**, 374–383 (2016)
13. Xu, X., Tang, M.: A new approach to the cloud-based heterogeneous MapReduce placement problem. IEEE **09**, 862–871 (2016)
14. Rattanaopas, K., Kaewkeeree, S.: Improving Hadoop MapReduce performance with data compression: a study using wordcount job. IEEE **03**, 564–567 (2017)
15. Merla, P.R., Liang, Y.: Data analysis using Hadoop MapReduce environment. IEEE **01**, 4783–4785 (2017)
16. Sheoran, A., Malathi, D., Senthil Kumar, K.: MapReduce scheduler: a bird eye view. IEEE **06**, 213–217 (2017)
17. Afzali, M., Singh, N., Kumar, S.: Hadoop-MapReduce: a platform for mining large datasets. IEEE **03**, 525–518 (2016)
18. Lin, J.-W., Selvi, A., Arul, J.: A performance-satisfied and affection-aware MapReduce allocation scheme for intelligent information applications. IEEE **01**, 46–51 (2016)
19. Ayub, M.S., Siddiqui, J.H.: Poster: efficiently finding minimal failing input in MapReduce programs. IEEE **01**, 177–178 (2018)
20. Malik, M., Tullsen, D.M., Homayoun, H.: Co-locating and concurrent fine-tuning MapReduce applications on microservers for energy efficiency. IEEE **01**, 22–31 (2017)

21. Chen, H.-L., Shen, Y.-S.: Reducing imbalance ratio in MapReduce. IEEE **01**, 279–282 (2017)
22. Jenifer, X.R., Lawrance, R.: Classification of microarray data using SVM MapReduce. IEEE **01**, 101–109 (2017)
23. Preet, S., Bagga, A.: Lempel–Ziv–Oberhumer: a critical evaluation of loss less algorithm and its applications. ICCS **01**, 175–185 (2018)

Measurement of Liquid Level Using Prediction Methodology

Trupti Nagrare(✉)

S.B. Jain Institute of Technology, Management and Research, Nagpur, India

Abstract. There are various applications in which it is required to measure level of liquid like chemical industry, food industry, medical tests etc. The methods for level measurement are also varies in every place. In this paper, work is done on video-based level measurement for scientific purpose where it requires accuracy in millimeter range. The camera for video capturing is placed at some distance and by image processing the level of liquid can be detected. Also, the time required to detect level is also a significant parameter over here because it has to read level in real time and it is changing continuous during calibration(experiment), so just after change in level, the system has to report that change. Here the level and volume are measured using image processing and for getting accuracy, curve fitting and machine algorithm is applied. With this, it achieves accuracy up to 98% in both measurements.

Keywords: Liquid level measurement · Image processing · Level and volume prediction machine learning theory · Algorithmic mechanism design · Computational pricing and auctions · Maximum likelihood estimation

1 Introduction

The liquid level measurement is required to explore more as there are various applications which need accuracy in the measurement. The accuracy required for such application is in micro range which is not possible to achieve using mechanical measurement devices like pressure sensors in that case. In most of the laboratory experiment and research when certain object is added or removed from the liquid, it changes the level. During this the liquid lost its stability and if reading is taken within few seconds it gives variable reading for level. The liquid takes some time to settle and this time depends on type of liquid, its viscosity and type of container. In most of the level measurement applications the accuracy is not so important at micro level like in dams, tank level measurement [1], bottle filling application [2], hydraulics and hydrology [3]. In pharma or chemical industry, an accuracy of liquid level measurement is required to a great extent. It's difficult to get such accuracy using existing technics, hence the need of new method which gives more accuracy in less time. This paper gives the computer vision-based method for liquid level measurement. The applications in which it requires visual level measurement, the container used must be transparent. Further, the paper is organized as: Sect. 2 gives the study about the previous work, Sect. 3 describes the method and system

M. Bhattacharya et al. (Eds.): ICICCT 2021, CCIS 1417, pp. 141–152, 2021.
https://doi.org/10.1007/978-3-030-88378-2_12

in detail, Sect. 4 explains the detail methodology for level and volume measurement, Sect. 5 shows the result and last Sect. 6 is about conclusion of the work.

2 Literature Review

Most of the work done in the level measurement is for water level measurement like in tank, rivers, bottles, dams. For measurement of water level in tank T. H. Wang uses digital camera and float method [1]. In this paper they used single digital camera placed on top of tank and 3 floats inside water level, images taken by camera and the pixel counts of the float in image were determine first for calculating the diameter, a subpixel resolution during the measurement was achieved. The mean square error was -/ + 1%. In [2] paper digital camera is used to detect the water level. In [7] paper also research was on water level measurement of industrial boiler, which uses image recognition algorithm such as improved BP neural network and feature matching. The complete system was based on CCD and FPGA. In [3] authors used image thresholding techniques and contour fitting algorithm to detect correct liquid level in bottle. There are two types of thresholding techniques adaptive thresholding and global thresholding techniques. The disadvantage of global thresholding technique is that it provides poor results under varying light condition. Under varying light condition Adaptive Thresholding technique provide better results. In paper [4] also liquid level was detected using camera as well as IR sensor for label information. The [9] paper recognizes the liquid surface and liquid level in various transparent container using computer vision for chemistry applications. The authors determine the curves of liquid using image recognition then used that to find liquid level, filled liquid and phase boundaries. The instrument designed in [5] was detected the water level and also able to compute the concentration of additives in the liquid. For this purpose, a multisensor model comprising a capacitive level sensor (CLS), ultrasonic level sensor (ULS), and capacitance pressure sensor is used to obtain information of the liquid. Fars Esmat Samann in his paper [11] works on liquid level as well as color detection using camera and image processing and also gets required accuracy. There are various methods for the level measurement of liquid but it has some drawbacks. first is, if some material added into liquid, it will take some time to settle down nearly 200–300 s (depends on liquid) and if readings taken before settling so it's not giving the accuracy. Second is, if an object inserted into the liquid and taken out, some liquid drops will also go out with the object. This will reduce the original liquid level. These issues are very important for level measurement but till now any of the researcher had not considered it in their research. In the level measurement area, most of the research is done on water level but other liquid level measurement is also very essential in many fields. There is some difference in measurement of water and other liquids. So here considering all these issues level measurement study is carried out but the prime goal is to measure the liquid level (Fig. 1).

3 Method and System Description

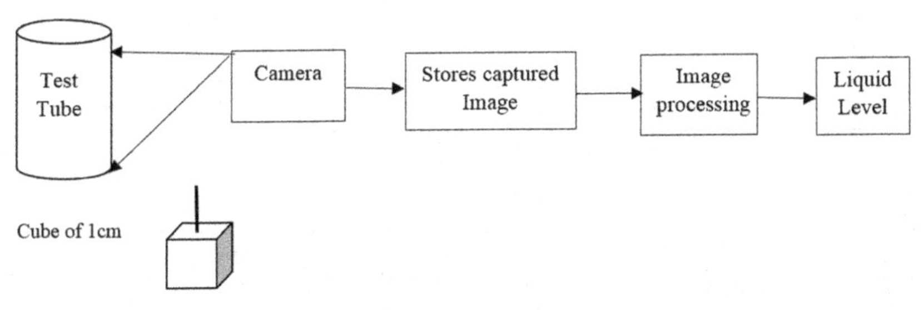

Fig. 1. Experimental setup for liquid level measurement

In the experimental setup web camera is placed in front of test tube which captures the continuous video for level detection. Liquid may be of any type. A cube of 1cm is taken to insert in test tube. With the help of thread, a cube is inserted and taken out from the tube. Repeat the process of cube insertion in the tube and every time the reading taken for liquid level must be same. But in experiment we get the reading for repetitions is less than the previous one as it is shown in Figs. 2 and 3. Suppose the original level of liquid is L1 and after inserting the cube, level changes by x. Take out the cube and insert again, now this time level changes which is less than x. It is continuing to decrease the level for each time the cube inserted. The graph in Fig. 2 shows the relation between number of time cube inserted and change in liquid level.

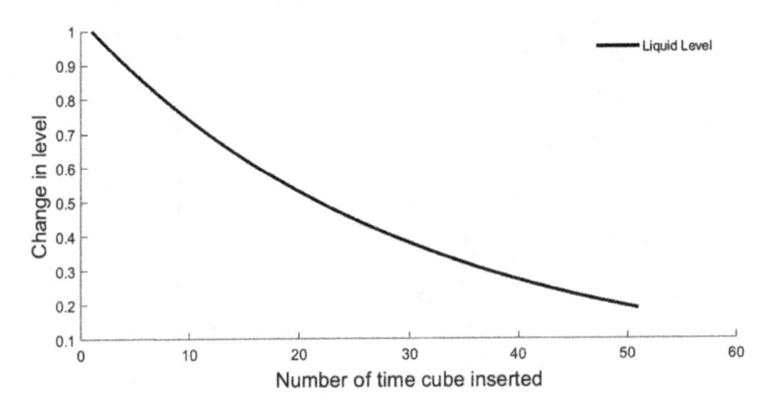

Fig. 2. Change in level with respect to number of time cube inserted

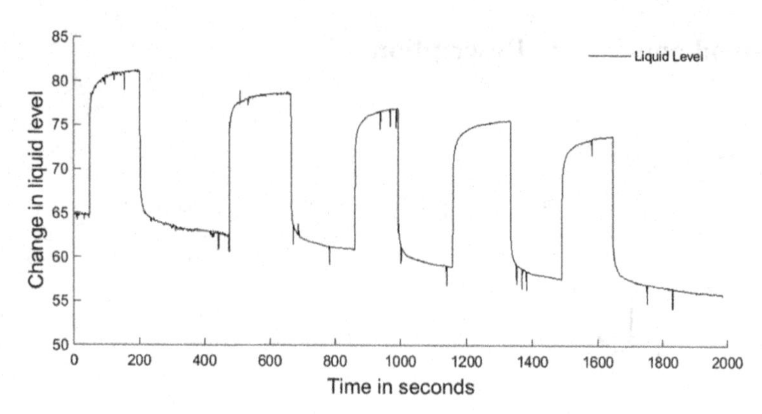

Fig. 3. Change in liquid level with respect to time in seconds

When cube is inserted and reading is taken after few seconds then the reading is more accurate. For initial time we always get the false reading and after few seconds the liquid get settle and the correct reading will be recorded. This settling time of liquid will be different for different type of liquid, it depends on the viscosity of the liquid. One more point need to consider here is that if the surface of test tube near the liquid is dry, then the reading is different than the reading when the surface is wet with the liquid filled. This is because of cohesive force by test tube, it affects only during the first reading, till the surface is dry.

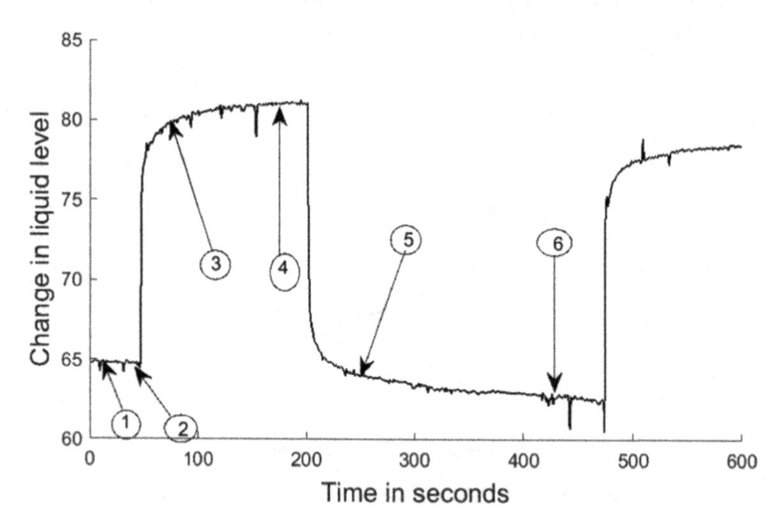

Fig. 4. Change in liquid level with respect to time

In Fig. 4, point 1 shows the initial level of liquid. At point 2 a cube of 1cm is inserted in the liquid which increases the level at point 3 at 100 s. The cube is kept inside for next 100 s, till this time level settle at point 4. In this period level varies from point 3 to 4. After this cube is removed from liquid due to which level drops down at point 5. Some time is given to settle the liquid nearly 150 to 200 s. In this period liquid level changes from point 5 to point 6. The cube material is also needed to consider here. If the cube material absorbs liquid or change its own shape then in this case the reading will be affected.

4 Methodology

When the process of level measurement started, liquid is filled in the tube at known level say L1, camera is turn on and start taking video. The 1cm cube is then inserted in the tube which change the liquid level, say x. after few seconds it is removed from tube, again the level is return to its original level. Then insert two 1cm cubes and take the reading for change in level. Repeat the same process for 3, 4, 5, 6 and 7 cubes.

4.1 Image Processing

The camera is capturing continues video for level measurement. This video is then process by converting it into frames of video. A frame is taken from the running video and stores it. This image is then converted to gray scale. This gray image is then converted to binary form that is in black & white mode. Then in this image black portion indicates the background whereas the white portion indicates the region of interest that is level in the liquid. After that, centroid is calculated for detecting the larger white portion in the image by eliminating the other areas. Liquid in the tube is increase in convex or concave shape according to the type of liquid. The following algorithm shows steps to convert image captured into the region of interest and detect its area which is the calibrated to get the liquid level.

4.2 Algorithm for Image Processing

- Start video streaming using camera
- Take frame from stream
- Apply preprocessing on frame
- Apply image processing algorithm for level detection

  - Normalize the image frame

- convert to gray image
- convert to black & white image
- Find maximum white area object in image
- Return centroid of area

4.3 Calibration for Measurement

For calibration we need standard scale value or distance between two points which known and by image process we also have distance between in pixel. Therefore, we can formulate the ration between actual scale distance & pixel distance for the given level as-

y known distance = x pixel distance.

Therefor any level we can find the unknown distance as.

Unknown distance = (y known distance)/(x pixel distance) x z pixels.

For calibration of the measurement, consider there are two levels in the tube, one is reference level which is taken at initial condition and second is the level which comes after some change in the liquid. The change in the liquid cause due to any reason like adding any material in the tube or removing it from liquid. So here we consider two conditions.

Framen- nth frame from video stream or live from sequence.
ΔL = Difference between initial level and current level.
L1 = Initial level.
Take next frame from stream.
Detect_level(frame).
Find ΔL = L0-L1.
if $\Delta L >$ threshold.
delay() // for stabilization of liquid level.
Detect_level(current frame).
If system is linear use calibration method to detect level.
If system is non linear use prediction mechanism to detect level.
Read new frame again from video stream.
Detect_level(frame).
{
Normalize the image frame.
Convert to gray image.
Convert to black & white image.
Find maximum white area object in image.

Return centroid of area.
}
The below Fig. 5 shows images for centroid detection.

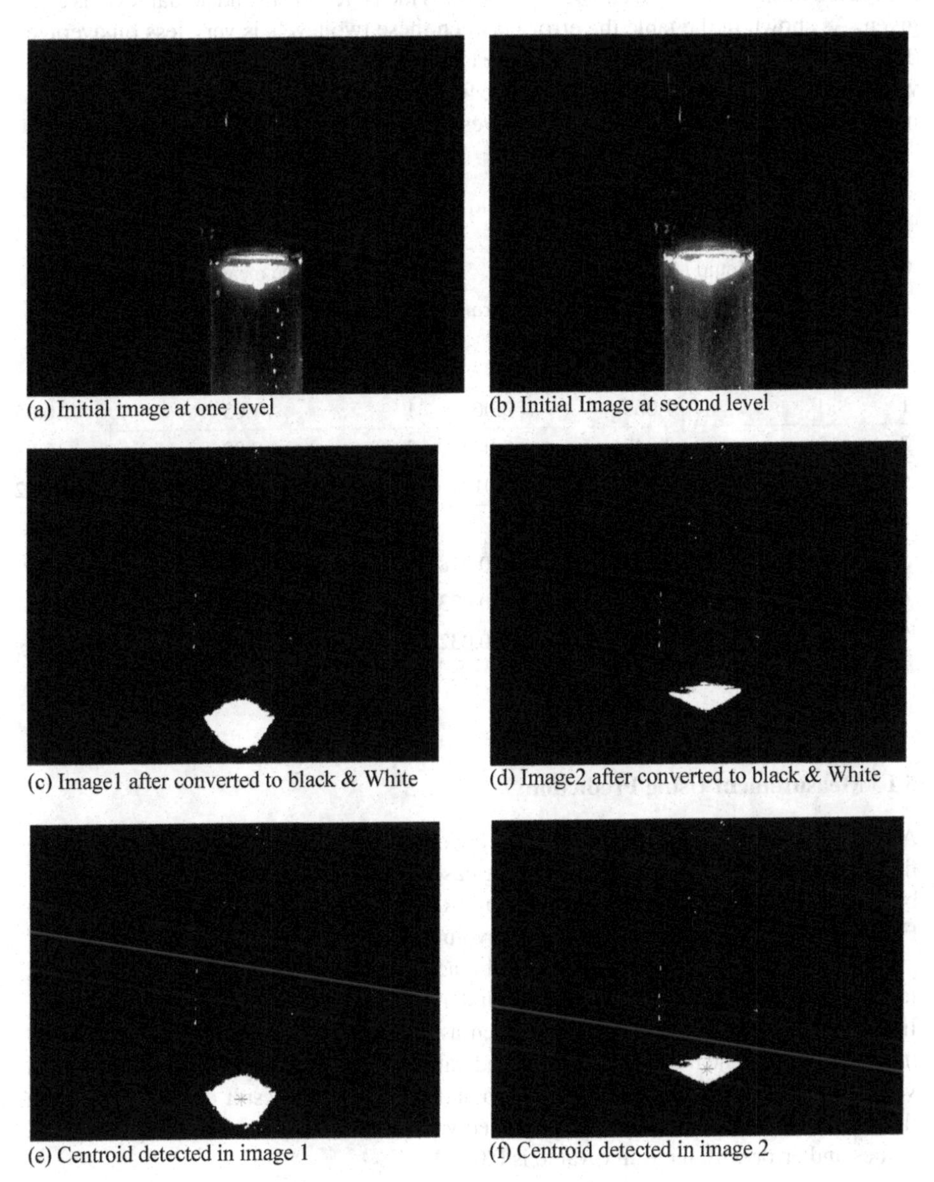

(a) Initial image at one level (b) Initial Image at second level
(c) Image1 after converted to black & White (d) Image2 after converted to black & White
(e) Centroid detected in image 1 (f) Centroid detected in image 2

Fig. 5. Images showing centroid detection

5 Result

The result of the experiment is given in this section. The Table 1 shows that two cases conducted in experiment. In case1, liquid level is measured in cm and actual level is also given. As shown in the table the error between these two levels is very less on average 0.043 cm. In case 2, volume inserted in the test tube is measured and it is also compared with actual volume inserted. In this case we found the error with average 0.03 cm^3. Thus, in both the cases, error is very less and gives more accuracy in the result.

Table 1. Readings for liquid level and volume measured.

Sr. No	Case 1- liquid level measurement			Case 2- volume measurement		
	Actual liquid Level (in cm)	Measured liquid level (in cm)	Error	Actual liquid level (in cm^3)	Measured liquid level (in cm)	Error
1	1	1.068	0.068	1	1.05	0.05
2	2	2.047	0.047	2	2.03	0.03
3	3	3.014	0.014	3	3.02	0.02
4	4	4.00	00	4	4.01	0.1
5	5	4.982	– 0.018	5	4.99	– 0.01
6	6	5.977	– 0.023	6	5.97	– 0.03
7	7	6.968	– 0.032	7	6.98	– 0.02

5.1 Measurement Using Prediction

After the experiment, we get two values for each case which are then used for curve fitting and prediction purpose. In both the cases, the reading obtain is linear, so it can be used for curve fitting for postprocessing. Results after curve fitting are given in the Fig. 6a and 7a for level measurement and volume measurement.

In other method of detection, we used machine leaning for prediction of the liquid level and volume inserted in the liquid. Under ML, by using the regression algorithm. In the algorithm, two parameters are taken as input that is early calibrated value and measured value. The output is the predicted value in both the cases that is for liquid and volume measurement. The plot in Fig. 6b and 7b shows the result of the same. From the result we can conclude that the predicted values are exactly same as early calibrated values and error with measured value is 0.0042.

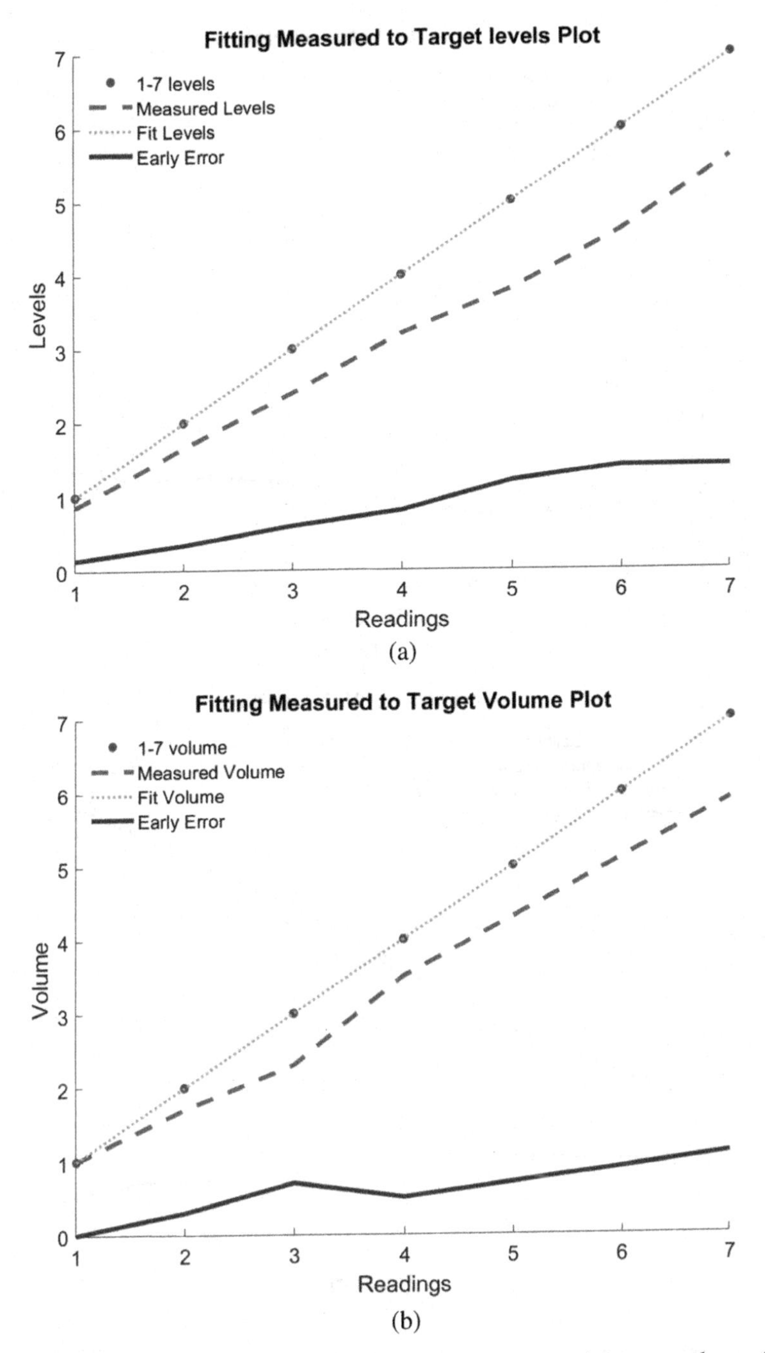

Fig. 6. a) Fitting measured to target level plot b) Fitting measured to target volume plot.

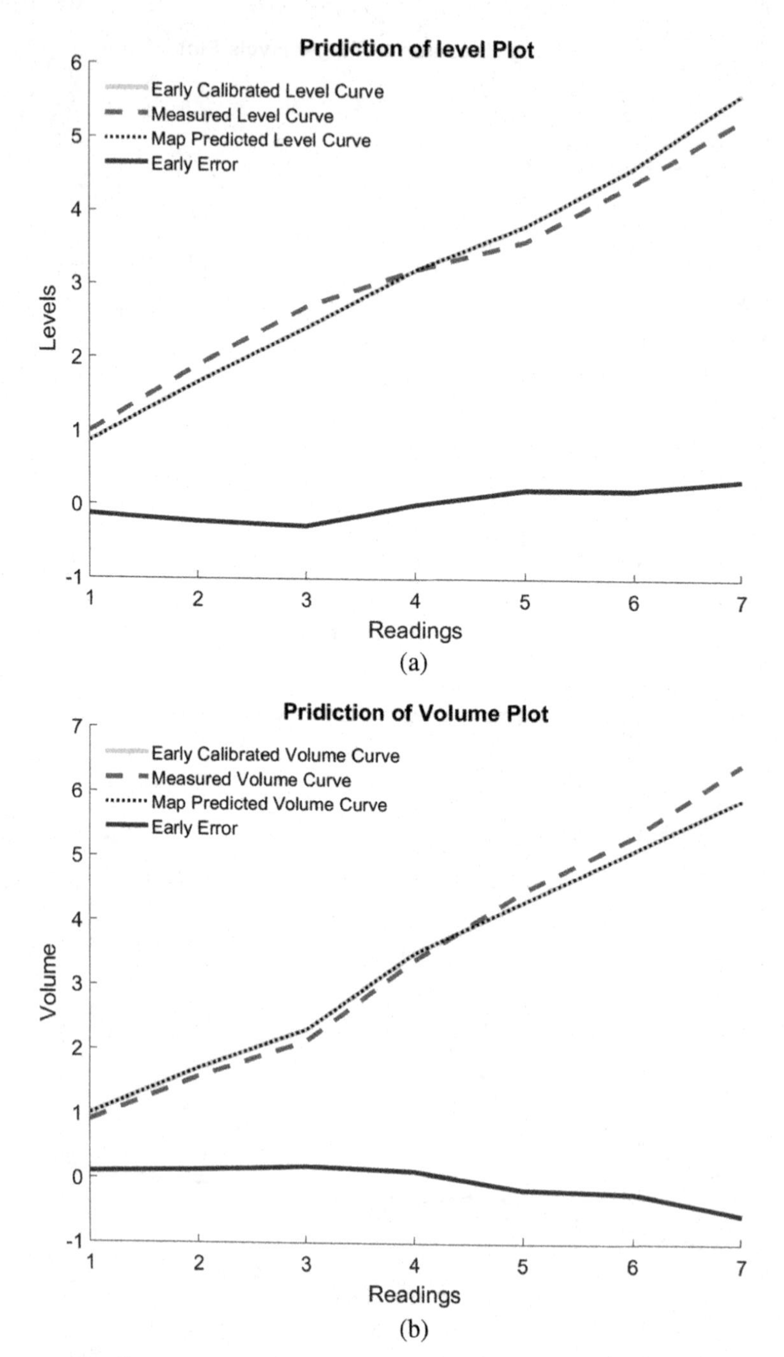

Fig. 7. a) Prediction of liquid level b) Prediction of volume in liquid.

6 Conclusion

The liquid level measurement is affected with different parameters as it is discussed in the paper. We propose the computer vision based liquid level measurement with predictor and curve fitting approach. The system is using a camera interface with PC or Laptop system. The video stream from camera is processed with frame by frame basis. The algorithm consists of preprocessing and postprocessing of images. Preprocessing consists mainly noise removal, normalization, conversion of color space, segmentation and morphological operation. Whereas post processing consists of formulization of level values from the raw centroid coordinates of the level segment portion of liquid. Mainly it includes filtering of signals [Level Values], appropriate delay for stabilization of liquid proportional to the its viscosity [affected by temperature], and at last curve fitting and prediction mechanism application. The application of liquid level can be a secondary indirect method for measurement of any other application like volume calculation etc. A proper calibration and proper test and prediction mechanism is needed for more accurate results. Here we discussed the two cases of such possibility.

We have studied about the viscosity of liquid, influencing the stability liquid level, for accurate reading. Liquid should be stable as much as possible, while in real time some addition of substances into or remove which results into a delay mechanism for proper valid readings.

References

1. Wang, T.H., Lu, M.C., Hsu, C.C., Chen, C.C., Tan, J.D.: Liquid-level measurement using a single digital camera. Measurement **42**(4), 604–610 (2009)
2. Chakravarthy, S., Sharma, R., Kasturi, R.: Noncontact level sensing technique using computer vision. Instrum. Measur. IEEE Trans. **51**(2), 353–361 (2002)
3. Gonzalez Ramirez, M.M., Villamizar Rincon, J.C., Lopez Parada, J.F.: Liquid level control of Coca-Cola bottles using an automated system. In: 2014 International Conference on Electronics, Communications and Computers (CONIELECOMP 2014), pp 148–154. IEEE (2014)
4. Shah, V., Bhatt, K.: Image processing based bottle filling and label checking using embedded system. Int. J. Innov. Res. Comput. Commun. Eng. **4**(5), 1–15 (2016)
5. Santhosh, K.V., Joy, B., Rao, S.: Design of an instrument for liquid level measurement and concentration analysis using Multisensor data fusion. J. Sensors **2020**, 1–13 (2020). https://doi.org/10.1155/2020/4259509
6. Majdalani, S., Chazarin, J.-P., Moussa, R.: A new water level measurement method combining infrared sensors and floats for applications on laboratory scale channel under unsteady flow regime. Sensors **19**, 1511 (2019). https://doi.org/10.3390/s19071511,MDPI
7. Bin, S., Zhang, C., Li, L., Wang, J.: Research on water level measurement based on image recognition for industrial boiler. In: 2014 26th Chinese Control and Decision Conference (CCDC) (2014)
8. Guo, J.C., Zhang, H.S.: Liquid level detection based on digital image processing. Tianjin University Electronic Information Engineering Institute (2010)
9. Eppela, S., Kachmanb, T.: Computer vision-based recognition of liquid surfaces and phase boundaries in transparent vessels, with emphasis on chemistry applications (2014). arXiv preprint: arXiv:1404.7174

10. Liu, X., Bamberg, S.J.M., Bamberg, E.: Increasing the accuracy of level-based volume detection of medical liquids in test tubes by including the optical effect of the meniscus. Measurement **44**(4), 750–761 (2011)
11. Samann, F.E.: Real-time liquid level and color detection system using image processing. Acad. J. Nawroz Univ. **7**(4), 223 (2018). https://doi.org/10.25007/ajnu.v7n4a293
12. Zhang, Z., Zhou, Y., Liu, H., Gao, H.: In-situ water level measurement using NIR-imaging video camera. Flow Measur. Instrum. **67**, 95–106 (2019)
13. Chandel, R., Gupta, G.: Image filtering algorithms and techniques: a review. Int. J. Adv. Res. Comput. Sci. Softw. Eng. **3**(10), 198–202 (2013)

Performance Analysis of Named Entity Recognition Approaches on Code-Mixed Data

Sreeja Gaddamidi and Rajendra Prasath(✉)

Indian Institute of Information Technology, Sri City, Chittoor, 630 Gnan Marg, Sri City 517646, Andhra Pradesh, India
{sreeja.g17,rajendra.prasath}@iiits.in
http://rajendra.2power3.com

Abstract. Online content is growing in multiple languages every day, and end-users share their opinions, reviews, blogs, tweets, etc., on Online Social Network (OSN) platforms. Such contents written by users having bi-lingual or multi-lingual abilities are harder to understand and analyze various aspects of user preferences and behaviors. In order to understand such code-mixed content, a family of tools such as language identification, named entity recognizer, parts-of-speech (POS) taggers, and other natural language tools have been developed. Named Entity Recognition (NER) is an essential task in text analytics and a prominent sub task of information extraction. This helps to identify, locate and classify entities into a fixed set of categories like name of a person or an organization or a location or time. It also helps to perform various NLP tasks such as role labeling, annotation and parse tree generation, etc. In this work, we present a study on the performance analysis of various approaches to the automatic NER of code-mixed data and propose an improved approach based on BiLSTM-CRF for NER. In this approach, we pass pre-trained word embeddings, trainable character embeddings and a few selected word structural features to BiLSTM-CRF network, which can make use of sentence level tag information along with context from both the directions in the sentence. We also compare various existing NER approaches for social media content, and present a detailed performance analysis of these approaches with the performance of our approach. It is observed that our proposed model gives comparable or outstanding results in the case of both the standard datasets [17, 18] used in our experiments.

Keywords: Classification · Code-mixed data · Named Entity Recognition · Natural Language Processing · Information extraction · Data mining · Decision trees · Information retrieval · Social networks · Social media · Ensemble methods · Neural networks · Deep Neural networks

1 Introduction

The number of users having fluency in multiple languages has been growing in the world wide web, and they tend to express their opinions, reviews, blogs, tweets and so on over online in mixed languages. More than one language script is used in such

M. Bhattacharya et al. (Eds.): ICICCT 2021, CCIS 1417, pp. 153–167, 2021.
https://doi.org/10.1007/978-3-030-88378-2_13

writings, and a specific monolingual system may not be able to process the content to arrive at some meaningful insights. Scripts used from multiple languages make it entirely different for traditional NLP tools to understand further and process it. This becomes more complicated when the words are not written in their native scripts and even more complex when characters from different languages are used together to write words in the given text document. So this emphasizes the need to develop systems that could find the languages used in the text content, specifically, to identify and recognize the named entities written in mixed-up languages.

The following are few example sentences taken from dataset [17] to illustrate some of the code-mixed data (Fig. 1).

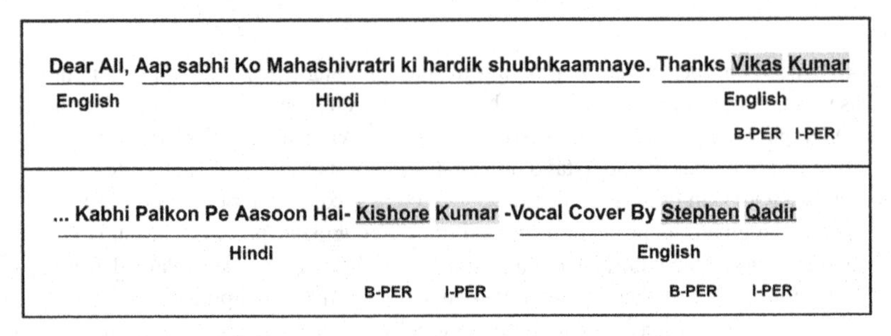

Fig. 1. Example sentences from dataset [17]

In the above examples, each sentence is written in multiple languages in a mixed-up manner (code-mixed data). Language Identification is one important task in such code-mixed data. Several approaches for Language Identification in code-mixed data were proposed in the literature [10, 17, 20]. Entity identification is even more challenging in such text and Srirangam et al. [24] recently proposed Named Entity Recognition approaches in Telugu-English Code-Mixed Data in Social Networks and Priyadharshini et al. [15] proposed entity recognition using meta embedding for Indian language corpus. In this research work, a hybrid approach that uses 4 steps including word and character embeddings, use of word-level features and sequence labelling using BiLSTM-CRF approach is proposed.

This paper is organized as follows: Sect. 2 describes works related to NER in code-mixed data. Different approaches explored here for NER in code-mixed data are given Sect. 3. Section 4 describes the proposed approach in detail. Experiments are described in Sect. 5. Finally Sect. 6 concludes the paper.

2 Related Work

In this section, we report the state-of-the-art research works related to the named entity recognition in code-mixed data. When lexical items from two different or semantically similar languages appear in one sentence, code-switching or code-mixing occurs [11], this is heavily seen in the communities that are fluent in more than one language. With the ever growing size of the world wide web, textual content on Online Social Network (OSN) increases every day. Many users, among several millions of active users on several OSNs, are fluent in bi-lingual or trilingual. Most of the available tools for understanding and analyzing online OSN text content fail to handle multilingual content. Several approaches/systems were proposed for handling code-mixed text including language identifiers[21], parts-of-speech taggers [27], and chunking [16]. Singh *et* al. [17] contributes to Named Entity Recognition in code-mixed data.

There are several approaches proposed for named entity recognition tasks in the code-mixed data including decision tree, conditional random fields (CRF) and long short term memory (LSTM) approach as mentioned in [18], and ensemble on random forest, decision tree and naive bayes approaches as suggested in [1]. Other approaches that use BiLSTM-CRF networks were also proposed, which allowed models to understand input features from both directions and make use of tag level information [17]. These approaches use only hand-crafted features, which are very expensive to develop, and fine-tuning these models to other tasks or domains is also not easy. [12, 15] focused on improving embedding input to models that improved overall performance of the models. Sravani [23] presented a comparison of word embeddings for named entity detection in codemixed Indian language data. Bhargava et al. [3] proposed a hybrid approach for recognizing named entities in Indian languages. Song et al. [22] proposed an approach in which characteristics based on characteristic templates are extracted by training the CRF models and then using them to recognize named entities in the monolingual text. [12] proposed BiLSTM trained character embeddings, for effective extraction of entities in Hindi-English code-mixed tweets which is said to improve BiLSTM-CRF model performance when passed along with glove word embeddings. [6] proposed BiLSTM-CRF NER model with CNN trained character features, and [9] suggested that adding Highway network on top of CNN character embeddings improves representation.

Inspired by these approaches, we propose a BiLSTM-CRF based approach that uses trainable character embeddings extracted from CNN-Highway network and pre-trained fasttext word embeddings along with a few manually picked up features that are related to social media code-Mixed text as inputs.

3 Approaches to NER

In this section, we describe different models explored in this research work:

3.1 Stanford NER Tagger

Stanford Named Entity Recognizer is one of the well proven systems in the NLP domain. Stanford NER, known as CRF (explained in Sect. 3.6) classifier, uses inference in linear chain conditional random field sequence models. CRF is primarily used as it could represent the sequence modeling very well allowing two facts: a) discriminative training and b) the flow of probabilistic information across the given sequence [8].

3.2 Decision Tree

In Machine Learning, Decision tree algorithm is the simplest form of supervised learning. As the name suggests, it tries to solve the feature to label mapping using tree representation. Each attribute is represented by a node in the decision tree and the class labels are represented by leaf nodes [18]. During the classification task, the problem of finding the label reduces to the problem of path finding with matching attributes till reaching the leaf node from the root.

3.3 Random Forest

A Random Forest consists of several tree-structured classifiers in which the model votes for the most popular class at the given input. The training set is used to grow the trees such that for each tree, a random feature vector is generated and the tree is grown using the training set and the random vector. After generating a large number of trees, the model is selected through a voting process for prediction of the labels [5].

3.4 Ensemble Learning Approach

Ensemble learning combines the prediction capabilities of multiple classifiers for extracting the named entities using features that could be based on the content and/or the context. The algorithm based on Ensemble Learning provides a better performance by combining the capabilities of multiple classification algorithms in predicting the labels [1]. In this work, we have used max voting and weighted voting ensemble learning techniques by leveraging the predictive capabilities of Decision Tree and Random Forest.

3.5 BERT

Bidirectional Encoder Representations and Transformers (BERT) model is designed as a transformer-based pre-trained bi-directional language representation model. This model considers unlabelled text data but conditions both contexts in all layers. Instead of predicting the next word given a context, BERT learns to predict the value of masked words (Masked Language Modelling) in a given sentence, as well as decide if two sentences are contiguous (next sentence prediction (NSP)) [7].

BERT Multilingual is pre-trained using Wikipedia text from the top 104 languages and has 110M parameters. It is trained with a shared word piece vocabulary and does not use any label indicating the input language. It, explicitly, does not support equivalent translation pairs to have similar representations. It is said to generalize across languages, handle transfer across scripts and to code-switching fairly well [14].

3.6 CRF with Fasttext Word Embeddings (CRF-FE)

Conditional Random Fields (CRF) [26] is a class of discriminative models used for prediction tasks where contextual information affects the current prediction. In case of named entity recognition, CRF helps to capture the dependency between tags. For a sequence $\mathbf{x} = (x_1,\ldots, x_T)$ of length T, the corresponding label sequence is $\mathbf{y} = (y_1,\ldots, y_T)$, and $Y(x)$ represents all possible label sequences. The probability of y is calculated by the following equation

$$P(y|x) = \frac{\sum_{t=1}^{T} exp\{f(y_{t-1}, y_t, x)\}}{\sum_{y'}^{Y(x)} \sum_{t=1}^{T} exp\{f(y'_{t-1}, y'_t, x)\}} \quad (1)$$

where $f(y_{t-1}, y_t, x)$ computes the transition score from y_{t-1} to y_t and the score for y_t. The optimization target is to maximize $P(y|x)$ value. When decoding for label sequence, the Viterbi Algorithm is used to find the maximum probability path. We conducted experiments by passing FastText word embeddings (will be elaborated in Sect. 4.1) and a few hand selected word features (will be elaborated in Sect. 4.3) as input to the CRF model.

3.7 BiLSTM-CRF with Fasttext Word Embeddings and BiLSTM Character Embeddings (WCE-BiLSTM-CRF)

In Bidirectional LSTM [2], the embeddings related to a specific word are given in direct order to a forward LSTM and in reverse order to a backward LSTM. Hence, it can understand the context in both the directions. We combine BiLSTM network and CRF network to form a BiLSTM-CRF network, which can use the sentence level tag information along with the past and future input features. For input, along with word embeddings, we also provide the model with character embeddings extracted from another BiLSTM model [12]. This character-level representation is generated by concatenating the output of both forward and backward LSTMs.

4 Proposed Methodology

Our proposed model is inspired from [6] and comprises bidirectional LSTM (Sect. 3.7) layer and ReLU activation, and a CRF (explained in Sect. 3.6) layer. The model takes three inputs, word embeddings, word features, and character embeddings for each token in a sentence. Instead of using only convolutional neural networks for character-level features, we added a highway network as suggested in [9].

4.1 Word Embeddings

We used FastText word embeddings [4] trained on Hindi-English codemixed tweets. FastText helps capture the synonyms and morphologically similar words [13] for each token present in a tweet. The word embeddings trained on the FastText architecture were not updated during the training of the model. It is observed that the FastText embeddings significantly improves the model's performance.

4.2 Character Embeddings

Character embedding layer is initialized randomly and updated during the model training. The output from the character embedding layer is fed to multiple 1D CNN layers with different kernel sizes. Depending on the size of the kernel, each CNN is suitably picking up a character n-gram. The concatenated output from CNN layers is passed to a highway network [9]. The Highway network [25] allows skip and nonlinear connections between its layers across the concatenated CNNs. They help to combine the local features detected by the individual filters adaptively. We observed that the model's performance worsened on removing the Highway Network layer. We used kernel sizes up to 4 as the average length of words is around 4 in both the datasets [19]. The model to learn the character embeddings is shown in Fig. 2.

4.3 Additional Word-Level Features

Our hand-picked features are described below. For each token, we pass a few structural features of current, previous and next tokens along with word and character embeddings as input.

- **Case based features:** These indicate whether a token is upper, lower, or title case. If a token is title cased, then the probability of it being a named entity increases, and if lowercase, the probability decreases.
- **Character based features:** These indicate whether a token has numerical characters or characters from the Roman alphabet. If a token has numerical characters, it can act as a good indicator to identify negative examples.
- **Word length:** This feature provides length of the token.
- **Affixes:** Prefixes and suffixes of length 1 to 5 extracted from token. If a word in English ends with eroring it is unlikely that the word is a named entity. Hence, these features help in identifying negative examples.

- **Hashtags and mentions:** Twitter users use # before a word to make it notable and @ to address a person or organization. Thus, features indicating whether a token starts with # or @ are considered, as the probability of the word to be an entity increases with these tags
- **BOS and EOS:** These indicate whether the current token is at the beginning or at the end of the sentence.

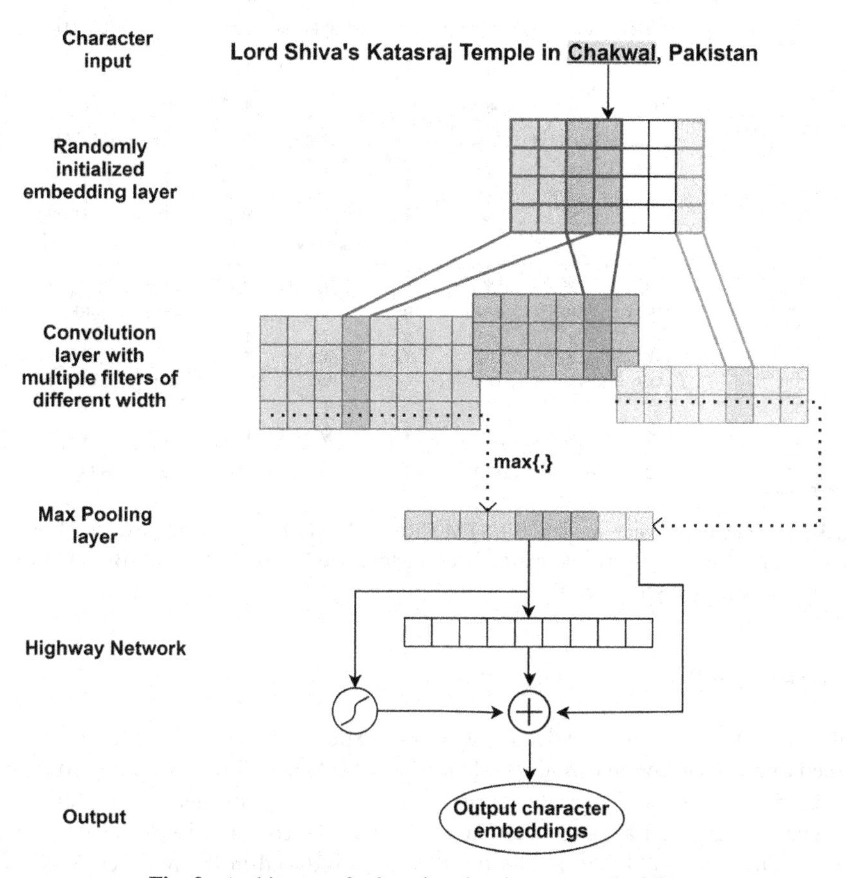

Fig. 2. Architecture for learning the character embeddings

4.4 Sequence Labelling Using BiLSTM-CRF

The extracted word and character level features of each word are concatenated and fed into a bidirectional LSTM layer. The output of the BiLSTM layer is fed to a fully connected layer with ReLU activation and then to a CRF layer. The best possible tag sequence is given as output. The architecture of our model has been portrayed in Fig. 3.

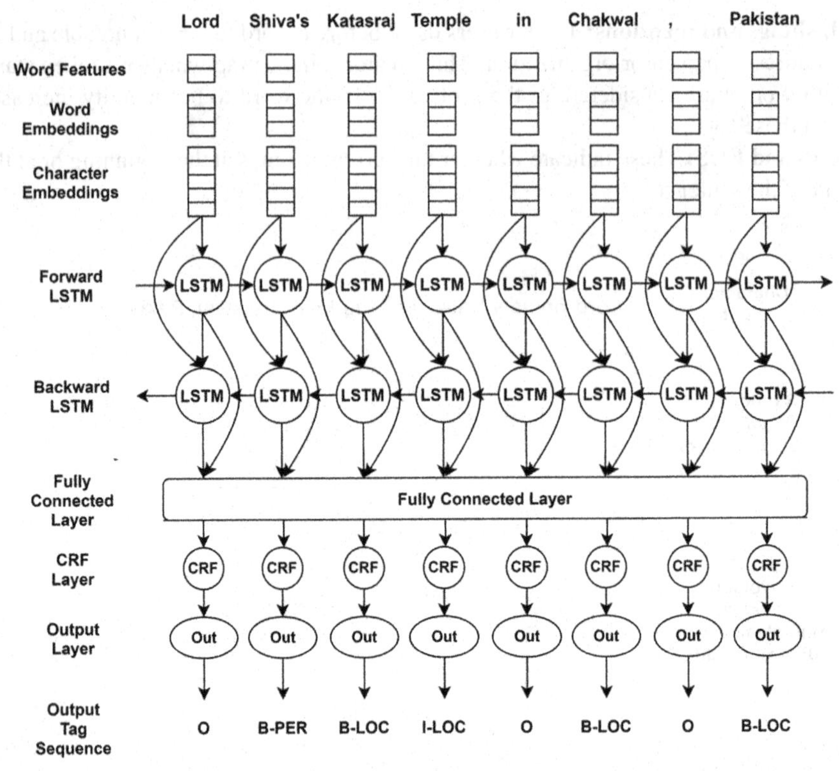

Fig. 3. Architecture of our model, BiLSTM-CRF with word features, fasttext word embeddings and CNN-Highway Networks extracted character embeddings (CNN-HWY-BiLSTM-CRF), applied to an example sentence.

5 Experiments

In this section, we discuss the details of various experiments carried out in recognizing Named Entities and summarize the evaluation results. We employed the standard evaluation metrics - precision (P), recall (R), and F1 measure. Experiments were performed on two datasets namely, Language and Named Entity Tagged Hindi-English Code Mixed Tweets dataset [17] (dataset 1) and the dataset introduced in the research work [18] (dataset 2), during our study. In both cases, various approaches were compared for NE recognition.

5.1 Dataset - 1: Hindi-English Code Mixed Tweets [17]

The first dataset is created by randomly sampling 50,000 tweets from the code mixed tweet dataset collected by (Patro et al., 2017). The researchers filtered this data for tweets with at least five Hindi tokens and removed tweets having words from languages other than Hindi and English (like transliterated Telugu). The final dataset comprises 35374 tokens, of which 13860 are English, 11391 are Hindi, and 10123 are the rest. Each token is annotated for three classical named entity types, Person (PER), Location (LOC), and Organisation (ORG), using the IOB format. The annotation process was carried out by three linguists proficient in both English and Hindi. There are 35,374 named entity tags distributed across various NE types in this dataset.

5.2 Dataset - 2: Code-Mixed Tweets Used in [18]

The second dataset used in [18], serves as a good baseline in code-mixed experiments and it contains 1,10,231 code-mixed tweets. These raw tweets collected to create this dataset consist of information specific to social events, politics, sports, and so on and these tweets are region specific, viz, it consists of Indian language code-mixed tweets. These raw tweets are then preprocessed and noisy and monolingual tweets are removed. The resulting corpus consists of 3,638 code-mixed tweets that are comparable to the baseline dataset used by the experiments FIRE 2016. The final data were annotated using the three named entity tags namely Person (PER), Organization (ORG), and Location (LOC) as in the first dataset. In this dataset, there are 68,506 named entity tags distributed under various NE types. The distribution of the different named entity tags in both datasets is shown in Fig. 4.

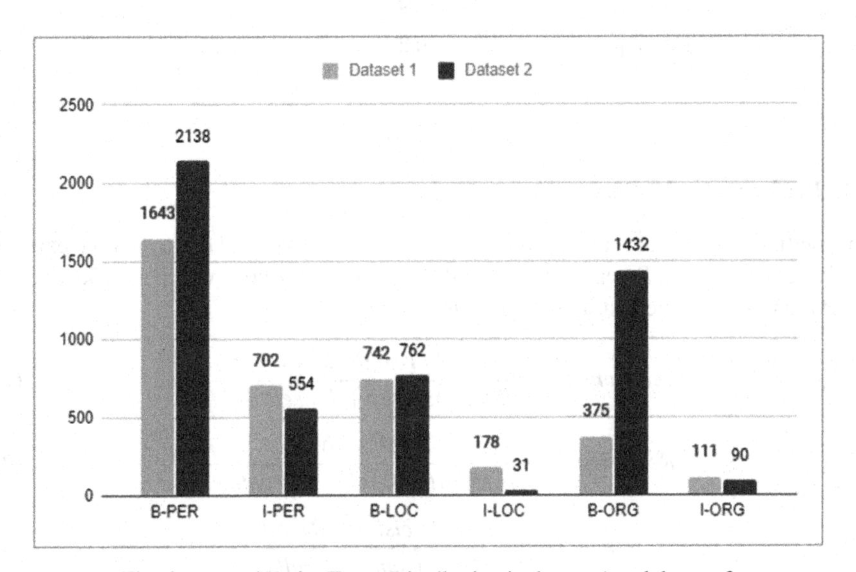

Fig. 4. Named Entity Tags - Distribution in dataset 1 and dataset 2

5.3 Experimental Settings

We compared the performance of various models namely Stanford NER, Decision Tree, Random Forest, Ensemble Max Voting, Ensemble Weighted Voting, BERT, CRF with fasttext word embeddings, BiLSTM-CRF with fasttext word and BiLSTM character embeddings, BiLSTM-CRF with word features, fasttext word embeddings and CNN-Highway Networks extracted character embeddings, on both the datasets.

We performed various experiments with different values of hyperparameters like word embedding size, character embedding size, number of CNN filters before choosing the best set. Table 1 shows some of the hyperparameters used in our final model.

Table 1. Model hyperparameter summary

Hyperparameter	Value
Size of Word Embeddings	300 (pre-trained, not trainable)
Size of Character Embeddings	64 (trainable)
Maximum sequence length	127
Maximum word length	66
Batch Size	32
Number of filters in 1-D CNN	32
Dropout	0.3
LSTM units	256
LSTM dropout	0.2

5.4 Performance Metrics

To quantitatively evaluate the performance of our model, we used the standard measures used in classification: Precision (P), Recall (R), and f_1 measures, which are widely used for evaluating in named entity recognition tasks.

$$Precision(P) = \frac{TruePositives}{(TruePositives + FalsePositives)} \tag{2}$$

$$Recall(R) = \frac{TruePositives}{(TruePositives + FalseNegatives)} \tag{2}$$

$$F_1Score = \frac{(2 * Precision * Recall)}{(Precision + Recall)} \tag{4}$$

5.5 Results and Discussion

Table 2 and Table 3 shows performances of different models that we implemented in this paper on various tags of both datasets. Initially, we implemented trained standard Stanford NER on the first dataset and achieved an outstanding precision, recall, and F1 Score. However, when we implemented the same on dataset 2, we found that the model performed poorly on all NE tags. Later we implemented other standard models like

Decision Tree, Random Forest and observed that unlike Stanford NER, they performed similarly on both datasets.

Even though the Decision Tree model can tag a single token most of the time, it gives low precision. Also, it can not understand continuation tags and can not differentiate between nouns and pronouns and named entity types. For example, "Ancient Mahakaleshwar Jyotirlinga" is tagged as (O, I-ORG, I-LOC) instead of (O, B-LOC, I-LOC). The Random Forest model is able to give good precision as it is averaging over many models and also is able to identify continuation tags. But it is unable to understand the context and is case sensitive. After obtaining good results from the above models, we worked with the ensemble learning approach, as suggested in [1], which combines predictions of Decision Tree and Random Forest. We used two techniques namely Max Voting and Weighted Average for the Ensemble Learning and found that the Weighted Average model performs better on both datasets (Fig. 5).

Table 2. *F*1 Scores of various approaches with dataset 1

Model	B-PER	I-PER	B-LOC	I-LOC	B-ORG	I-ORG	Other
Stanford NER	**0.9**	**0.89**	**0.89**	**0.81**	**0.87**	**0.74**	0.99
Decision Tree	0.54	0.54	0.58	0.39	0.45	0.22	0.92
Random Forest	0.65	0.61	0.69	0.5	0.64	0.41	0.97
Max Voting	0.57	0.66	0.62	0.48	0.41	0.18	0.96
Weighted Voting	0.66	0.65	0.71	0.37	0.65	0.47	0.97
BERT	0.79	0.79	0.78	0.58	0.68	0.40	0.99
CRF-FE	0.7	0.69	0.7	0.59	0.65	0.48	0.97
WCE-BiLSTM-CRF	0.62	0.65	0.5	0.31	0.52	0.15	0.99
CNN-HWY-BiLSTM-CRF	0.75	0.75	0.74	0.51	0.6	0.27	0.99

We also experimented with the BERT model. BERT is pre-trained in language model tasks on raw unlabeled texts. This pre-trained deep bidirectional model can reach state-of-the-art results in many tasks such as question answering and Natural Language Inference. We used the BERT multilingual cased model, which is trained on multiple languages, including Hindi and English. From our experiments, we found that the BERT model performs well on the first dataset; however, its performance is not significant on a few tags of the second dataset. We observed that the model is able to give a high recall value and understand the context better, but it is case sensitive as it was pre-trained on cased data. This might actually be problematic in the case of social media text where users might not follow grammar. We also observed that in the case of phrases like "Sadanandan Master" and "Rawat Ji", the model is tagging them as (B-PER, I-PER) instead of (B-PER, O). We also transliterated Hindi tokens of the first dataset (language tags were not available for the second dataset) from Roman to Devanagari and passed them to our model as suggested in [14]. We observed an increase in the F1-Score, but the increase was not significant.

Table 3. $F1$ Scores of various approaches with dataset 2

Model	B-PER	I-PER	B-LOC	I-LOC	B-ORG	I-ORG	Other
Stanford NER	0.07	0.04	0.05	0.04	0.08	0.04	0.93
Decision Tree	0.34	0.46	0.33	0.18	0.4	0	0.97
Random Forest	0.46	0.59	0.53	0.25	0.50	0.08	0.98
Max Voting	0.38	0.56	0.47	0.20	0.39	0.04	0.97
Weighted Voting	0.45	0.57	0.53	**0.67**	0.5	0.0	0.98
BERT	0.63	0.55	0.59	0.0	0.49	0.0	0.97
CRF-FE	0.69	0.71	0.58	0.4	0.58	**0.67**	0.98
WCE-BiLSTM-CRF	0.74	0.65	0.64	0.2	0.7	0.39	0.99
CNN-HWY-BiLSTM-CRF	**0.8**	**0.77**	**0.77**	**0.67**	**0.76**	0.54	0.99

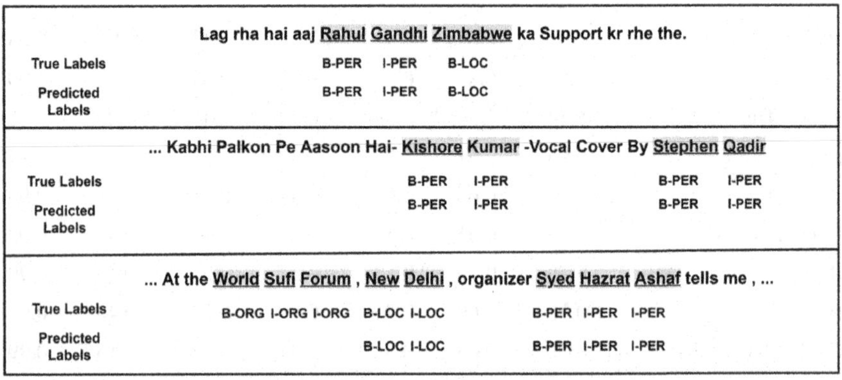

Lag rha hai aaj Rahul Gandhi Zimbabwe ka Support kr rhe the.
True Labels: B-PER I-PER B-LOC
Predicted Labels: B-PER I-PER B-LOC

... Kabhi Palkon Pe Aasoon Hai- Kishore Kumar -Vocal Cover By Stephen Qadir
True Labels: B-PER I-PER B-PER I-PER
Predicted Labels: B-PER I-PER B-PER I-PER

... At the World Sufi Forum , New Delhi , organizer Syed Hazrat Ashaf tells me , ...
True Labels: B-ORG I-ORG I-ORG B-LOC I-LOC B-PER I-PER I-PER
Predicted Labels: B-LOC I-LOC B-PER I-PER I-PER

Fig. 5. Sample output using BiLSTM-CRF with word features, fasttext word embeddings and CNN-Highway Networks extracted character embeddings model

In order to capture the dependency between named entity tags, we implemented CRF with pre-trained FastText word embeddings as input. We observed that the model very well handled continuation tags. But, it is case sensitive and does not understand the sentence; for example, it tagged "Buy Sabudana @ 64/-" as (B-PER, I-PER, O, O) instead of (O, O, O, O). To obtain the context from both the directions in a sequence, we added a BiLSTM layer on top of the CRF model and observed a considerable improvement in the model's performance. Passing character embeddings to the model along with word embeddings increased the precision value. The BiLSTM-CRF with FastText word embeddings and BiLSTM character embeddings model achieved good precision and recall value. However, for words with more length, BiLSTM character embeddings tend to make more mistakes [29].

From Table 4 and Table 5 we can observe that our model gives a fair precision and recall compared to the other models on dataset 1 with the average F1-score of 0.7 and performs better than all the other models on dataset 2 with an average F1 score of 0.77.

Table 4. Weighted averages of precision, recall and $F1$ Scores for different tags (excluding Other tags) using various approaches on dataset 1

Model	Avg. Precision	Avg. Recall	Avg. F1-Score
Stanford NER	**0.95**	**0.82**	**0.88**
Decision Tree	0.53	0.54	0.54
Random Forest	0.83	0.52	0.64
Max Voting	0.61	0.56	0.56
Weighted Voting	0.82	0.54	0.65
BERT	0.76	0.78	0.75
CRF-FE	0.79	0.61	0.69
WCE-BiLSTM-CRF	0.59	0.54	0.56
CNN-HWY-BiLSTM-CRF	0.72	0.69	0.7

Table 5. Weighted averages of precision, recall and $F1$ Scores for different tags (excluding Other tags) using various approaches on dataset 2

Model	Avg. Precision	Avg. Recall	Avg. F1-Score
Stanford NER	0.07	0.06	0.06
Decision Tree	0.33	0.54	0.4
Random Forest	0.48	0.52	0.49
Max Voting	0.34	0.56	0.41
Weighted Voting	0.47	0.51	0.48
BERT	0.45	0.76	0.56
CRF-FE	0.58	**0.76**	0.65
WCE-BiLSTM-CRF	0.73	0.66	0.69
CNN-HWY-BiLSTM-CRF	**0.8**	**0.76**	**0.77**

CNN-Highway Network character embeddings are capturing spelling and morphological variations due to the N-gram features captured by the convolution layers. Here, the length of the words will not affect the model's performance. The obtained results show that this model performs better than all the other models mentioned in the paper.

6 Conclusion

In this paper, we have presented a study on the performance analysis of various approaches on the automatic Named Entity Recognition (NER) of codemixed data. We also proposed an approach based on the hybrid model that uses BiLSTM-CRF with various textual features. We have performed a comparison of various Named Entity

Recognition approaches and shown that our BiLSTM-CRF with word features, fasttext word embeddings and CNN-Highway Networks extracted character embeddings model performed better than the existing approaches. We have also observed that there is a significant improvement in the performance of the proposed model on different datasets. Subsequently, we plan to explore Transformer-CRF models as suggested in [28] and improve word embeddings by pre-training them on a larger corpus.

References

1. Arya, S., Majumder, A., Majumder, S., Kundu, A., Shamim, N., Rakshit, P.: Ensemble learning approach for named entity recognition from Hindi-English tweets. CSI J. Comput. **3**(2), 47–51 (2020)
2. Bengio, Y., Simard, P., Frasconi, P.: Learning long-term dependencies with gradient descent is difficult. Trans. Neur. Netw. **5**(2), 157–166 (1994). https://doi.org/10.1109/72.279181
3. Bhargava, R., Tadikonda, B.V., Sharma, Y.: Named entity recognition for code mixing in Indian languages using hybrid approach. In: FIRE (2016)
4. Bojanowski, P., Grave, E., Joulin, A., Mikolov, T.: Enriching word vectors with subword information. Trans. ACL **5**, 135–146 (2017)
5. Breiman, L.: Random forests. Mach. Learn. **45**(1), 5–32 (2001). https://doi.org/10.1023/A:1010933404324
6. Chiu, J.P., Nichols, E.: Named entity recognition with bidirectional LSTM-CNNs. Trans. Assoc. Comput. Linguist. **4**, 357–370 (2016)
7. Devlin, J., Chang, M.W., Lee, K., Toutanova, K.: BERT: pre-training of deep bidirectional transformers for language understanding. arXiv preprint: 1810.04805 (2018)
8. Finkel, J.R., Grenager, T., Manning, C.: Incorporating non-local information into information extraction systems by Gibbs sampling. In: ACL 2015, ACL, USA, pp. 363–370. (2005). https://doi.org/10.3115/1219840.1219885
9. Kim, Y., Jernite, Y., Sontag, D., Rush, A.: Character-aware neural language models. In: Proceedings of the AAAI Conference on Artificial Intelligence, vol. 30 (2016)
10. Mandal, S., Singh, A.K.: Language identification in code-mixed data using multichannel neural networks and context capture. In: Proceedings of the 2018 EMNLP Workshop W-NUT: The 4th Workshop on Noisy User-generated Text, pp. 116–120. ACL, Brussels, Belgium, November 2018. https://doi.org/10.18653/v1/W18-6116
11. Moyer, M.G.: Pieter Muysken, Bilingual Speech: A Typology of Code-Mixing, pp. xvi, 306. Cambridge University Press, Cambridge (2002). Lang. Soc. **31**(4), 621–624
12. Narayanan, A., Rao, A., Prasad, A., Das, B.: Character level neural architectures for boosting named entity recognition in code mixed tweets. In: 2020 International Conference on Emerging Trends in IT and Engineering (ic-ETITE), pp. 1–6. IEEE (2020)
13. Pilán, I., Volodina, E.: Exploring word embeddings and phonological similarity for the unsupervised correction of language learner errors. In: Proceedings of the Second Joint SIGHUM Workshop on Computational Linguistics for Cultural Heritage, Social Sciences, Humanities and Literature, pp. 119–128 (2018)
14. Pires, T., Schlinger, E., Garrette, D.: How multilingual is multilingual BERT? arXiv preprint arXiv:1906.01502 (2019)
15. Priyadharshini, R., Chakravarthi, B.R., Vegupatti, M., McCrae, J.P.: Named entity recognition for code-mixed Indian corpus using meta embedding. In: 2020 6th International Conference on Advanced Computing and Communication Systems (ICACCS), pp. 68– 72. IEEE (2020)

16. Sharma, A., Gupta, S., Motlani, R., Bansal, P., Shrivastava, M., Mamidi, R., Sharma, D.M.: Shallow parsing pipeline - Hindi-English code-mixed social media text. In: Knight, K., Nenkova, A., Rambow, O. (eds.) NAACL HLT 2016, San Diego California, USA, 12–17 June 2016, pp. 1340–1345. ACL (2016)
17. Singh, K., Sen, I., Kumaraguru, P.: Language identification and named entity recognition in Hindi-English code mixed tweets. In: Proceedings of ACL 2018, Student Research Workshop, pp. 52–58 (2018)
18. Singh, V., Vijay, D., Akhtar, S.S., Shrivastava, M.: Named entity recognition for Hindi-English code-mixed social media text. In: Proceedings of the Seventh Named Entities Workshop, pp. 27–35 (2018)
19. SNUDerek: multiLSTM for Joint NLU (2018). https://github.com/SNUDerek/multiLSTM
20. Solorio, T., Liu, Y.: Learning to predict code-switching points. In: Proceedings of the Conference on Empirical Methods in Natural Language Processing, pp. 973–981. ACL, USA (2008)
21. Solorio, T., Liu, Y.: Learning to predict code-switching points. In: EMNLP 2008, pp. 973–981. ACL, Hawaii, October 2008. https://www.aclweb.org/anthology/D08-1102
22. Song, S., Zhang, N., Huang, H.: Named entity recognition based on conditional random fields. Clust. Comput. **22**(3), 5195–5206 (2017). https://doi.org/10.1007/s10586-017-1146-3
23. Sravani, L., Reddy, A.S., Thara, S.: A comparison study of word embedding for detecting named entities of code-mixed data in Indian language. In: 2018 International Conference on Advances in Computing, Communications and Informatics (ICACCI), pp. 2375–2381 (2018). https://doi.org/10.1109/ICACCI.2018.8554918
24. Srirangam, V.K., Reddy, A.A., Singh, V., Shrivastava, M.: Corpus creation and analysis for named entity recognition in Telugu-English code-mixed social media data. In: Proceedings of the 57th Conference of the ACL 2019, Florence, Italy, pp. 183–189 (2019)
25. Srivastava, R.K., Greff, K., Schmidhuber, J.: Highway networks. Preprint arXiv:1505.00387 (2015)
26. Sutton, C., McCallum, A.: An introduction to conditional random fields for relational learning. Introduction Stat. Relat. Learn. **2**, 93–128 (2006)
27. Vyas, Y., Gella, S., Sharma, J., Bali, K., Choudhury, M.: POS tagging of English-Hindi code-mixed social media content. In: Proceedings of the 2014 Conference on Empirical Methods in Natural Language Processing (EMNLP), pp. 974–979. ACL, October 2014
28. Yan, H., Deng, B., Li, X., Qiu, X.: TENER: adapting transformer encoder for named entity recognition. arXiv preprint arXiv:1911.04474 (2019)
29. Zhai, Z., Nguyen, D.Q., Verspoor, K.: Comparing CNN and LSTM character-level embeddings in BILSTM-CRF models for chemical and disease named entity recognition. arXiv preprint arXiv:1808.08450 (2018)

Quantum Based Deep Learning Models for Pattern Recognition

Prakhar Shrivastava(✉), Kapil Kumar Soni, and Akhtar Rasool

Department of Computer Science Engineering, MANIT, Bhopal, Madhya Pradesh, India

Abstract. The machine learning model influences the pattern recognition based on extraction of relative patterns, but not enough capable of efficiently processing the data set that needs layered interaction. So, deep learning model takes the advantage of artificial neural network for processing the data in layered abstraction by exploring massive parallelism, although the classical implementations of such model may not be competent due to the processing and storage of large neural networks. Current research explores quantum computation potentials and utilizes its significance for supporting inherent parallelism, as the machine perform exponential operations in single step of execution. The possible classical designs of pattern recognition using deep learning are Hopfield network and Boltzmann machine and their equivalent quantum models can remain effective and overcome the processing limitations of classical model by incorporating Grover's search method and quantum Hebbian learning. The aim of writing this article is to introduce the necessity of deep learning model, an emergence of quantum computations, discussion over requisites of data preprocessing techniques, then to propose the classical equivalent quantum deep learning model along with the algorithms, complexity comparison and speedup analysis followed by conclusive aspects that proves effectiveness of quantum deep learning model and future works.

Keywords: Quantum pattern recognition · Machine learning · Deep learning · Grover's search · Neural networks · Superposition · Quantum Hebbian learning

1 Introduction

1.1 Pattern Recognition and Machine Learning

The problem for searching a pattern in a tremendously huge data set is foundational aspect for evolution of pattern recognition. The field of pattern recognition is unfolded due to the recent developments and use of regularities in data. This discovery of regularities can be obtained with the use of machine learning algorithms, the association rules are set to figure out the pattern in these regularities and classify them into different clusters. There are some heuristics easily available to find out patterns e.g. handcrafted rules for finding alphabet based on shapes available but practically it will have some exceptions and provide inefficient results. Exceptionally good results can be found by using machine learning concepts where huge set of $N = \{a1, a2, \ldots, an\}$ alphabets called the training set is used to configure the parameters of a dynamically adaptive model. The different

M. Bhattacharya et al. (Eds.): ICICCT 2021, CCIS 1417, pp. 168–183, 2021.
https://doi.org/10.1007/978-3-030-88378-2_14

classes of alphabets in training set are known prior to applying it to model. The class of alphabet can be expressed with a vector $\boldsymbol{V}$ [4, 15].

Machine learning algorithm yields the output which is a function and can be represented as $(\boldsymbol{a_i}) = \boldsymbol{v}$ where a_i is the input alphabet passed to the function and function will map it to one of the vector $\boldsymbol{v} \in \boldsymbol{V}$. The behavior of function ***Y(a)*** is decided during training phase. Once the model is trained then it can map the new input alphabet to its corresponding output class. The ability to categorize the new input correctly is known as generalization. This property is the main goal of the pattern recognition [1].

1.2 Machine Learning and Deep Learning

In 1959, Arthur Samuel introduced this term and defined it as field of study that enables computer to learn without being explicitly programmed [1]. However, a lot of development in this domain happened over the time. So, machine learning can be redefined as an automated process that gives machine the ability to analyze the huge data sets, identify patterns and learn from the data to concretize predictions and evolve decision making. Currently, the drawback of machine learning is the inner working of these machines is a black box, which makes it hard to understand and explain. As a result, it will become tough to completely rely on the output of these machines [16, 15].

Neural Networks take huge amount of time for training which is considered as a major issue and blocked the development of neural networks with several hidden layers for a long time. Deep learning unblocked this issue by putting computational tricks and massive parallel architectures on the plate. In deep learning, depiction of information is distributed i.e. layers in the network provides different levels of abstraction and is directly proportional to each other. Deep learning model have the capability to generalize better supporting automation, automatically building the layers of abstraction that are necessary for critical learning tasks. Deep Learning habitually combines with supervised and unsupervised learning. Unsupervised learning supports the training process by acting as a regularizer and help achieving optimization. Deep learning is prone to over fitting problem because a greater number of layers can capture the rare dependencies in training data. Using unsupervised learning for preprocessing and post processing the data can come suppress the problem [4, 12].

1.3 Pattern Recognition Using Deep Learning

Recently, deep learning techniques have accomplished huge progress in the domain of pattern recognition such as object recognition. In object recognition also known as object classification, deep learning models have achieved supreme growth in terms of performance as compared with classical classification algorithms. For German traffic sign recognition, the multi-column DNN has been proposed. To study neuropsychiatric conditions based on functional connectivity patterns, standard classifiers like the SVM have been used frequently and broadly. Recently, the DNNs have been employed to classify the whole-brain resting-state FC patterns of schizophrenia. To improve the efficiency of classification in terms of performance, a novel *maximum margin multimodal deep neural network (3mDNN)* was introduced to take benefit of the multiple local descriptors of an image. Compared with standard algorithms, this method, considering the information of multiple descriptors, can achieve discriminative power. DNNs can also be used for the wind speed patterns classification and the supervised multispectral land-use classification. It should be considered that deep learning techniques can also be applied to hand posture recognition. Since the features generated by traditional algorithms are limited, and it is difficult to detect and track hands with normal cameras, the DNNs are employed to produce enhanced features. Based on *functional near infrared spectroscopy (FNIRS)*, deep learning techniques have achieved promising results in classifying brain activation patterns for BCI [4, 13, 14].

1.4 Classical and Quantum Aspects of Deep Learning

There are several of classes of deep learning architecture such as restricted Boltzmann machine, the convolution neural networks etc. These different verticals are used for different categories of problems and provide a better performance compared to their conventional algorithms. But as name suggests, these are deep neural networks which requires tremendous amount of data to work on and processing at each neuron take significant amount of time to execute. If overall time complexity is to be considered, then it would raise exponentially. And, it would require huge space to store these networks and related data set which makes space complexity also heavier. As amount of data is increasing on a per day basis it becomes crucial to reduce the complexities and increase the performance of the system [4, 11].

Quantum computing provides an insightful approach to decrease the efforts and complexities by using quantum fundamentals such as superposition, entanglement, etc. It just not only provides the significant speed up but also reduces the space complexity. As with the recent developments in field of quantum machine learning, researchers now are exploring quantum deep learning and trying to come up with more optimized approaches. One of the proposed approaches is to put the neurons of the giant neural network into the superposition and then try to figure out the path in the neural network which leads to the desired output for a particular training instance. As a result, all the neurons of the network are not exposed. In this way, reduction in time complexity can be achieved. And to store this deep learning architecture quantum random access memory (qRAM) model can be used which will also help providing significant mitigation in usage of memory. As a result, a huge speed up can be observed in efficiency, carrying out of operations in the network as well as the storage [1, 16].

1.5 Motivation and Organization of Paper

If deep learning and quantum computing are taken into consideration and explore the problems in different verticals of deep learning, it can be concluded that it requires huge amount of time and space to execute its operation. On the other hand, quantum computing was on the top and was successfully improving machine learning algorithms by providing significant amount of speedup. The main objective of this article to gain insight on what effect quantum computing do when deep learning models are concerned. Is it possible to apply quantum algorithms on deep learning and obtain an observable speedup as compared to its classical variant? The organization of paper revolves around deep learning architecture and quantum computing. Firstly, introduction regarding various fields like machine learning, deep learning, pattern recognition and quantum computing has been provided. Deep learning models like Hopfield networks and Boltzmann machine and quantum fundamentals like quantum gates, superposition, entanglement, Grover's algorithm etc. are discussed. After that classical equivalent quantum deep learning models are described. Comparative analysis between classical and quantum deep learning algorithm are analyzed and depicted. At the end, conclusion and future work are also discussed (Fig. 1).

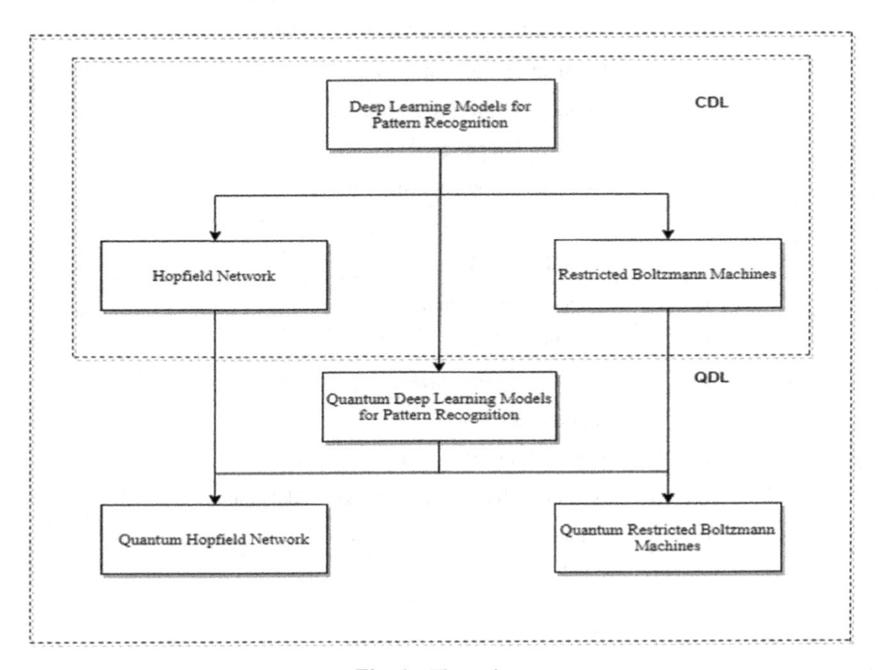

Fig. 1. Flow chart

2 Classical Deep Learning (CDL)

As the name suggests that it consists of the deep neural network where deep means the depth of the hidden layers used for the given problem. The deep neural networks used for the pattern recognition gives feasible result and is successfully applicable in pattern recognition domain is due to single reason which is deep neural networks are good in recognizing things. This is because of the property of neural networks i.e. a neural network is computationally universal which means any logic can be implemented in the nodes of neural network and they will start behaving accordingly. For example, neural network can be configured to behave like any logic gates i.e. NOT gate, OR gate, NAND gate etc. Even they can be shaped as to behave like a copy operator [4] (Table 1 and Fig. 2).

Table 1. Symbols table CDL (Symbol, Description)

Symbols	Description
n	# Artificial binary neurons
x_i	Artificial binary neuron
	'i' ∈ {1, 2, …, n}
x^t	Transpose of x (artificial binary neuron)
W	(n × n) dimensional Weighting Matrix
w_{ij}	Element of weighting matrix W
$T^{(k)}$	Training set, where k ∈ {1, 2, …, K}
K	# Activation patterns
$x^{(n)}$	New activation pattern
θ	User defined threshold vector for each neuron
E	Neural network energy
H	Overall Hamiltonian
a_i and b_{ij}	Tunable weights
c	Output/classifier layer
d	Input/data layer
P(c, d)	Probabilities of classifier-data pair
q(c, d)	Actual probability q of classifier-data pair
d	Data d ∈ {0, 1} at all nodes

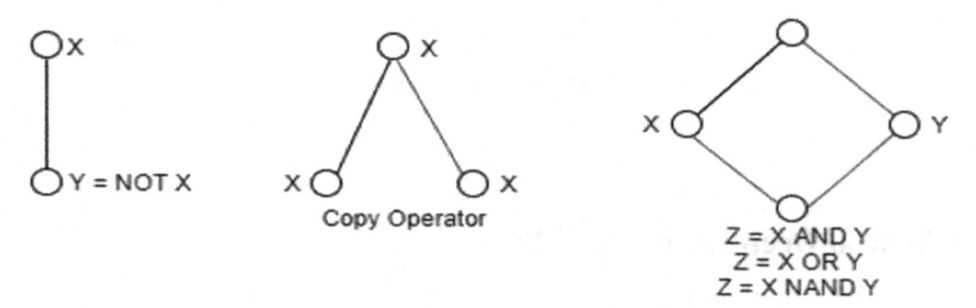

Fig. 2. Basic operations on combination of neurons on basis of input to output flow

If ferromagnetic and anti ferromagnetic coupling is considered to implement these logic circuits, then this will enforce low energy states to implement arbitrary flow through the logic circuits from input layer to output layer and vice-versa.

2.1 Hopfield Neural Network

Let us first consider some fundamentals of neural network. Let's have a collection of n artificial binary neurons $x_i \in \{-1, 1\}$ with 'i' ∈ {1, 2, …, n}, which are depicted by the activation pattern vector $x = \{x_1, x_2, \ldots, x_n\}^t$ where x^t is known as transpose of x. The neural network can be simply represented as connected graph, which can be stored in a (n x n) dimensional weighting matrix W. Its element is denoted by w_{ij} which specify connection between neuron i and j. It is to be noted that there is no self- loop that's why $w_{ii} = 0$. Filling the weighting matrix W can be done by teaching network on a set of training data. These sets can contain known activation pattern for input and output neurons. There are various learning tools available like backpropagation, gradient descent, and Hebbian learning [3, 7].

The Hopfield network is a fully visible, single-layered, and undirected neural network. It can be trained using Hebbian learning rule. It sets the weighting matrix elements w_{ij} according to the number of occurrences in the training set that the neuron 'i' and 'j' comes together. Consider a training set $T^{(k)}$ having 'K' activation patterns where $k \in \{1, 2, \ldots, K\}$. The normalized weighting matrix will be represented as –

$$W = \frac{1}{n}\left[\frac{1}{K}\sum\nolimits_{k=1}^{K} x^{(k)}\left(x^{(k)}\right)^t - I_n\right] \tag{1}$$

Where I_n is n dimensional identity matrix [3].

Now let us assume that K activation patterns are stored in the weight matrix W. Consider a new activation pattern $x^{(n)}$ is being supplied to the training set which will be compared with the stored patterns. The traditional method of running Hopfield network is to initialize it in the activation $x^{(n)}$ and then operating it iteratively whereas neuron 'i' is selected at random and updated according to the below setting –

$$x_i \leftarrow \begin{cases} +1 \; if \; \sum_{j=1}^{n} w_{ij}x_j \geq \theta_j \\ -1 \quad Otherwise, \end{cases} \tag{2}$$

Whereas $\theta = [\theta_i]_{1=1}^{n}$ user defined threshold vector which determines the switching threshold for each neuron. Each member θ_i should be triggered such that its magnitude becomes order of at most 1. The result for every update will try to stabilize the network energy as

$$E = -\frac{1}{2}x^t W x + \theta^t x \tag{3}$$

Eventually network will converge to local minimum of E after certain number of iterations [3, 7].

As weight matrix W has been kept steady due to learning Hebbian rule local minimum for every $x^{(k)}$ can be obtained. As a result, the output of the Hopfield network is ideally one of the trained activation patterns. The utilization of this memory architecture is also clear and can be used in any kind of binary pattern recognition.

2.2 Restricted Boltzmann Machine

RBMs are broadly used in deep learning networks because of their historical importance and simplicity compared to other architectures. RBMs are used to produce stochastic models of ANNs which can learn the probability distribution based on the input. RBMs come from a family hierarchy of Boltzmann machines. BMs can be decoded as neural networks with stochastic processing units interconnected bidirectional. As it is slightly difficult to learn the properties of an unknown probability distribution, RBMs have been developed to simplify the structure of the network and increase the efficiency of model. It is also well observed that RBMs is a special type of Markov random field having stochastic visible layer in one layer and stochastic hidden units in another layer [10].

In RBMs, neurons form a bipartite graph as every neuron is connected to the neuron from other layer but not from the neuron of the same layer. To teach an RBM, the Gibbs sampling is adopted. Data can be produced by RBM, by starting with an arbitrary node in one layer and run Gibbs sampling. All the nodes in hidden layer will be eventually get updated once the state of the nodes in input layer is given. This updating node is an iterative process and will carry on until network stabilizes. After that weights within the RBM are obtained by maximizing the likelihood probability of this RBM. The basic idea here is that having a neural network and a visible layer that contains data d ∈ {0, 1} at all nodes. Every node also has σ_z terms which penalizes 0 and enhances 1 or ferromagnetic coupling and anti ferromagnetic coupling with respect to all nodes. So, the overall Hamiltonian comes out to be an Ising model -

$$H = \sum_{ij} a_i \sigma_z^2 + b_{ij} \sigma_z^i * \sigma_z^j \tag{4}$$

And a_i and b_{ij} are called weights which are supposed to be tunable. The basic working is to tune these weights in such a way that if visible layer is biased energetically to obtain a piece of data then low energy states of the system will automatically makes this classifier to predict the output correctly [10].

As discussed in previous section that neural networks are good at recognizing things that applies with RBM also as RBM can enforce any computational relationship between classes of output and data. So, the main objective is to tune the weights a_i and b_{ij} so that the probability is given by a Boltzmann distribution

$$P(c, d) = \frac{1}{z} \sum_h e^{-\beta H(c,b,d)} \tag{5}$$

where 'c' and 'd' are output and input layer encoded as classifier and data layer. It means that the probabilities for classifier-data pair (c, d) match the actual frequency of (c, d) in the data q(c, d) which is nothing but the actual probability q of the (c, d). So, the main objective is to try and learn a physical system whose Boltzmann distribution matches the probabilities of data. More precisely, adjust these weights ai and bij to minimize the relative energy between the actual data distribution and current grasp of the data distribution. This is nothing but a minimization of effective free energy of Boltzmann machine which can be formulated as

$$\sum_{c,d} q(c, d) \log \frac{p(c, d)}{q(c, d)} \tag{6}$$

Now, the next step is to identify the learning methods for RBM. There are several learning mechanisms that can be implemented like Gibbs sampling in which you sample the Boltzmann distribution with the help of a known algorithm known as contrastive divergence. Another mechanism is back propagation developed by Hinton in which weights are adjusted starting from data layer to increase the probability of matching the pattern more accurately in the upcoming hidden layers and then errors obtained at classifier levels are back propagated to generate more accurate sequences of data. Another most used notion is stochastic gradient descent on weight space in which we let computer find the sets of weights which takes the system to right direction. Another most interesting algorithm is wake-sleep algorithm which is based on the working on biological brain. Firstly, neural network is tuned with inputs so that it sets up the flow of the network and then the productive connection which back propagates the information is set which as a result will try to predict the flow from input to output based on the network configurations. Now a new data set is created and is trained on these productive connections which exhibit much better results than compared to get it trained with real world [5, 6].

3 Quantum Computing Methodologies

3.1 Basic Fundamentals

3.1.1 Quantum Bits

As classical computers are concerned, digital signals are depicted either in 0 or in 1. One bit provides two states as 0 and 1. Two bits provide states as 00, 01, 10, and 11. Therefore, n bits provide 2^n states. Now, in quantum computers, data is represented in quantum bits also known as qubits which is nothing but a two-state system. A Qubit can form two states at a time which leads to perform exponential computations in limited steps and hence exploits intrinsic parallelism [17].

3.1.2 Quantum Superposition

Qubits reveals a quantum physical property known as superposition. Qubit can describe superposition of 0 and 1 i.e. it can be in both the states simultaneously whereas in classical bits are represented as either 0 or 1.

$$A = x|0\rangle + \mathrm{y}|1\rangle \tag{7}$$

"$|\text{Ket}\rangle$" is described as "Ket Vector" and coefficients 'x' and 'y' are known as probabilistic amplitudes. $|x|^2$ and $|y|^2$ gives the probability of qubits which is –

$$|x|^2 + |y|^2 = 1 \tag{8}$$

It means Qubit A is measured in ket 0 with probability $|x|^2$ and it is measured in ket 1 with probability $|y|^2$ [17].

3.1.3 Quantum Entanglement

Entanglement is a characteristic manifest by qubits which states that if first qubit is calculated then it will say something about the second qubit. This predicts that if the bits are entangled then they must be dependent to each other in some sense. This feature can be used to generate the huge computing power. For example, when n-bit register is given classical computer will store one of the 2^n possible outcomes. But the quantum computer can store all the possibilities simultaneously [17].

$$A = \frac{1}{\sqrt{2}}(|01\rangle + |10\rangle) \tag{9}$$

Equation 9 shows that second qubit can be in either 0 or 1 until first qubit is measured. After that the probability of second qubit will become either 0% or complete 100% to be in a specific state [17].

3.1.4 Quantum Gates

In classical computers, digital information is processed with the help of logic gates. Similarly, quantum computers have quantum gates. Quantum gates are the foundational units for any machine to carry forward their intermediate results. These gates are used to carry in commute qubits values from one state to other. These different states will depend on the type of gate is being used. Quantum gates are mathematically represented as linear operators or in the form of transpose matrix. These gates show an interesting property that they are reversible in nature. This property provides energy efficiency and nurture entanglement. Few crucial universal quantum gates are Hadamard gate, C-NOT gate, Walsh Hadamard gate, etc. [17].

3.2 Grover's Search

An unstructured database contains N records and it becomes critical to figure out the required entry. A classical algorithm will take O(N) time in worst case. And, on average it is required to search a large fraction of database. As quantum system provides intrinsic parallelism it will take much less time as compared to classical counterpart. Grover's search will find such an element in $(\sqrt{N})$ time. Grover's search mainly deals with searching of an element which covers major section of problems related to deep learning framework. This is the reason when performing deep learning model on quantum architecture Grover's search is preferred [17].

3.3 Complexity Analysis of Grover's Search

The beginning phase of Grover's search involves quantum registers of 'k' qubits where 'k' is the search space of size $N = 2^k$ and all are initialized to $|0\rangle$. The whole system is put into equal superposition of states by using one of the quantum gate known as Hadamard gate. Next Grover Iteration is applied which provides amplitude magnification. Grover stated that the best probability of the outcome to be true, phase rotation is required which can only be achieved after $\left(\frac{\pi}{4}\sqrt{2^n}\right)$ iterations. Initially, oracle call is required to establish the searching configuration. After that it matches the system state each time it wants to rotate the phase. If condition satisfies it rotates the phase by π radians otherwise it will do nothing. This will negate the phase but does not change the probability. The crucial part is the next step which is diffusion. This step will perform mean inversion across mean. The running time complexity of this algorithm can be defined as the cumulative sum of oracle function, two Hadamard gate and a conditional phase shift. The single operation of Grover iteration take O(2k) time explicitly including the O(k) time from phase shift which leads to total time complexity of Grover's algorithm to be $O\left(2^{\frac{k}{2}}\right)$. It covers $O\left(\sqrt{N}\right) = O\left(\sqrt{2^k}\right) = O\left(2^{\frac{k}{2}}\right)$ iterations each with a runtime of O(k).

4 Classical Equivalent Quantum Deep Learning Models (QDL)

Quantum methodology can be implemented with deep learning models to give it a quantum touch and achieve some significant speedup in terms of runtime complexity. If pattern recognition is considered, then the above discussed classical deep learning models can be implemented by applying quantum fundamentals to extract the same result in more efficient way. Quantum computing can help train restricted Boltzmann machine in much less time and provides more dealing framework for deep learning that classical and as a result provide improvements. There are certain quantum algorithm exists which can perform better than the conventional algorithms in terms of both training and model quality. The quantum algorithms like Gradient *Estimation via Quantum Sampling (GEQS)* and *Gradient Estimation via Quantum Amplitude Estimation (GEQAE)* works to implement a coherent analog of Gibbs sample state for Boltzmann machine. Similarly, quantum computing can be employed on Hopfield networks also and provide some improvements specifically for pattern recognition application. Some of the quantum coped models are discussed below [11] (Table 2).

4.1 Quantum Hopfield Network for Pattern Recognition

As discussed in the Sect. 2, there are various learning tools available. One of which is Hebbian learning. Quantum Hebbian learning can be designed to provide some efficient working neural network. Quantum Hebbian learning depends on two important factors one is that quantum algorithms can be performed to extract the knowledge from the weighting matrix W. Second W can directly be associated with mixed state τ of a memory register of M Qubits. It is given as –

$$\tau = W + \frac{I_n}{n} = \frac{1}{M}\sum\nolimits_{m=1}^{M} |x^m\rangle\langle x^m| \tag{10}$$

But there encounters a problem for efficient preparation of state $|x^m\rangle$ which can be resolved by various techniques available like qRAM. Let's say P_{req} is the time required to prepare each $|x^m\rangle$. The time complexity required to compute this will be $O(\log n)$ [3].

Consider that τ has been prepared and knowledge must be extracted. It would be difficult to recover the training states $|x^m\rangle$ if τ is an output from an unknown quantum system because τ splits to any random set of pure states as it is not always unique. Although, important information retrieval such as eigen values and eigen vectors can be done with the scan of quantum state tomography of τ. τ can be used as a quantum state to efficiently simulate $e^{i\rho t}$. After simulating, we can use τ to figure out eigen values and

Table 2. Symbols table QDL (Symbol, Description)

Symbols	Description
W	(n x n) Weighting matrix
τ	Mixed state of a memory register
M	#Qubits
$\vert x^m\rangle$	Quantum training state
P_{req}	Time required to prepare each $\vert x^m\rangle$
I_n	n-Dimensional identity matrix
N	Total number of unitary operators
$\{Un\}^N\ n = 1$	Set of N unitary operators
$\vert x(n)\rangle\langle x(n)\vert$	Memory projectors
S	Swap matrix
$e - H$	Diagonal matrix developed after matrix exponentiation of H
φ	Partition function
ρ	Density matrix
Pc	Marginal Boltzmann probability
ω_c	Diagonal matrix containing diagonal elements to be 1 when in $\vert c\rangle$ state, otherwise 0
I_D	Identity matrix on data layer
σ_y	Non diagonal elements matrices
Γ	Represents negative log-likelihood

Note: Some of the symbols are mentioned in CDL

eigen vectors with the help of quantum phase estimation algorithm needing the overall time as $P_{eigenvalues}$ $O(poly(\log n.\frac{1}{\varepsilon}, M))$ [3].

Consider that there is a set of N unitary operators $\{U_n\}$ $n = 1$ applied on M + 1 registers of qubit and can be defined according to

$$U_n = |0\rangle\langle 0| \otimes I + |1\rangle\langle 1| \otimes e^{-i|x^{(n)}\rangle\langle x^{(n)}|\Delta t} \tag{11}$$

These unitary impose the different memory pattern projectors $|x^{(n)}\rangle\langle x^{(n)}|$ on the basis of condition applied and for a small time change Δt. With the help of these unitary simulations on $e^{i\rho t}$ can be done by applying them for a large number of times. Let S be the swap matrix for subsystem σ and $|x^n\rangle$. It can be referred as –

$$\begin{aligned} U_S &= e^{-i|1\rangle\langle 1|\otimes S\Delta t} \\ &= |0\rangle\langle 0| \otimes I + |1\rangle\langle 1| \otimes e^{-i|1\rangle\langle 1|\otimes S\Delta t} \end{aligned} \tag{12}$$

Where $|1\rangle\langle 1|\otimes S$ is 1 - Sparse and can be effectively simulate. Now for sparse Hamiltonian simulation, quantum algorithms with constant oracle calls can be used with run time complexity of O (log n). This can be mathematically represented as

$$\mathrm{tr}_2\left\{U_S\left(|q\rangle\langle q| \otimes |x^n\rangle\langle x^n| \otimes \sigma\right)U_S^\dagger\right\}$$
$$= U_n(|q\rangle\langle q| \otimes \sigma)U_n^\dagger + O\left(\Delta t^2\right) \tag{13}$$

The trace is put to second subsystem which contains the state $|x^n\rangle$.

Now, N unitary operations can be applied sequentially for 'd' repetitions. That is

$$U_p = \left(\prod_{n=1}^{N} U_n\right)^d \tag{14}$$

where $\Delta t = \frac{t}{dN}$.

Now if standard Suzuki-trotter is applied then,

$$\gamma = \left(e^{-i|1\rangle\langle 1|\frac{t}{(nN)}} \bullet \bullet \bullet \bullet\ e^{-i|N\rangle\langle N|\frac{t}{(nN)}}\right)^d - e^{-i\rho t} \tag{15}$$

This can be approximately equal to $O\left(\frac{t^2}{d}\right)$. Hence $d \in O(t^2/\gamma)$ repetitions are required cascaded with N sparse Hamiltonian simulations per repetitions and takes total runtime to $O(Mt^2/\gamma)$. This allows us to use multiple copies of $|x^m\rangle$ as quantum states and these training states need not to be in superposition. Conclusively ρ can be simulated to a precision γ with the number of exercises of U_n with run time of O (t^2/γ). Each U_n can be run with logarithmic time using sparse matrix simulations resulting total time complexity of quantum Hebbian learning to $O(\text{poly}(\log n, t, N, \frac{1}{\gamma}))$. This working of quantum Hebbian learning shows that by implementing weight matrix from training data set as a arbitrary mixed quantum state and then exploit the density matrix and use it in quantum algorithms for high level machine oriented functions that is eigen values and eigen vectors [7].

This quantum learning algorithm can be used in quantum Hopfield neural network for pattern recognition. First step would be to configure the input state $|\theta\rangle$. This will contain the incomplete activation pattern and the threshold data. Quantum Hebbian learning will be imposed with sparse Hamiltonian simulation to stage quantum phase estimation. Conditional rotation of an ancilla would help availing eigen values larger or equal a fixed number α set by user. This is then joined by an uncalculating of the first register of F qubits which can be done by same quantum phase estimation protocol. Beyond the calculation of ancilla one can find the pure quantum state to work with. To recreate the new input state, new runs of quantum algorithms are required. This is known as cloning theorem of states. It would be inappropriate to again use the in-transit data of one run of quantum algorithm in other input patterns. Although, multiple input patterns can simultaneously be recreated with the help of superposition of input patterns. It means information about G patterns can be extracted from the resulting state where g = 1, 2 ..., G be G patterns [3, 7].

4.2 Quantum Boltzmann Machine for Pattern Recognition

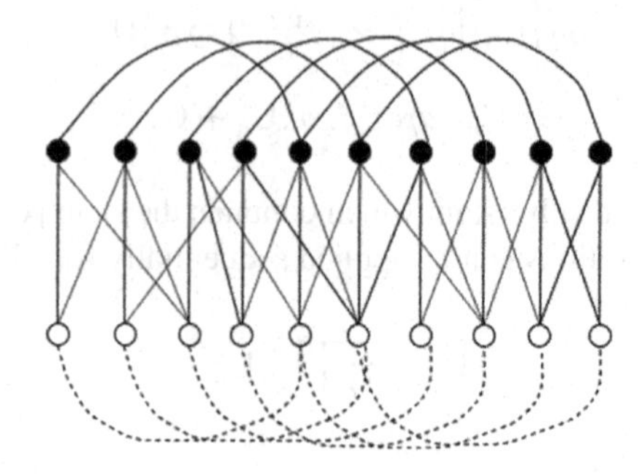

The mathematics behind quantum mechanics is Fourier transforms matrices (operations and inversions) and vector calculus majorly. Classical spins or bits can be replaced by quantum bits (qubits) in order to achieve the hidden complexity improvements. Let us take a matrix with 2^M possible number of states i.e. $2^M \times 2^M$ diagonal matrix called as Hamiltonian –

$$H = \sum_{ij} a_i \sigma_z^2 + b_{ij} \sigma_z^i * \sigma_z^j \tag{16}$$

This Hamiltonian contains all the diagonals values as energy elements corresponding to all 2^M binary states z ordered lexicographically. The eigen states of the Hamiltonian is represented by $|c, d\rangle$ where again 'c' and 'd' are output and input layer respectively and encoded as classifier and data layer. Matrix exponentiation can also be taken out easily by Taylor expansion

$$e^{-H} = \sum_{i=0}^{\infty} \left(\frac{1}{i!}\right)(-H)^{-i} \tag{17}$$

e^{-H} is the diagonal matrix contain 2^M diagonal elements. Now, density matrix is defined with the help of partition function $\varphi = \mathrm{Tr}[e^{-H}]$ [5, 6].

$$\rho = \varphi^{-1} e^{-H} \tag{18}$$

Clearly, the diagonal elements of density matrix would be the Boltzmann probabilities of all 2^M states. For a provided state $|c\rangle$, marginal Boltzmann probability can be defined as

$$P_c = Tr[\omega_c \rho] \tag{19}$$

where ω_c set the threshold only to the diagonal elements corresponding to the classifier points being in state $|c\rangle$. In other words, ω_c is a diagonal matrix containing diagonal elements to be 1 when in $|c\rangle$ state, otherwise 0. Quantum mechanically, it can be written as

$$\omega_c = |c\rangle\langle c| \otimes I_D \tag{20}$$

where I_D is identity matrix on data layer [6].

Additionally, work can be done on non-commutative effects of quantum mechanics by adding off-diagonal elements to the Hamiltonian. Transverse field to Ising Hamiltonian is an example which can be used for this purpose. This transverse field will help is setting up non-diagonal matrices. So the transverse Ising Hamiltonian can be represented as

$$H = \sum_{i} C_i\sigma_y + \sum_{ij} a_i\sigma_z^2 + b_{ij}\sigma_z^i * \sigma_z^j \tag{21}$$

where, σ_y is the non-diagonal elements matrices. All eigen states of H are now in superposition in the calculation basis created with the help of classical states $|d, c\rangle$. In QBM, quantum probability distribution with density matrix is used which contains off-diagonal elements also. For every calculation qubits are read out in σ_z basis and output is classical value $+1$ or -1. Due to this outstanding property of quantum mechanics, after every calculation an output for classifier layer will comes with the probability of P_c [6].

To train the QBM, Shifting of Hamiltonian parameters θ is applied in such a way that the probability distribution P_c moves close to the probability of data in data layer. This can be done by minimizing the log-likelihood which can be represented as

$$\Gamma = -\sum_{c} p^{data} \log \frac{Tr[\omega_c e^{-H}]}{Tr[e^{-H}]} \tag{22}$$

Differentiation of above equation is performed with respect to the Hamiltonian parameters to calculate the gradient efficiently using sampling. Now the matrices obtained are static in nature and also gradient cannot be found trivially as in case of classical Boltzmann machines. The terms need to be permuted and traced to find the expression needed to establish the optimal gradient. Gibbs sampling is then applied to fetch out the desired results. Hence, training of QBM is achieved in this manner. However, this is not the optimized approach as separate calculation is required for each data vector and process becomes inefficient basically for large data sets. Various workarounds are available to deal with this problem. These workarounds can be integrated with this structural algorithm to address various functional requirements. The time required to train is $O(IME\sqrt{\eta})$ where 'I' is number of iterations, 'M' is training data set, 'E' is number of edges in model and η is a normalizing constant required to provide a probabilistic view of quantum algorithm used [5, 6, 12].

5 Comparative Analysis Between Quantum and Classical Models

Deep learning models	Training complexity	Model complexity
Classical Hopfield network	$O(KNn)$	$O(KNn)$
Quantum Hopfield network	$O(\text{poly}(\log n, t, N, \frac{1}{\gamma}))$	$O(\text{poly}(\log n, \frac{1}{\upsilon}, N, \frac{1}{\gamma}))$
Classical RBM	$O(IME\eta)$	$O(IME\eta)$
Quantum RBM	$O(IME\sqrt{\eta})$	$O(IME\sqrt{\eta})$

6 Conclusion and Future Work

This paper provides the insight about pattern recognition, classical deep learning, and quantum counterpart of classical deep learning. As discussed, quantum computing can provide an appreciable speed up via various approaches like quantum Hebbian learning used in Hopfield networks, quantum Boltzmann machines, quantum associative memory used to store input patterns and activation patterns in terms of speed and memory. The time complexity analysis is also compared and discussed. There are certain opens which needs to be addressed to achieve the purest meaning of quantum computing. Deep quantum learning nowadays means having a classical model and quantum algorithms like Grover's search, quantum annealing is being used for data preprocessing and weight discovery process on it. But to make a universal deep quantum learning which means network should be capable of universal quantum computing itself to perform quantum version of deep learning models on it. Research is also going around finding out whether quantum Fourier transform, boson sampling can be performed by universal deep quantum learning.

References

1. Biamonte, J., Wittek, P., Pancotti, N., Rebentrost, P., Wiebe, N., Lloyd, S.: Quantum machine learning. Nature **549**, 195–202 (2018)
2. Trugenberger, C.A.: Quantum Pattern Recognition. Quantum Inf. Process. **1**, 471–493 (2002)
3. Rebentrost, P., Bromley, T.R., Weedbrook, C., Lloyd, S.: Quantum Hopfield neural network. Phys. Rev. A **98**, 042308 (2018)
4. Liu, W., Wang, Z., Liu, X., Zeng, N., Liu, Y., Alsaadi, F.E.: A survey of deep neural network architectures and their applications. Neurocomputing **234**, 11–26 (2017)
5. Wiebe, N., Kapoor, A., Granade, C., Svore, K.M., Microsoft: Quantum inspired training for Boltzmann machines. arXiv:1507.02642 (2015)
6. Amin, M.H., Andriyash, E., Rolfe, J., Kulchytskyy, B., Melko, R.: Quantum Boltzmann machine. Phys. Rev. X **8**, 021050 (2018)
7. Widrow, B., Kim, Y., Park, D., Perin, J.K.: Nature's learning rule: the Hebbian-LMS algorithm. In: Artificial Intelligence in the Age of Neural Networks and Brain Computing, pp. 1–30 (2019)
8. Schutzhold, R.: Pattern recognition on a quantum computer. Phys. Rev. A **67**(6), 062311 (2002)
9. Prousalis, K., Konofaos, N.: A quantum pattern recognition method for improving pairwise sequence alignment. Sci. Rep. **9**, 7226 (2019)
10. Fischer, A., Igel, C.: Training restricted Boltzmann machines: an introduction. Pattern Recogn. **47**(1), 25–39 (2014)
11. Wiebe, N., Kapoor, A., Svore, K.M.: Quantum deep learning. Quantum Physics (quant-ph); Machine Learning (cs.LG); Neural and Evolutionary Computing (cs.NE) (2015)
12. Beer, K., Bondarenko, D., Farrelly, T., et al.: Training deep quantum neural networks. Nat. Commun. **11**, 808 (2020)
13. Pinkse, P.W.H., Goorden, S.A., Horstmann, M., Škorić, B., Mosk, A.P.: Quantum pattern recognition. In: 2013 Conference on Lasers & Electro-Optics Europe & International Quantum Electronics Conference CLEO EUROPE/IQEC, Munich, p. 1 (2013)
14. Prousalis, K., Konofaos, N.: Quantum pattern recognition for local sequence alignment. In: 2017 IEEE Globecom Workshops (GC Wkshps), Singapore, pp. 1–5 (2017)

15. Dunjko, V., Briege, H.J.: Machine learning & artificial intelligence in the quantum domain: a review of recent progress. Rep. Prog. Phys. **81**, 074001 (2018)
16. Shrivastava, P., Soni, K.K., Rasool, A.: Classical equivalent quantum unsupervised learning algorithms. Procedia Comput. Sci. **167**, 1849–1860 (2020). ISSN 1877-0509
17. Shrivastava, P., Soni, K.K., Rasool, A.: Evolution of quantum computing based on Grover's search algorithm. In: 2019 10th International Conference on Computing, Communication and Networking Technologies (ICCCNT), pp. 1–6 (2019)

Empirical Laws of Natural Language Processing for Neural Language Generated Text

Sumedha(✉) and Rajesh Rohilla

Department of Electronics and Communications, Delhi Technological University, Delhi, India
rajesh@dce.ac.in

Abstract. In the domain of Natural Language Generation and Processing, a lot of work is being done for text generation. As the machines become able to understand the text and language, it leads to a significant reduction in human involvement. Many sequence models show great work in generating human like text, but the amount of research work done to check the extent up to which their results match the man-made texts are limited in number. In this paper, the text is generated using Long Short Term Memory networks (LSTMs) and Generative Pretrained Transformer-2 (GPT-2). The text by neural language models based on LSTMs and GPT-2 follows Zipf's law and Heap's law, two statistical representations followed by every natural language generated text. One of the main findings is about the influence of parameter Temperature on the text produced. The LSTM generated text improves as the value of Temperature increases. The comparison between GPT-2 and LSTM generated text also shows that text generated using GPT-2 is more similar to natural text than that generated by LSTMs.

Keywords: Long Short Term Memory networks (LSTMs) · Transformers · Generative Pretrained Transformer 2 (GPT-2) · Text Generation · Zipf's Law · Heap's Law

1 Introduction

Language is a very important part of our lives since it is something we use to communicate our thoughts. At an abstract level, language can be explained as a collection of alphabets and characters which can be accessed through some rules known as grammar. Even for humans, it takes years to learn a language because of its complexity. As the rules are not standard in every language, the existence of sarcasm and context adds extra fuzziness to the process of understanding it by the system. For example, the word duck in terms of cricket and general terms hold completely different meanings [10]. Due to all these challenges, natural language generation and processing have for quite some time been considered as among the most testing computational assignments.

Standard Neural Networks are not used for text generation because the length of input and output is not fixed, it may vary with each sentence. For this sequential data Recurrent Neural Networks (RNNs), Gated Recurrent Units (GRUs), Long Short Term Memory Networks (LSTMs), Encoder-Decoder Models, or some type of Transformer is

M. Bhattacharya et al. (Eds.): ICICCT 2021, CCIS 1417, pp. 184–197, 2021.
https://doi.org/10.1007/978-3-030-88378-2_15

used. Since these are predictive models as well as generative models i.e., they can learn from sequence and generate a completely new sequence based on training data fed to it [10]. In this paper, we have generated text using LSTMs and Generative Pretrained Transformer 2 (GPT-2). This process consists of two parts first being preparing the data. In the preparation step various processes such as data cleaning and exploration, tokenization, character to integer mapping, applying sliding window technique, etc. are done, so that data can be fed to the machine learning model. The next step is training the model which includes defining the type of model, adding embedding, neural network and dropout layers, deciding values and types of various hyper-parameters for the model, etc.

After generating text, we have evaluated how close the neural language generated text is to human-generated text by checking if it follows Zipf's law and Heap's law, two empirical laws followed by every man-made text in every language. Along with this, we have seen the dependence of text generated on a parameter called Temperature and compared text generated by LSTMs to that generated using GPT-2.

2 Related Work

The task of text being generated by a machine was first seen in mid the 1960s with the emergence of ELIZA, a chatbot-like computer program built at MIT Lab for Artificial Intelligence. ELIZA breaks the input into sentences, parses it based on a simple pattern, and searches for keywords. Based on that keyword it generates a generalized response and if no keyword is found it asks the user to elaborate the input [6]. Although it works well, the amount of intelligence involved is very little.

Comparatively more commercial and intelligent text generation systems using Markov Chains first appeared in March 2017. Markov Chains predict the next word using the current word, as the output depends only on the current word, text loses semantics and context [11]. To overcome this problem, using Recurrent Neural Networks (RNNs) is recommended. In RNNs the current output is a function of present input as well as output of the previous neural network, hence it recollects the past and, conclusions it makes are also influenced by what it has learned from the past as well as current data being fed to it. As we go on with training RNNs, weights try to adjust themselves to minimize the error. In this process weights at the end will have more influence on the text generated as compared to weights at beginning of the network and the weights at the start slowly become zero. This process is known as vanishing gradients and is solved by using Long Short Term Memory networks (LSTMs) [10]. LSTMs are an exceptional sort of RNNs, equipped for learning comprehensive dependencies. They have an extra input known as cell state. The cell state is updated in each step such that if weights are too high it lowers them, if they are much less it increases them, hence avoiding the problem of vanishing or exploding gradients [12]. Along with research in the field of text generation using LSTMs, possibilities of using Generative Adversarial Networks (GANs) for text generation are also being explored [5].

The most recent and well-performing machine learning models for text generation are Transformers. These models work on self-attention mechanism and are developed using encoders and decoders. All the best performing neural network architectures in the field of Natural Language Processing are found to be variants of transformers e.g.: BERT, GPT-2, etc., and the recent researches are focused on improving these models [9].

3 Methods

3.1 Zipf's Law and Heap's Law

All human-generated texts in all the languages follow Zipf's law and Heap's law. If our machine-generated text also follows these laws, we say that this neural language generated text is close to the natural language generated text.

According to Zipf's law, for all human-generated texts, the rank-frequency distribution is an inverse relation (frequency is proportional to the inverse of rank) [4]. In simple terms it can be explained as, the word that occurs the greatest number of times occurs two times more than the word that occurs the second most number of times, three times more than the third most frequently occurring word, and so on [11]. In natural language processing, Zipf's law is mostly used in text compression algorithms, so that we know about the frequency of words and algorithm compresses words that occur very frequently and not the rarely occurring ones. Along with text compressors it is also used in various NLP Algorithms and text generators [12]. Although it is an empirical law, many explanations have been attributed to its occurrence such as the Principle of least effort which says that, people tend to read and write easily so they use the same words more and more and it also gets passed to future generations.

Mathematically, Zipf's law can be represented as:

$$f(r) = Cr^{-k} \tag{1}$$

where f(r) is the frequency of term, r is its rank, C is a constant and k has value approximately equal to 1.

Heap's law states that as the size of document increases, the rate at which the number of distinct words increase in it takes a downturn e.g.: Suppose in a document with 1000 words no. of unique words are 100, then for a document with 2000 words no. of unique words will be less than 200, for a document with 3000 words no. of unique words will be much less than 300 and so on [8].

The archived meaning of Heaps' law says that the quantity of exceptional words in content of n words is approximated by:

$$Q(n) = An^{k} \tag{2}$$

where A is a positive constant which is usually up to 100, k is between 0 and 1, most frequently in range the 0.4 to 0.6 [11].

3.2 Long Short Term Memory Networks

Recurrent Neural Networks (RNNs) are the neurons which receive input from the previous cell as well as present cell. This can be seen in Fig. 1, where $y_2(t)$ =

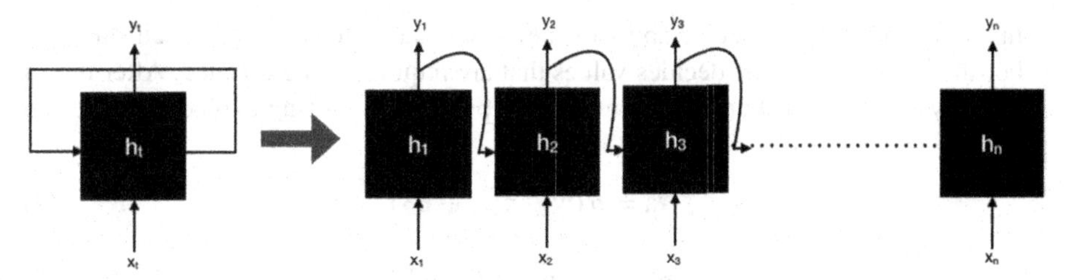

Fig. 1. Unrolled RNN cell

$f(x_2(t) + y_1(t-1))$. Cells that are the function of input from previous cell are known as memory cells. As a result, RNNs remember the past. These are designed to work with sequential data like sentences, audio, video, etc. [10].

While training the network on large sequences RNNs begin to forget the starting part due to vanishing gradient issue which is solved utilizing LSTMs. Remembering details for a long time comes naturally to LSTMs because of their structure and design.

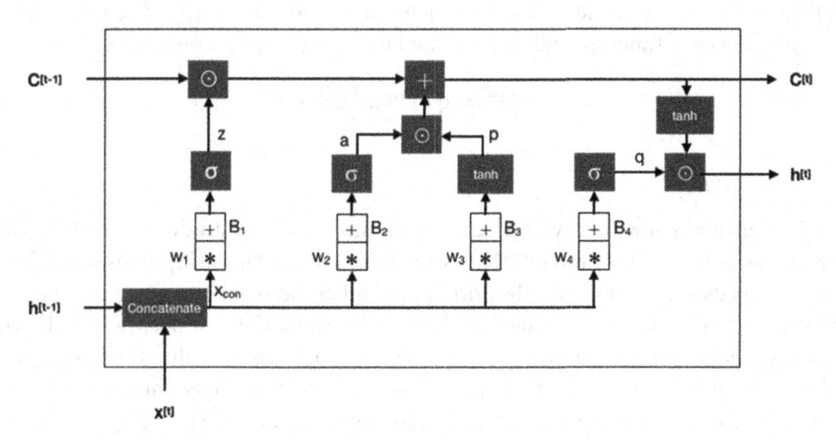

Fig. 2. LSTM cell

In Fig. 2, $C^{[t]}$ represents the memory cell i.e., it provides the memory feature to LSTM cells, $h^{[t-1]}$ is the output of the previous cell. The LSTM generates $h^{[t]}$ and $c^{[t]}$ as its outputs [13].

Its functioning can be explained in four steps or layers:

- First, a forget Layer makes a decision on information to be kept and thrown away. In this step, we pass $h^{[t-1]}$ and $x^{[t]}$ into a sigmoid function, which generates an output between 0 and 1. Output 0 represents forgetting that information and an output of 1 represents keeping that information. e.g.: When working with a subject we may need to remember its gender while selecting that pronoun but may not need to remember that feature for the next subjects. Hence, this information can be forgotten.

$$z = \sigma(w_1 \cdot x_{con} + B_1) \tag{3}$$

- In the second step, we decide about the new information to be stored in cell state $c^{[t]}$. For this, a sigmoid layer decides values that are required to be updated. After this, a tanh layer (as it distributes gradients, hence prevents vanishing/exploding) gives an array of new values.

$$a = \sigma(w_2 \cdot x_{con} + B_2) \tag{4}$$

$$p = \tanh(w_3 \cdot x_{con} + B_3) \tag{5}$$

$$q = \sigma(w_4 \cdot x_{con} + B_4) \tag{6}$$

- In this step we update the old state cell by deciding the output for $c^{[t]}$ and discarding information about old subject and adding new details.

$$C^{[t]} = z \odot C^{[t-1]} + p \odot a \tag{7}$$

- For the last step, we update the output H(t). First, a sigmoid layer is present which tells about the part of the cell state which will contribute to h[t]. Then, its result is given to the tanh function and is multiplied by sigmoid gate's output.

$$h^{[t]} = q \odot \tanh\left(C^{[t]}\right) \tag{8}$$

3.3 Transformers

Transformers are a new family of Machine Learning models introduced in 2017. These work on the self-attention model, that is we look for a relationship between different input sequences [2]. This means they don't remember the whole sentence at once, but a perimeter α is assigned. $\alpha_{<1,1>}$ decides how much value the first network holds while generating the first word, similarly $\alpha_{<2,1>}$ decides how much value the first network holds while generating the second word and so on. In transformers, this process is done multiple times, so this is known as multi-headed self-attention [3].

As, can be seen in Fig. 3, a transformer cell consists of encoder and decoders blocks. An encoder block is further a collection of many encoders connected serially, similarly, a decoder block is also a collection of connected decoders. An encoder converts text to word embeddings which are fed to the decoder, which as output generates text. Many latest models based on the transformer architecture are BERT, GPT-2, GPT-3, etc. In this paper, text is generated using Generative Pretrained Transformer 2 (GPT-2), which is based on a transformer model and uses decoder blocks. These have shown remarkable performance in every field of Natural Language Processing including text generation [9].

4 Experiments and Results

4.1 Dataset Used

As dataset, the famous book "The Adventures of Sherlock Holmes" by Sir Arthur Conan Doyle is utilized. The book is made available through Project Gutenberg and contains

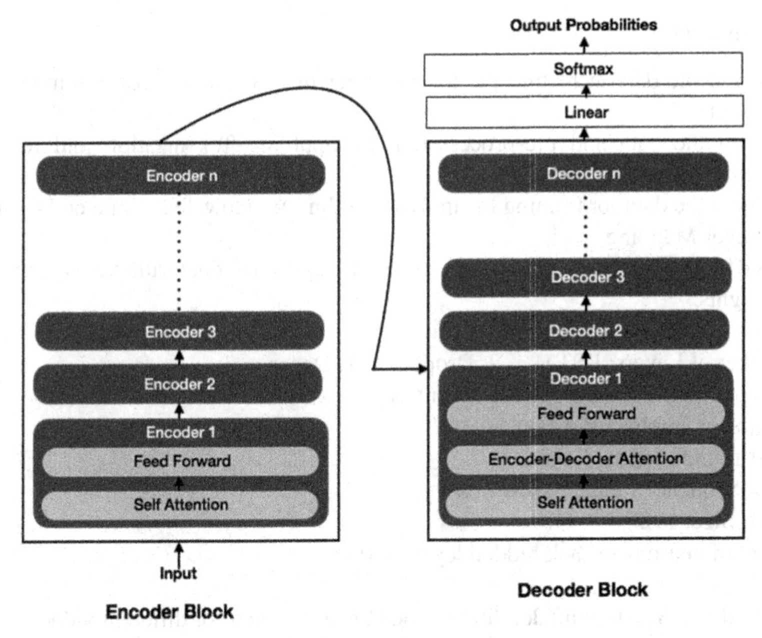

Fig. 3. Transformer cell

a total of 594,197 characters. Most ordinarily happening words in the dataset and their recurrence can be seen in Fig. 4.

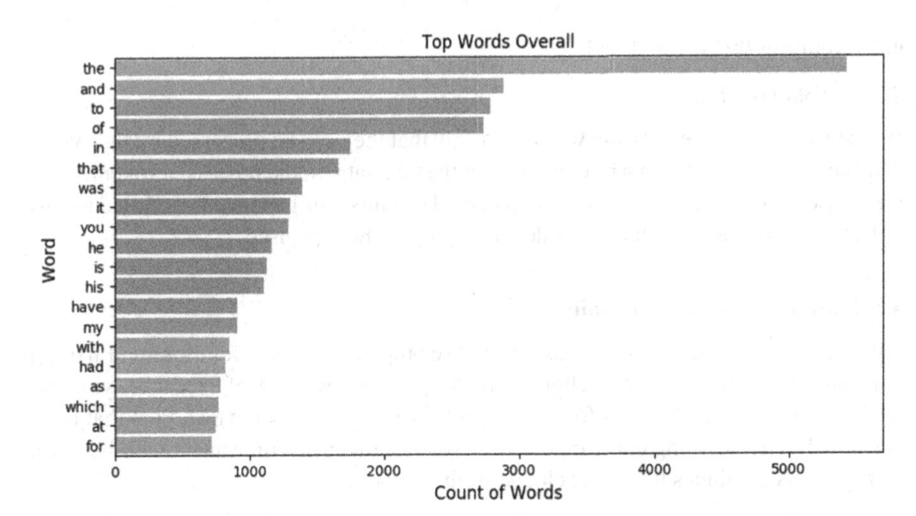

Fig. 4. Most ordinarily happening words in the dataset and their recurrence

4.2 Approach

- Explored the dataset to find the total number of characters, most commonly used words, etc.
- Cleaned the data and pre-processed it by applying Tokenization and removing punctuations and stop words
- Prepared the data for training by applying Sliding Window Technique and Character to Integer Mapping
- Tuned hyperparameters and finally generated sequential model with the accompanying highlights:

  Number of Layers: LSTM = 2, Dropout = 1, Dense = 2
  Loss computed: Categorical cross-entropy
  Optimizer employed: Adam
  Activation function: ReLU
  Input Sequence Length: 25 words
  Batch Size: 192
  Count of neurons in each hidden layer: 250

- Created LSTM based model, fitted it, and generated text for different values of word length and hyper-parameter Temperature in range 0.1 to 3
- Generated text using GPT-2 for different word lengths
- Compared how well LSTM and GPT-2 generated text follows Zipf's law and Heap's law
- Compared the texts generated using LSTMs and GPT-2

4.3 Defining the Terms Used

4.3.1 Tokenization

It is a process of expressing the words in a way that they can be processed by the system and is usually the first step while processing the text data. In the process of tokenization, we divide the original text into smaller pieces. The units thus generated are called tokens. Tokens can be words or characters depending upon the type used.

4.3.2 Sliding Window Technique

After obtaining an array of words as tokens, we prepare the model for training. First step in it being the sliding window technique. In this process a window of fixed size (sequence length is taken to be 25 words for this project) is operated on some text, then that tssext is learned. After learning that window is moved one token at a time and remembers that. This process continues till we reach end of the sequence.

4.3.3 Character to Integer Mapping

In this process, each character is assigned a unique integer. It is done because for machines it is easier to read numbers as compared to text data. This process reduces complexity and increases speed of the model [8].

4.4 Results Obtained

4.4.1 Variation of Text Generated Using LSTMs with Parameter Temperature

Table 1. Text generated using LSTMs keeping document length = 50 words and varying Temperature in range 0.1 to 3

(Temperature, document length)	Text generated
(0.1, 50)	Each when we followed the back from some mark the openly preceded in spite of the slight man in the man work and the heading gave the north taken and dealings respectable 4 and of an attacked of the drug out of the help and with a small public face
(0.5, 50)	Is used living and violates support in the dead of reading the work seen of which was whine beside the edge of the great house which had been hereditary sense hands across to its crop to explain you through the first hand of ballarat in the fancier faced hat which
(1, 50)	All machinery is given you through hand where is very such so long that miss remark is its laughed or coupled' judgment from all the house was the disposition of the city branch of it and save the full which he had promised to have spoken to us to
(1.5, 50)	Across together the show little gloom of burning in the u.s. and accordance heard repeated sign brown and distributed project gutenberg tm electronic works and you do not narrow to be
(2, 50)	Robberies three of a shriek in disappearing acid with the inspector tragedy of the spellbound sun were inches of hercules by the manager of those gaiters was of the same residence the light men and the help like being waistcoat yet in the aperture his is was as his sleeping
(2.5, 50)	Behind the white curling edge which are thrown at a manner bow on peeped on a shoulders face which is over the house a show plain man hair crystallised the smaller observe of the front of his lantern and left the date of the sea indexing i do not touch
(3, 50)	The edge there was breaches crushed and violates support in a poisoning this and door to the same corridor by the odessa vault of water which trafalgar him to come investment we had just a very asked witted carried under the irene world against me by order to prevent the

Temperature is a hyperparameter in neural networks which controls the randomness of text generated or any output by scaling the un-normalized predictions before applying activation function.

In Table 1, we have measured the effect of variation of hyperparameter Temperature on the text generated, by keeping the word length in document as constant and we can see that as temperature (randomness) increases grammatical errors decrease and sentences start making comparatively more sense.

4.4.2 Text Generated Using GPT-2 for Different Word Lengths

We implemented a pre-trained transformer model using Hugging Face and generated a text of lengths = 25, 50, 75 and 100 with random seed text [10]. The results obtained can be seen from Figs. 5, 6, 7 and 8.

```
0: Watson you are in an accident?

Pam: Oh yes, and I would tell you, but I'm
```

Fig. 5. Text generated using GPT-2 for word length = 25

```
1: Watson you are the greatest football fan ever, what can you say in your defence?

"I could probably give you one reason why I love being a fan of rugby union so much.

"I'm not going to pretend to
```

Fig. 6. Text generated using GPT-2 for word length = 50

```
2: Watson you are no longer allowed to leave your job but your job does not matter. He tells me that his only aim is
to get people to know and respect his intelligence; this is important for a successful company, otherwise he doesn't
understand his customers. You are a smart man and I applaud your intelligence, but the rest is a complete waste of th
e time, energy
```

Fig. 7. Text generated using GPT-2 for word length = 75

```
3: Watson you are very well. Now you are saying that you would not get away with not getting away with the stuff. Are
you really implying that you were in a position where you knew that you were not making it a crime to say "fuck you"
in public to the police officer? Is that what you are implying?

Hmmm. Is that what you are implying? Is that what you are implying? It certainly strikes me as a strong indication of
some sort of crime of verbal
```

Fig. 8. Text generated using GPT-2 for word length = 100

From the obtained text we can see that the text generated by GPT-2 is understandable to great extent, but sudden change in topic can be observed for each word length. Comparing this to LSTMs generated text we can say that text by GPT-2 makes more sense and is grammatically better. Also, the process of text generation by GPT-2 is more computationally expensive than the process involving LSTMs.

Table 2. Word rank vs. frequency plot and number of unique words for text generated using LSTMs for different values of Temperature and Document Length in words

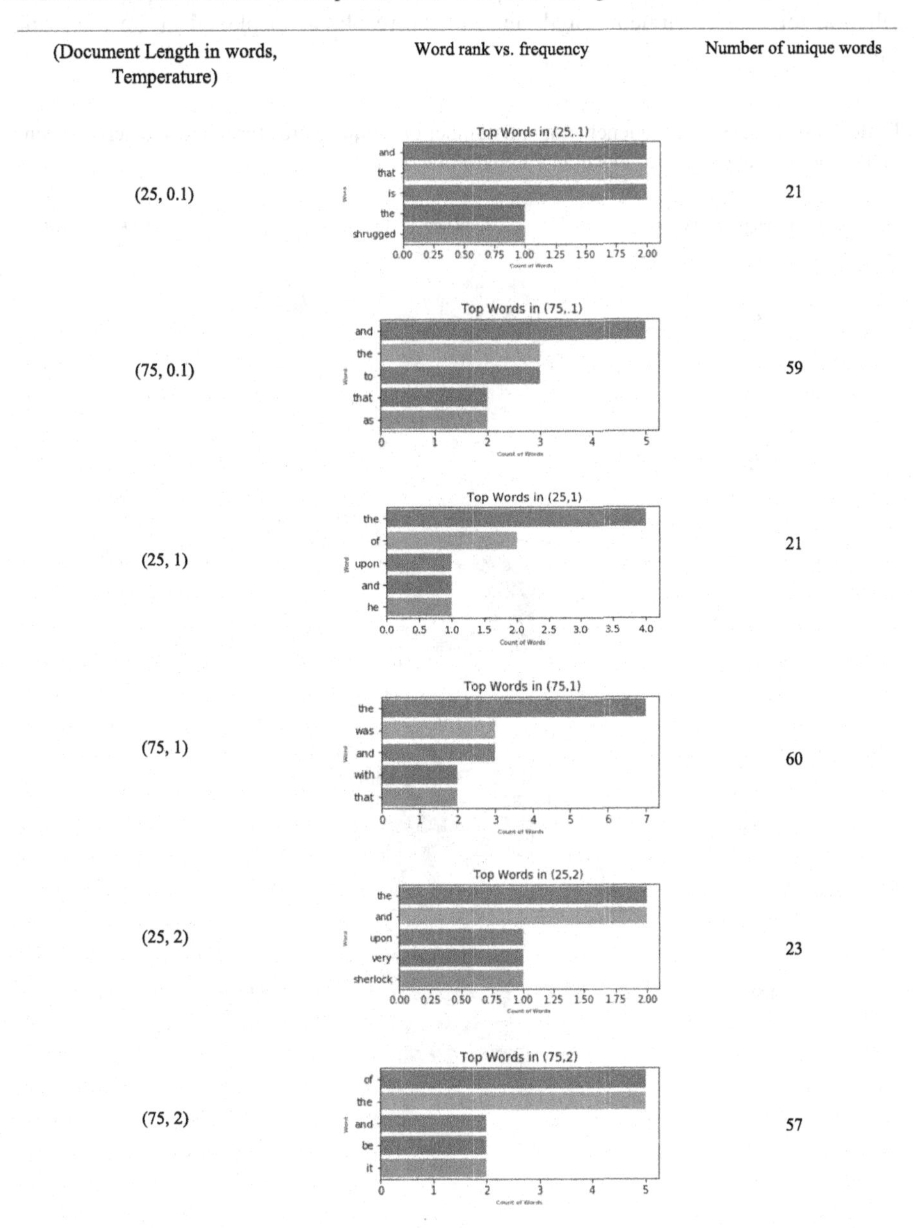

(Document Length in words, Temperature)	Word rank vs. frequency	Number of unique words
(25, 0.1)	Top Words in (25,.1)	21
(75, 0.1)	Top Words in (75,.1)	59
(25, 1)	Top Words in (25,1)	21
(75, 1)	Top Words in (75,1)	60
(25, 2)	Top Words in (25,2)	23
(75, 2)	Top Words in (75,2)	57

4.4.3 Verifying Empirical Laws for Text Generated Using LSTMs and GPT-2

From Table 2, we can see that for all temperatures and document lengths word rank and frequency are inversely proportional to each other.

For verifying these two power laws for GPT-2 we generated text with various word lengths and calculated number of unique words and word rank vs its frequency distribution for each document length and got the results as displayed in the following table.

Table 3. Word rank vs. frequency plot and number of unique words for the text generated using GPT-2 for different values of Document Length in words

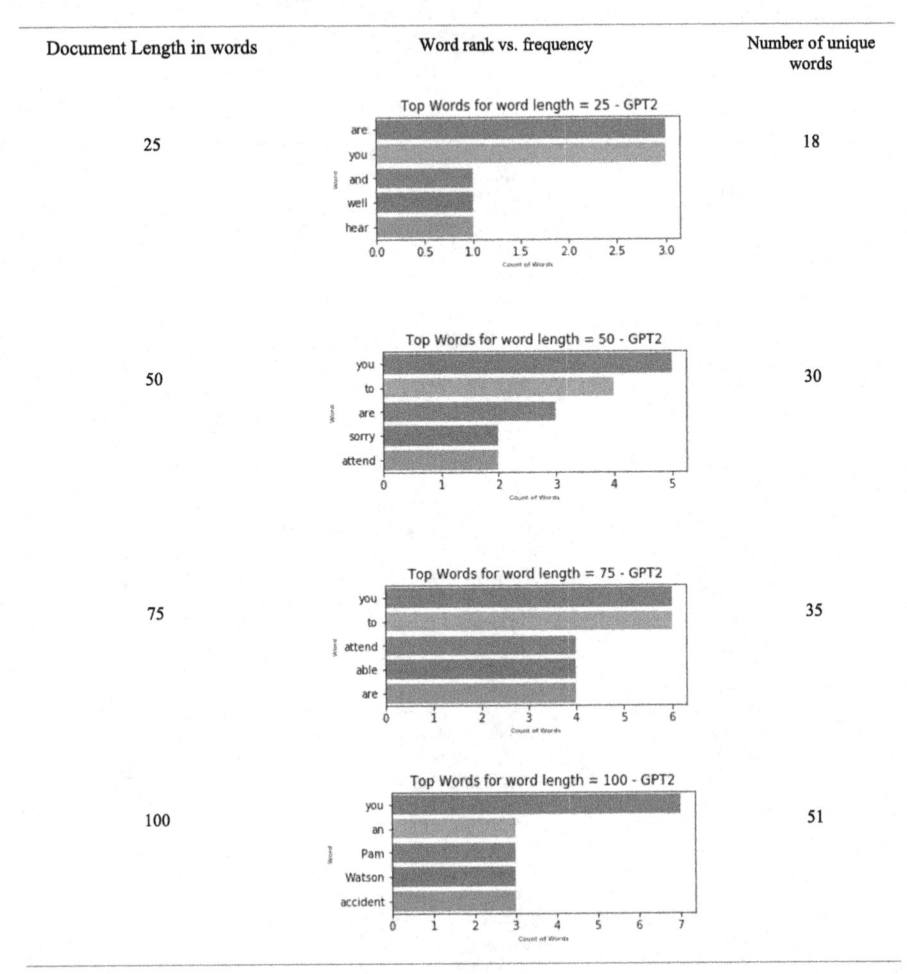

Document Length in words	Word rank vs. frequency	Number of unique words
25	Top Words for word length = 25 - GPT2	18
50	Top Words for word length = 50 - GPT2	30
75	Top Words for word length = 75 - GPT2	35
100	Top Words for word length = 100 - GPT2	51

From Table 3, we can see that for all document lengths word rank and frequency are inversely proportional to each other for GPT-2 generated text as well. Hence, we can say that Zipf's law is being followed by text generated by LSTMs and GPT-2.

On plotting Document length vs. Number of unique words for LSTMs and GPT-2 we get the plot as shown below:

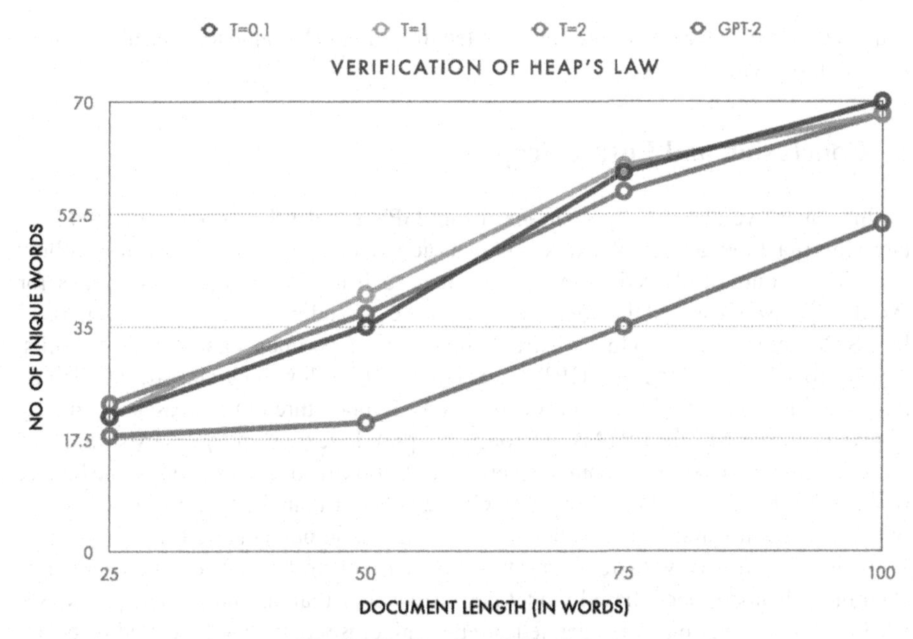

Fig. 9. Verification of Heap's Law

From Fig. 9, we can see that as document size is increasing, the rate at which the number of distinct words is increasing has dropped. This tells us that Heap's law is also being followed for the text generated by the neural language models (both LSTM networks and GPT-2).

4.5 Comparison with Existing Works

M. Lippi, M. A. Montemurro [1] generated text with LSTMs using torch-rnn package, with 2 LSTM layers and each layer consisting of 1024 cells. For training purposes, they split training data such that 100 characters are processed at a time and used dropout = 0.7. Text generated using these parameters showed the most amount of similarity to natural text, for the value of Temperature = 1 [1]. We generated LSTM based text using Keras library, with 2 LSTM layers, each layer containing 250 cells and a dropout = 0.8. For training the LSTMs we used word based approach instead of character based approach, and used 25 words to be processed at a time. Text generated using these parameters can be seen in Table 1 and it shows that as the value of hyper-parameter T increases, the quality of text and its similarity to human generated text increases.

For generating text using OpenAI GPT-2 Y. Qu, P.Liu [2] used a BERT tokenizer and trained the GPT-2 model. The generated model gives accurate readability but generates duplicates [2]. We utilized sample decoding, GPT-2 tokenizer and gpt2-medium, a pre-trained model with 1024 hidden layers, 16 heads and 345 million parameters for OpenAI GPT-2 based text generation. The text generated using these parameters can be seen in Sect. 4.4.2. The text is readable to great extent and unique upto word length = 75, but some repetitions can be observed for word length = 100. The same text can also be seen to randomly switch topics.

5 Conclusion and Future Scope

In this paper, we generated pseudo text using GPT-2 and LSTMs and evaluated how close this machine-generated text is to human-generated text by checking if they follow statistical features followed by man-made text such as Zipf's and Heap's Laws for Words. We also measured the effect of temperature or randomness on the text generated by LSTM networks. To do this, we tried out various experiments and obtained results, which proved that LSTM and GPT-2 generated texts follow both statistical laws i.e., Zipf's and Heap's law. We also observed that as Temperature (T) increases the quality of text also improves for LSTMs. Although the pseudo text given by LSTMs improves with T, it consists of some grammatical errors and stops making sense after some length, on the other hand, the text generated by GPT-2 is better than the text given by LSTMs in terms of grammar and quality, but lags in terms of computational cost.

This study opens ways for many types of research in the future. The Generative Pretrained Transformer-2 based model performs better than all the models previously used for text generation and generated samples are close to human-generated texts but, it also has some limitations such as because of its huge size it is more computationally expensive compared to previous models, this model gives good performance on generalized topics but performs poorly on scientific or technical data and abrupt changes in the topics can also be noticed. Future researches in the natural language processing community aim to reduce the computational cost of Transformers and LSTMs and to remove the limitations discussed to as much extent as possible.

References

1. Lippi, M., Montemurro, M.A., Degli Esposti, M., Cristadoro, G.: Natural language statistical features of LSTM-generated texts. IEEE Trans. Neural Netw. Learn. Syst. **30**(11), 3326–3337 (2019)
2. Qu, Y., Liu, P., Song, W., Liu, L., Cheng, M.: A text generation and prediction system: pretraining on new corpora using BERT and GPT-2. In: 10th International Conference on Electronics Information and Emergency Communication (ICEIEC), Beijing, China, pp. 323–326. IEEE (2020)
3. Santillan, M.C., Azcarraga, A.P.: Poem generation using transformers and Doc2Vec embeddings. In: 2020 International Joint Conference on Neural Networks (IJCNN), Glasgow, UK, pp. 1–7. IEEE (2020)
4. Wang, D., Cheng, H., Wang, P., Huang, X., Jian, G: Zipf's law in passwords IEEE Trans. Inf. Forensics Secur. **12**(11), 2776–2791 (2017)
5. Li, C., Su, Y., Liu, W.: Text-to-text generative adversarial networks. In: International Joint Conference on Neural Networks (IJCNN), pp. 1–7. IEEE, Rio de Janeiro (2018)
6. Weizenbaum, J.: ELIZA – a computer program for the study of natural language communication between man and machine. Commun. ACM **9**, 36–45 (1966)
7. Gatt, A., Krahmer, E.: Survey of the state of the art in natural language generation: core tasks applications and evaluation. JAIR **61** 65–170

8. Godor, B.: World-wide user identification in seven characters with unique number mapping. In: 12th International Telecommunications Network Strategy and Planning Symposium, New Delhi, India, pp. 1–5. IEEE (2006)
9. Vaswani, A., Shazeer, N., Parmar, N., Uszkoreit, J., Jones, L., Gomez, A.N.: Attention is all you need. NIPS (2017). https://arxiv.org/abs/1706.03762. Accessed 14 Feb 2021
10. Raghav, B.: Text Generation in NLP - Springboard India. https://in.springboard.com/blog/text-generation-using-recurrent-neural-networks/. Accessed 22 Dec 2020
11. Lü, L., Zhang, Z.K., Zhou, T.: Zipf's law leads to Heaps' law: analyzing their relation in finite-size systems. PLoS ONE **5** 0014139 (2010)
12. Li, W.: Random texts exhibit Zipf's-law-like word frequency distribution. IEEE Trans. Inf. Theory **38**(6), 1842–1845 (1992)
13. Otter, D.W., Medina, J.R., Kalita, J.K.: A survey of the usages of deep learning for natural language processing. IEEE Trans. Neural Netw. Learn. Syst. **32**(2), 604–624 (2020)

Story Generation from Images Using Deep Learning

Abrar Alnami(✉), Miada Almasre(✉), and Norah Al-Malki(✉)

King Abdulaziz University, Jeddah, Saudi Arabia
aalnami0033@stu.kau.edu.sa, {malmasre,nasalmalki2}@kau.edu.sa

Abstract. Recently, the problem of creating descriptive captions for images became a significant one. However, human languages' expressivity had been among the challenges that hindered researchers from widely experimenting with creating linguistically rich captions for images. That motivated us to utilize advanced deep learning algorithms to generate captions for images. The researchers proposed an AI model utilizing deep learning and natural language processing algorithms, which has two main components, an image-feature extractor, and a story generator. The researchers trained the first component (image-feature extractor) of the model to predict object names in images. The second component (story-generator) was trained on a custom short descriptive sentence which considered short stories. So, the output from the first component (list of words) will be entered into the second component to generate stories on input images. Thus, when testing the model's performance, a list of names will be entered from the first component so that the second generator arranges them and generates a short story from them. The proposed model developed could generate a short story expressive of an input image as shown by the results of a logical value used on the BLEU scale of 0.59, which further research is planned to improve.

Keywords: Convolutional neural network · Deep learning · Object detection · Image captioning · Long short-term memory · Neural networks

1 Introduction

In Computer Vision (CV) image captioning is a widely researched topic that deals with more than detecting and recognizing an image's components. It involves as well identifying the connection between the objects represented in an image and the narrative generated to describe these objects which brings in the role of Natural Language Processing (NLP) and Machine Learning (ML) [7]. Various applications have benefited from advancements in this field like ones facilitating the sharing and captioning of visual information (images, or videos) via social media platforms, or applications that are assisting blind and visually impaired individuals in making sense of the visual data surrounding them.

Earlier CV research has investigated image recognition and classification tasks, which are basic problems in the area of image captioning [23]. However, recent studies

M. Bhattacharya et al. (Eds.): ICICCT 2021, CCIS 1417, pp. 198–208, 2021.
https://doi.org/10.1007/978-3-030-88378-2_16

attempted to go beyond these simple recognition and classification tasks to automatically generate natural language that describes the content of images in ways that imitate humans. This kind of research demands the development of advanced language models and the implementation of advanced machine learning techniques and algorithms [1, 14].

The persistent problem while researching image captioning is that natural languages are characterized by many stylistic features that enrich how humans describe the world around them. One major aspect of language composition is "expressivity," often found in literary artifacts, which highlights elements like "intent", "emotion", and "idea". Until recently, research has focused primarily on either producing accurate descriptions of objects in an image or generating syntactically or semantically accurate texts. It is then required that machines be taught to mimic the human expressive style to generate coherent image captions. This is not an easy task to accomplish, and research in computer vision is currently interested in this problem. Researchers have started implementing Deep Learning (DL) techniques and Artificial Intelligence (AI) to effectively generate natural language descriptions of images. In such research, various language models are created using RNN (recurrent neural network), LSTM (Long short-term memory), CNN (convolutional neural network), cGRU (Gated Recurrent Unit), and TPGN (Tensor Product Generation Network). As far as methods are concerned, image captioning research utilizes Deep Learning techniques, used: encoder-decoder method, Attention Mechanism, Novel Objects, or Semantics [6, 17].

The lacking focus on generating expressive (stylistically rich) captions for images in earlier research has motivated the researcher to investigate this issue. Therefore, this paper introduces a combination of deep learning algorithms (RNNs, LSTMs, etc.) and NLP techniques to create an AI system prototype. Such a prototype can detect the content of images (encoder) and then produce a sequence of words that appropriately describe these images. In this context, the method of choice is the Encoder-Decoder architecture used widely in image captioning research [27]. Figure 1 generally demonstrates the proposed model's architecture.

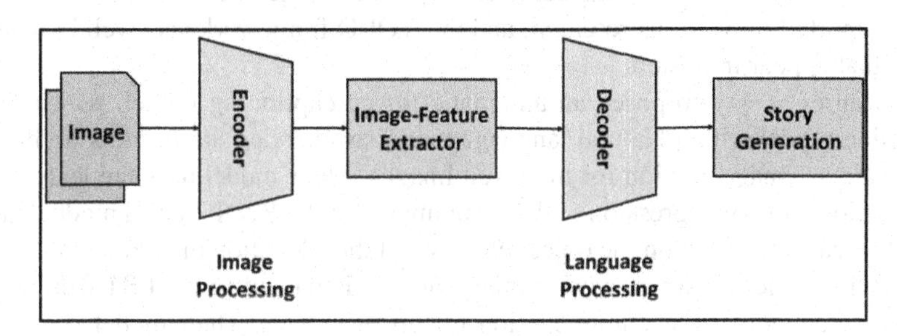

Fig. 1. General architecture of proposed model.

2 Related Work

Image classification, detection, and localization have been a subject of extensive studies in the field of CV in the last few decades [20]. This domain has been researched widely and notably enhanced due to the emergent development in the field of DL and neural networks. DL techniques have recently become popular in CV due to the results achieved in image classification, natural language processing (NLP), and object detection. DL models developed for object detection, inspired by image classification findings, and object detection based on deep learning have also accomplished advanced results [1]. This section briefly reviews the related literature done in the image feature extraction and text generation fields.

2.1 Image Feature Extraction

Object Detection. Object detection is a subject of interest in Machine Learning (ML) due to its potential applications in image captioning, video monitoring, and robotics, which utilizes advanced DL algorithms to teach computers how to detect and extract objects from images. Object detection is used to identify objects according to their class and predict their location by showing bounding boxes around the objects. Various DL methods have been significant developed in the object detection field to achieve useful results [20].

Recently, CNNs have been trained obtain advanced and accurate results in several applications such as speech recognition, object detection, and NLP. For example, DL techniques can take an input image regarding object detection and assign importance to several objects in that image. Lately, research was done in utilizing deep learning algorithms as CNN has significantly improved the accuracy of object detection. Based on CNN's, popular DL techniques automatically perform object detection, such as R-CNN, YOLOv2, YOLOv3, YOLOV4, SSD, RCNNS, etc. [25].

W.-T. Chu et al. [2] a deep model of the neural network that relies on movie poster images was proposed to detect visual object information in posters to classify movie genre. They pretrained a CNN to extract visual features and adopt a YOLO framework trained on the MSCOCO. They tested their object detection model on a large movie poster dataset. The evaluation of results demonstrate that the outcome of model is superior to models reported in previous research, and the YOLO framework acts well in detecting objects that appear in posters.

M. Han et al. [4] proposed an automatic image captioning model, which utilizes a detection model with a Natural language model as encoder and decoder architecture to generate an image caption for the given images. Their model uses the learned class classification and box regression of the input image for the RNN. Such a model learns to generate a caption based on the object classes and their location that are being detected using YOLO. The YOLO network's final outcome is passed to the LSTM network as an input to generate an accurate caption for given images. Their model showed that it generates an excellent image description and understands image relation more than image classification. However, their model has limitations because it only used 20 classes for detection, reflecting on the diversity of generating captions.

X. Yin et al. [29] the authors proposed the OBJ2TEXT model, which concentrates on representing the relationships between objects and their locations in the given image, which can generate a coherent and semantically associated caption. Therefore, they proposed a model (OBJ2TEXT) combined with YOLO's object detector to extract an object and its location. This OBJ2TEXT-YOLO trained on the MSCOCO dataset consists of object annotations for the object detection process. After evaluating the proposed model on the MS-COCO dataset, the results show their model can acquire valuable information of objects from images it helps generate image captions.

Image Classification. Image classification is an interesting research area, which refers to image labeling as being associated with one of the predefined categories [12]. Several image classification techniques have been developed, including CNN, Support Vector Machine (SVM), Nearest Neighbor (NN), and Gradient Boosting (GB) [22]. However, there has been a lot of significant research about image classification based on CNN networks that can efficiently classify images [5, 10, 15].

According to their type, the authors proposed a novel model in [10] to classify Bangladeshi birds. They created a dataset containing 1600 images of 26 various Bangladeshi bird types. To extract the features from bird images, they utilized a VGG-16 network. They have implemented four various ML algorithms to classify the type of birds. The proposed model evaluated on their dataset and the SVM and kernel method get the highest accuracy of 89%.

C. Lee et al. [15] proposed a novel model for extracting features from images using VGG16 with pre-trained weights, which are necessary to generate a sentence using the LSTM technique Flicker8k dataset. They implemented a VGG16 with predefined weights on the Keras application module. For extracting the features from images, they remove the last layer of the VGG16 instead of the object classification process. A major part of their model is the feature extraction from images based on VGG16, and extracted features fed to the LSTM with the training captions. After evaluating the model on the Fliker8k dataset, the accuracy of generated sentences, obtained using a BLEU metric, scored 0.43.

M. F. Haque et al. [5] an improved object detection method was proposed construct on a very deep convolutional network (DCNN) for accurate and important object detection. They have constructed a VGG network based on the very DCNN, which extracts notable features for detecting objects from images that help to achieve significant accuracy. High accuracy in large-scale image classification achieved by the VGG network and accuracy conspicuously for object detection.

2.2 Text Generation

Automatic text generation does not always make sense as human-generated ones do. Therefore, it requires applying natural language processing techniques to analyze and generate meaningful human language sentences [25]. This section briefly reviews the related literature in the text generation and NLP fields, focusing on deep learning techniques.

Dipti Pawade et al. [19] in their work, a story scrambler system has been proposed using RNN and LSTM, which takes sequences of various short stories as input with different or same storylines to generate a new story like human thought. They have trained their model by varying the values of the number of layers, sequence length, and batch size. Initially, the input file is processed to build a dictionary of unique words and divided them based on the sequence length fed into the model. Then, select the highest probability of the next word, and the window is shifted by one word. Continues this process until it reaches the number of words demanded in the story. Humans evaluate the generated stories based on event linkage, grammar correctness, degree of importance, and uniqueness. The evaluated result of their model has obtained 63% accuracy.

Thomaidou et al. [28] have designed a model for generating commercial text snippets of advertising material in an automated manner that utilizes the product's landing page as input. Firstly, it extracts the related and significant keywords from the landing page. Then, have been applied the sentiment analysis on the keywords reserved that have a positive meaning. Finally, built a natural language generation model to create the best form of the final advertisement text. The authors experimented with their model on 100 products from eBay and BestBuy, which are the major online catalog aggregators. They show that the proposed model can automatically generate compact and comprehensive advertisement sentences, surpassing all the baseline approaches.

Similarly, Chandra et al. [13] have proposed an algorithmic architecture to generate the context for e-commerce products based on NLP and DL. Both abstraction and extraction are the most common methods of using an unsupervised manner to summarize the text or generate content. The extraction phase is a sentence retrieval for extracting keywords and objects from a specific product. The abstraction phase is a sentence generation that utilized the RNN network beside LSTM to generate the product descriptions. After sentence generation, they used Text Rank to order sentences contextually based on their importance to obtain a coherent read. These two methods are analyzed, and it is shown the proposed architecture can fully automate the generation of the content without any manual interference, and it is decreasing the cost of manual-curation.

Parag Jain et al. [11] proposed a model that automatically generates coherent narratives based on a sequence of independent short narrations. They have used the most common text-generation frameworks: Statistical Machine Translation (SMT) and Deep Learning (DL). The incoherent input text has to be translated into stories using phrase-based and syntax-based SMT. A deep RNN is then implemented to encode the sequence of input descriptions that have a different length and decode them to generate a meaningful and comprehensive story.

3 Methodology

This proposed model adopts the encoder-decoder architecture, where an image-feature extractor and a story generator are linked together to generate English short stories about input images. Figure 2 illustrates the two main components of the proposed model.

3.1 Image-Feature Extractor (Encoder)

A CNN is a neural network capable of extracting information from an image and have obtained good accuracy in CV applications such as object detection, image classification, and OCR in natural images, etc. [9]. The encoder takes the image as an input and trains it is using a deep learning algorithm. The decoder generates text seed by integrating two model object detection and image classification. This phase can be broken down into two components:

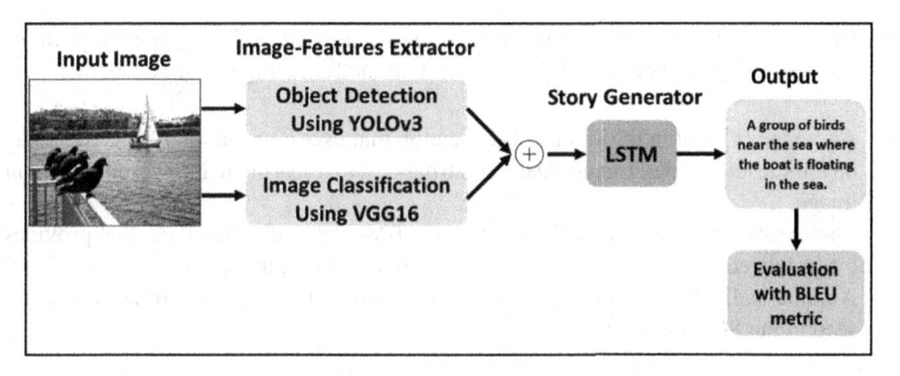

Fig. 2. The proposed model's architecture.

Object Detection. An extensive study has been conducted on various object detection models, which resulted in the researcher's current interest in the fastest one, the YOLO model. YOLO is an object detector model that takes an image and detects objects within it rather than classifying it. YOLO utilizes a single neural network that is pre-trained to predict bounding boxes that define the object location in the image, and class probabilities [21].

The current research uses YOLOv3, an enhanced version of the original YOLO that better detects small objects and runs faster. YOLOv3 is a neural network that uses CNN in the backend and is pre-trained on the MSCOCO dataset, including 328,000 images in 91 object classes [16]. It uses the learned box regression and class classification of the given image as an input for the LSTM that learns later to generate a caption based on the detected objects and their location [8]. The process of object detection using YOLOv3 can be broken down into the following steps:

1. An image is an input in the image processor.
2. A DL algorithm is implemented on a pre-trained model of YOLOv3 to perform object detection on an unseen image.
3. Extracts the features from the overall image to predict all objects present in the image.
4. The network output will return all detected bounding boxes, and the corresponding predicted class name and its confidence scores for all objects present in the image.
5. Automatically, the previous step's output is a text seed that would be used as input in the Text generator model.

Spatial Classification. To enhance the descriptive nature of the proposed model, spatial classification is used VGG16 (Visual Geometry Group), which is a CNN architecture developed for image classification tasks, which has 16 layers to recognize an input image of 224 × 224. It has three fully connected layers, and there are 4096 channels in the first two layers, and 1000 channels in the third layer that is called a SoftMax. It was pre-trained on the ImageNet dataset for feature extraction [8]. This stage gives the spatial features of images to be used in the story generation's subsequent process. The trained model has been prepared through this process, which is used to classify the image in the test dataset. The process can be broken down into the following steps:

1. An image is input in the spatial classification that uses a pre-trained VGG16 Model on the subset of Open Images Dataset V6 dataset, extracts the features from the input image.
2. These extracted features will be used in a SoftMax classification layer that provides a normalized probability distribution over predicted output classes.
3. Automatically, the output merges with the previous stage output and feeds as a text seed in the Sory Generator part.

3.2 Story Generator (Decoder)

The story generation component takes seed words as input and then attempts to generate a new story. It utilizes RNN-LSTM to generate a new story based on the given seed words. RNN network dominants the complex ML issues that require sequences as inputs. RNN is a form of artificial neural network (ANN) architecture that can take arbitrary length sequences as input. As well as long-term dependencies can be learned from sequential data features [24]. LSTM is a form architecture of RNN that has been introduced to address gradient vanishing problems in RNN models, which is an extended version of RNNs that facilitates keeping track of past data in memory more accurately. It comprises three gates to manage the passing of information: input gate, forget gate, and output gate. As well as, LSTM has a cell that keeps and holds the previous values until a forgotten gate instructs the cell to forget certain values. A new input adds to the cell using an input gate, while an output gate determines when the vectors have transferred to the next hidden state [26].

4 Experiment and Evaluation

4.1 Dataset

Two datasets will be used in the experiment, one for the image Feature Extraction specifically on the spatial classification mechanism (Open Images Dataset V6), which has 9 million annotated images for object detection, image classification, and segmentation. The images were chosen from six different groups; each group has two thousand images. They were manually chosen to demonstrate a spatial classification of images such as (sea, sky, mountains, forest, etc.). And another for the text generator that the researchers prepared custom short descriptive sentences that considered short stories.

4.2 Experimental Setup

This proposed model will be built using the python language. NLP processing can utilize several neural network models in a Python programming environment such as Chainer, Deeplearning4j, Dynet, Keras, Nlpent, etc. However, the most popular one is the TensorFlow framework, which the Google Brain Team created and made open source in 2015 [3]. TensorFlow API has two approaches to recognize images which are, Classification and Object Detection. CNN is trained to recognize sets of entities such as humans, plants, objects, etc. Subsequently, images are classified as belonging to either of these sets or not. The object detection approach is more effective than classification since multiple objects in the same image can be detected, as well as tag them based on their location inside the image. However, TensorFlow needs a High-level API such as Keras to simplify DL algorithms' complexity run on TensorFlow's top that can be easier to use.

4.3 Evaluation Metrics

In this research, however, what is needed is the generation of natural language that mimics literary texts' style but do not reproduce chunks of the original dataset. To evaluate the final product, i.e., the generated stories, I implement automatic and human-based metrics. BLEU, or the Bilingual Evaluation Understudy, is an automatic method for scoring the accuracy of a translation compared to reference translations [18]. Though used to evaluate machine translations' accuracy, BLEU is also implemented to evaluate textual content generated for many NLP tasks.

BLEU employs an adapted method of precision to compare a translated (or generated) sentence with several sentences that belong to an original reference, depending on the amount of matching n-grams between them. The n-gram precision value is calculated by summing up the count of all clipped translated (generated) sentences' n-grams and dividing them by the total count of these n-grams.

5 Results

This section reports the findings from proposed models' performance and accuracy, which evaluate the generated story by using the BLEU metric. The AI system was implemented that able to generate stories of a given image. Firstly, the Image features extraction model identifies probabilities to all the objects that are inside the image. Then, it converts the image objects into a word vector as a text seed. This text seed is given as input to LSTM in the story generator part to form the story's sentences. Some of the generated results are shown in Fig. 3. Using testing images available on Flickr8k dataset, the BLEU scores of the proposed model achieved 0.59, and on Flickr30k the BLEU score is 0.63, and for MSCOCO, the BLEU score is 0.52.

Results outlined above show it is obvious the features of images extracted for most cases are relevant to the content that exists in the pictures. The results on BLEU-1,2,3,4 metrics of three public datasets illustrate in Table 1. A perfect match results in a 1.0 BLEU score while a mismatch results in a 0.0 score. Therefore, one can say that our BLEU scores do not represent a perfect match which we do not want to achieve because it would mean an exact reproduction of the original captions.

Fig. 3. A some of the generated results.

Table 1. The BLEU results on Flickr8K, Flickr30K, and MSCOCO datasets.

Dataset	BLEU-4	BLEU-3	BLEU-2	BLEU-1
Flickr8k	0.64	0.61	0.59	0.56
Flickr30k	0.76	0.67	0.63	0.6
MSCOCO	0.56	0.53	0.51	0.49

6 Conclusion

A novel AI system for generating an expressive story in English based on one image is proposed, which combined the DL algorithms (LSTMs) and NLP techniques. The image-features extractor (encoder) and text-generator (decoder) successfully learned the proposed model, generating an expressive short story of the input image. It was noticeable during the evaluation that BLEU scores that demonstrate its reasonable functionality. To enhance the feature extraction of the model for future work, we will concentrate on alternating pre-trained image models. Also, to enhance the model efficiency on text generator will use a large corpus dataset like news articles and other data sources online.

References

1. Amritkar, C., Jabade, V.: Image caption generation using deep learning technique. In: 2018 Fourth International Conference on Computing Communication Control and Automation (ICCUBEA), pp. 1–4. IEEE (2018)
2. Chu, W.T., Guo, H.J.: Movie genre classification based on poster images with deep neural networks. In: Proceedings of the Workshop on Multimodal Understanding of Social, Affective and Subjective Attributes, pp. 39–45 (2017)
3. Ganegedara, T.: Natural Language Processing with TensorFlow: Teach Language to Machines Using Python's Deep Learning Library. Packt Publishing Ltd. (2018)
4. Han, M., Chen, W., Moges, A.D.: Fast image captioning using LSTM. Cluster Comput. **22**(3), 6143–6155 (2019)
5. Haque, M.F., Lim, H.Y., Kang, D.S.: Object detection based on VGG with ResNet network. In: 2019 International Conference on Electronics, Information, and Communication (ICEIC), pp. 1–3. IEEE (2019)
6. Hays, J., Efros, A.A.: IM2GPS: estimating geographic information from a single image. In: 2008 IEEE Conference on Computer Vision and Pattern Recognition, pp. 1–8. IEEE (2008)
7. He, X., Deng, L.: Deep learning for image-to-text generation: a technical overview. IEEE Signal Process. Mag. **34**(6), 109–116 (2017)
8. Hoang, L.: An Evaluation of VGG16 and Yolo V3 on Hand-Drawn Images. University Honors These (2019)
9. Hossain, M.A., Sajib, M.S.A.: Classification of image using convolutional neural network (CNN). Glob. J. Comput. Sci. Technol. (2019)
10. Islam, S., Khan, S.I.A., Abedin, M.M., Habibullah, K.M., Das, A.K.: Bird species classification from an image using VGG-16 network. In: Proceedings of the 2019 7th International Conference on Computer and Communications Management, pp. 38–42 (2019)
11. Jain, P., Agrawal, P., Mishra, A., Sukhwani, M., Laha, A., Sankaranarayanan, K.: Story generation from sequence of independent short descriptions. arXiv preprint arXiv:1707.05501 (2017)
12. Kamavisdar, P., Saluja, S., Agrawal, S.: A survey on image classification approaches and techniques. Int. J. Adv. Res. Comput. Commun. Eng. **2**(1), 1005–1009 (2013)
13. Khatri, C., et al.: Algorithmic content generation for products. In: 2015 IEEE International Conference on Big Data (Big Data), pp. 2945–2947. IEEE (2015)
14. Lakshminarasimhan Srinivasan, D.S., Amutha, A.: Image captioning-a deep learning approach. Int. J. Appl. Eng. Res. **13**(9), 7239–7242 (2018)
15. Lee, C.: Image caption generation using recurrent neural network. J. KIISE **43**(8), 878–882 (2016)
16. Lin, T.-Y., et al.: Microsoft COCO: common objects in context. In: Fleet, D., Pajdla, T., Schiele, B., Tuytelaars, T. (eds.) ECCV 2014. LNCS, vol. 8693, pp. 740–755. Springer, Cham (2014). https://doi.org/10.1007/978-3-319-10602-1_48
17. Ordonez, V., et al.: Large scale retrieval and generation of image descriptions. Int. J. Comput. Vis. **119**(1), 46–59 (2016)
18. Papineni, K., Roukos, S., Ward, T., Zhu, W.J.: Bleu: a method for automatic evaluation of machine translation. In: Proceedings of the 40th Annual Meeting of the Association for Computational Linguistics, pp. 311–318 (2002). Story Generation from Images using Deep Learning1118
19. Pawade, D., Sakhapara, A., Jain, M., Jain, N., Gada, K.: Story scrambler-automatic text generation using word level RNN-LSTM. Int. J. Inf. Technol. Comput. Sci. (IJITCS) **10**(6), 44–53 (2018)

20. Rashid, M., Khan, M.A., Sharif, M., Raza, M., Sarfraz, M.M., Afza, F.: Object detection and classification: a joint selection and fusion strategy of deep convolutional neural network and sift point features. Multimed. Tools Appl. **78**(12), 15751–15777 (2019)
21. Redmon, J., Divvala, S., Girshick, R., Farhadi, A.: You only look once: unified, real-time object detection. In: Proceedings of the IEEE Conference on Computer Vision and Pattern Recognition, pp. 779–788 (2016)
22. Ren, X., Guo, H., Li, S., Wang, S., Li, J.: A novel image classification method with CNN-XGBoost model. In: Kraetzer, C., Shi, Y.-Q., Dittmann, J., Kim, H.J. (eds.) IWDW 2017. LNCS, vol. 10431, pp. 378–390. Springer, Cham (2017). https://doi.org/10.1007/978-3-319-64185-0_28
23. Russakovsky, O., et al.: ImageNet large scale visual recognition challenge. Int. J. Comput. Vis. **115**(3), 211–252 (2015)
24. Sainath, T.N., Vinyals, O., Senior, A., Sak, H.: Convolutional, long short-term memory, fully connected deep neural networks. In: 2015 IEEE International Conference on Acoustics, Speech, and Signal Processing (ICASSP), pp. 4580–4584. IEEE (2015)
25. Salehinejad, H., Sankar, S., Barfett, J., Colak, E., Valaee, S.: Recent advances in recurrent neural networks. arXiv preprint arXiv:1801.01078 (2017)
26. Skovajsová, L.: Long short-term memory description and its application in text processing. In: 2017 Communication and Information Technologies (KIT), pp. 1–4. IEEE (2017)
27. Staniūtė, R., Šešok, D.: A systematic literature review on image captioning. Appl. Sci. **9**(10), 2024 (2019). https://doi.org/10.3390/app9102024
28. Thomaidou, S., Lourentzou, I., Katsivelis-Perakis, P., Vazirgiannis, M.: Automated snippet generation for online advertising. In: Proceedings of the 22nd ACM International Conference on Information & Knowledge Management, pp. 1841–1844 (2013)
29. Yin, X., Ordonez, V.: Obj2text: generating visually descriptive language from object layouts. arXiv preprint arXiv:1707.07102 (2017)

Detecting Text-Bullying on Twitter Using Machine Learning Algorithms

Abdullah Yahya Abdullah Amer(✉) and Tamanna Siddiqui

Department of Computer Science, Aligarh Muslim University, Aligarh, India

Abstract. The increasing use of social media leads to a large number of users-created unstructured text data. Due to the prevalence of social media, cyberbullying has become the main problem. Cyberbullying may cause many serious consequences on a person's life and community. That is because of various cultures, intellectual and educational backgrounds. The distinction between offensive language and hate speech is an essential Challenge in Detecting noxious textual content. In our work, we proposed a method in order to automatically analyze that tweet on Twitter within binary labels: Offensive and non-offensive. Utilizing the tweets data set, then we implement analyses granting N-grams as a feature and comparative between Term Frequency-Inverse Document Frequencies (TFIDF), and we achieve in, Using Decision Tree Classifier, Multinomial NB (Naïve Bayes), Linear SVC (Support Vector Classifier), and AdaBoost Classifier, K-neighbors Classifier, and Logistic Regression machine learning models. Then tuning a model fitting the best results we get in (TFIDF). We achieve the best accuracy, 0.924 in the Linear SVC classifier, best F1 Score 0.942% in the same classifier, the best Precision 0.975 in the AdaBoost classifier, and Best Recall 0.977 in Multinomial NB. In the other model (Count-Vectorizer), we achieve Best accuracy 0.925 in Logistic Regression classifier, Best F1 Score 0.942% in the same classifier, best Precision 0.976 in AdaBoost classifier, and Best Recall 0.941 in Multinomial NB.

Keywords: Cyberbullying detection · Feature extraction · Machine learning algorithms · Twitter · Hate speech

1 Introduction

Online Social media has become hugely attractive in the current years. Subscribers or users have used this one is websites as real-time and new communication tools; powerful data sources anywhere can build their profiles own, including comment the public and chat with some friends, notwithstanding physical limitations and geographical location who belong to it. By this consideration, these websites have become universal communication [1, 2]. The data from online social tools can give us unique shrewdness in the development of societies and social networks, which are considered problematic in phases of extent and scale. Besides, those digital programs can exceed the bounds of a physical [3]. Cyberbullying is an ultimately steadfast version of popular styles of bullying, including negative impacts onto the sufferer. A cyberbully can harry or

M. Bhattacharya et al. (Eds.): ICICCT 2021, CCIS 1417, pp. 209–222, 2021.
https://doi.org/10.1007/978-3-030-88378-2_17

harry his/her victims a website community. Online tools, like different sites, Social networks (e.g., Twitter, Facebook) have become grown essential parts of a user lifetime. Hence, those websites have grown then produce popular platforms for cyberbullying deception on the negative side [4]; also, their proliferation and popularity have increased cyberbullying events [5]. Such an addition is usually connected to an event that common cyberbullying is challenging to use method. A criminal bullying them, victims, and not any personal encounter uses tools such as a cellphone or a laptop connected to the Internet [6]. The features of social media have too extended the ability of the cyberbully to before unreachable countries also the location. Social networks Twitter is the famous platform social media service that allows users to read, send 140-characters' per Time. Twitter covers more than 500 million users and generates at least around five hundred million tweets per day. Nearly 80% of active user Twitter users tweet's utilizing those devices or mobile phones. Here the social media platform has grown a powerful, similar real-time news channel [7]. In this research article, we propose to use data tweets to promote cyberbullying detection performance results. A specific classification of our dataset includes several columns on Twitter, such as df_index, label, and full text, to train and test our model detection and improve its performance results. Implementing machine learning may give Offensive and Non-Offensive cyberbullying prediction events to build a mastering machine learning model approach depending on several features. The common signs of certain factors are features that are doing then the participation of features extraction into a model which interacts great within this class. We then selected the best features at the highest special powers with cyberbullying, and non-cyber-bullying tweet is the difficult task [8], which wants much work to create a Classifier machine learning model [9]. Therefore, we aim to get the best outcomes by comparing two vectors Tf-idf Vectorizer and count_vector, to predict a cyberbullying detection process with classifying features that follow be applied into machine learning plans in order to define bullying tweets dataset from no cyberbullying.

2 Literature Review

Several machine learning methods have been composed to take the issue of cyberbullying detection. The preponderance of the ways deals with the feature extraction of the text data. Sanchez, H et al. [10] involved recognizing bullying in social media, particularly on Twitter. There is not a past trained assignment of utilizing text classification to detect bullying to the best of our opinion. They train work datasets from Twitter to generally use phrases of damage, which are analyzed by noisy labels. The text data are openly accessible and can get directly from the Twitter streaming API. To classification of tweet comments, more identified as tweets, they used the Naïve Bayes classifier. Its accuracy was nearby on 70% during Training within standard terms of insult data.

K. Nalini et al. [11] Cyberbullying internet predation abuse minors, special teens who did not have sufficient supervision while they used the machine. The huge volume of data collected in unstructured records cannot just be applied for additional processing by networks, which typically text as a more series of characters string. Hence, particular processing techniques, algorithms are needed in order to obtain helpful patterns. Text Mining is the process of necessary, yet unknown, knowledge from the text data. Text classification is a major research problem in the area of natural language process NLP. Proposed an efficient method to detect cyberbullying text from Twitter within a weighting scheme of feature extraction.

Chatzakou, D. et al. [12] Work on takes a primary concrete stage to know the abusive act features on Twitter and become today's most prominent social communications platforms. And they classified 1.2 million users, including 2.1 million tweets, matching users competing into conversations about obviously common topics same the NBA, these also likely to be hate-relevant, so because of the Gamergate discussion. Also, search for particular demonstrations of offensive behavior, i.e., cyberbullying, is a hate-related community area. Then presented a solid methodology for define aggressors and bullies from the regular Twitter users, also serial-based characteristics. Used several cases ML algorithm and analyzed those accounts with around 90% accuracy.

A. Saravanaraj et al. [13] suggested work the detection concerning cyberbully news and rumor tweets on Twitter are combined into a single application, forward with those the cyberbully contents in the tweet text detected used Random Forest and Naïve Bayes classifier. The name, age, and gender of the cyberbully tweeted people detected used feature extraction methods. The rumor tweets were detected in the proposed work with used type and topic particular classification and Twitter speech-act classifier.

Reynolds, K. et al. [14], Cyberbullying used technology as a mechanism to bully somebody. Becomes an issue increase in current years; social media sites are rich in the medium for bullies, young and teenage adults who used the sites are exposed to attacks. Inside machine learning, used detect text models applied on bullies their victims, then improve controls into automatically identify cyberbully content text data. Depreciated data they work from the website Form spring. The text data held a label used by the Amazon Mechanical Turk web service. Applying machine learning methods, his work done by the Weka tools, trained the machine to identify cyberbullying contents. They used each of the C4.5 decision tree student obtained to recognize the right positives with around 78.5% accuracy.

3 Proposed Approach

This part will discuss our approach work and how the process data flow and by a method is illustrated in the figure below. Here we used model TFIDF in order to train workers; then, we used features extracting N-gram of text data to provide more reliable results, then we applied with the same last step but with the Count-Vectorizer method [15]. Moreover, the weight them according to their Term Frequency – Inverse Document Frequency TFIDF approach also shows encouraging results [16]. Both of them give us different results. Included the analysis concerning features including the pre-eminent classifiers for text model, we proposed extricating n-grams from the tweet. For two, our mothed then weigh them according to the TFIDF Count-Vectorizer values. Then we did feed those features to ML algorithms to text Classifiers. Specified a collection of the tweet, our task's purpose is to classify as them into two classes: offensive or non-offensive. Finally, we compare finding to our model evaluation (Fig. 1).

When we perform classification, the initial step is to understand the difficult issue and recognize potential extraction features and labels. Features are these properties or

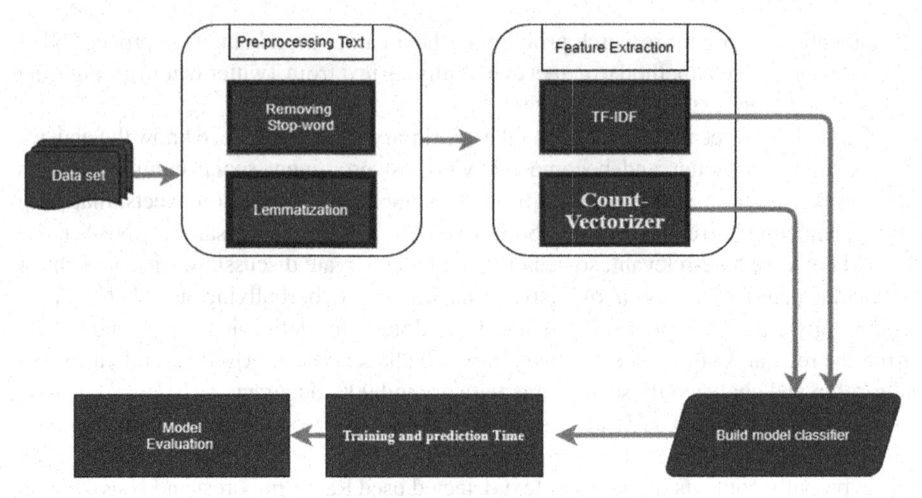

Fig. 1. Proposed approach

features which affect the on our results of the labels. These attributes are known as features that help the text model classify. Each classification has a pair stage, a learning stage, and the evaluation stage. In the learning stage, the classifier train and test it is model on a given dataset, and in last the evaluation stage, it tests the classifier performance model. The performance result is evaluated on the foundation of different parameters like accuracy, F-score, precision, and recall, as shown in Fig. 2 below.

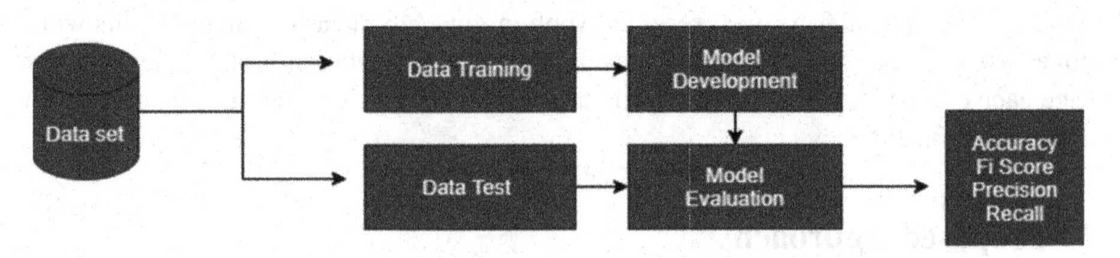

Fig. 2. Perform classification

3.1 Data Collection

The dataset that we have used is a tweet download from Kaggle, its binary text data consisting of two labels Detecting Cyberbullying in Tweets datasets. It consists of three columns: df_index, full-text, and label. In this dataset, tweets corresponding to the df_index, full-text, and label are divided into the following two labels: "offensive" and "Non-offensive."

3.2 Data Preprocessing

In this part of the text preprocessing here, we need the three tasks involved in dataset handling correctly. The first task involves Stop words and exceptional character removal, removing unnecessary from the datasets, Tokenization, Stemming, and Normalization. These steps are needed for transferring text data of human to machine language-readable frame format during the additional process. We will also present text preprocessing tools. Next, a text is taken; we begin within text normalization, which involves: converting each letter to upper case or lower, expanding abbreviations, removing accent marks, punctuations, and other diacritics, converting numbers in words or remove characters, removing stop words, particular words, and sparse terms, removing white spaces. After that, we split the text dataset into two parts: the training dataset to include 70%, and the testing dataset contains 30% of the data set [17, 33].

3.3 Text Classification

Text classification or text mining is part of the natural language process. NLP is a field where analysis algorithms are used to text. The task is to select a document within one (or more) classes based on its content. Sometimes, those classes are manual with humans. A classifier to determine how to classify the text documents requires any ground fact [18]. For this design, the input targets are divided into train and test datasets. Train datasets are these wherever the text documents are already labeled. Test datasets are where the text records are unlabeled. The purpose is to learn the already labeled text data train data and applied that to the test data, and predict the class label for the testing data set correctly. Therefore, the classifier is the building of an actual and a Learner Classifier. The classifier then follows that classification function to classify the unlabeled set of text records documents. Here, learning is described as supervised learning for a supervisor who works as a teacher Training or directing the learning method [19]. The selection of the measurement of the train and test dataset is significant. Classification analyzed tweets Applying Text Classifier to Detect Cyber Bullying.

3.4 Feature Extraction

The feature extraction or feature engineering Concerned by block helps convert the regulated text data into categorical/numeric characteristics that are used for learning mothed by the supervised learning models. That process is further named Vectorization as we convert each text document toward feature vectors to be feed within the supervised classification of the models. Unusual standard methods that are regularly used for Vectorization are the Term Frequency – Inverse Document Frequency Model [20].

3.5 N-gram

N-grams Concerned by used unigram, trigram, and bigram, as binary, features of texts are extensively used in natural language processing NLP and text mining tasks. The use of n-grams for improving NLP and text mining features to supervised Machine Learning models like SVMs, Naive Bayes, models, etc. To use various N-gram into a word, If X = Num of terms into the specific document d, the numbers of n-gram to document d that is: N-grams d = X – (N – 1) [21].

3.6 Count-Vectorizer

The Count-Vectorizer gives easy access to both tokenize a collection set of text documents and build a dictionary of well-known words and encode new records applying that vocabulary. We can do it as follow: Create a case of the Count-Vectorizer class, training vector, build the classifier, training the classifier, get the test vector, predict and score the vector. An encoded vector is replaced, including a length of the entire vocabulary and an entity count for the whole Time per word that appeared in the text record. That is useful when applying text data and transform each term into each text in vectors [22].

3.7 TF-IDF

TF-IDF is for "Term Frequency, Inverse Document Frequency. " It's a method to evaluate the effect of terms (or "words") in a document based on wherewith frequently they appear across different text documents. The score means the weight of that term in association with the original train dataset.

$$\text{TF-IDF score is provided with} : TF - IDF = tfij * idfi \tag{1}$$

Numerically, number frequency tfij defines the sense of a word i into tweet j. It is defined as:

$$tf_{ij} = \frac{N_{ij}}{\Sigma N_j}$$

Where N_ijis the number frequency of words i in tweets j and $\sum$N_i is the number frequency of all words in tweet j. (IDF) idfi assign the importance of terms or word i in the entire train data set. It is set as $Idf_{i=} \frac{log|t|}{|t_j W_i \in t_j|}$.

Where |t| is the total amount of tweets, the number where the words Wi appear. All tweets contain a vector of words, and all words are indicated in the vector by its TF-IDF score [23, 24].

4 Machine Learning Algorithms

Feature engineering of the tweet was applied to build models to detect cyberbully. Here we tested different machine learning ML methods to extract those fittest classifiers. Then we tested the decision tree classifier, MultinomialNB, linear SVC, AdaBoost classifier, Neighbors classifier, and logistic regression. Certain techniques are discussed as follows [25].

4.1 Multinomial NB Classifier

One of the most popular classification models on machine learning is categorical text data analysis, particularly text data. Works similar to Gaussian naive Bayes; however, the features are assumed to be multinomial distributed. In practice, this means that this classifier is commonly used when we have discrete data [26].

4.2 Linear SVC Classifier

Linear SVC (Support Vector Classifier) concerned with fit data that present returns a "best fit" hyper-plane to categorize divides and the data. Wherever next to get a hyper-plane, after that can feed characteristics into the classifier in order to get what "predicted" class. That performs that particular algorithm rather than fit our needs; however, it can use that during many situations [27].

4.3 AdaBoost Classifier

The common thought behind boost classification is to Training predicted sequence, all Training to improve it. The two generally applied boost model algorithm is Gradient Boost also AdaBoost. In the proceeding algorithm, cover AdaBoost. At a high level [28].

4.4 Kneighbors Classifier

Concerned by organized and classified a preponderance vote, it is neighbors, within an aim being selected into a class multiple commons between its k nearest neighbors. If k = 1, later, the objects are directly assigned into the class from the single nearest neighbor [29–32].

4.5 Logistic Regression (LR) Classifier

Concerned by classification algorithms to utilize to predicts the likelihood of definite dependents variable during logistic regression LR, the dependents variable is a binary classification while includes data records coded as 1 (yes, true, etc.) or 0 (no, false, etc.). Moreover, the RL design model predicts P(Y = 1) being a function of X [30, 31].

5 Results and Discussions

There are several methods for machine learning and feature extraction that can be applied to detect text-Bullying. It is not evident which of the techniques would work the best. Accordingly, we carried out an experiment where we tested various combinations of machine learning and feature extraction methods and evaluated their performance [34, 35]. We run a collection of experiments tests in order to measure achievement for five Classifiers such as Decision Tree, Multinomial Naive Bayes, Linear SVC, AdaBoost, Kneighbors, and Logistic Regression). All five classifiers did run to apply every proposed feature based on cross-validation. Fig. 3 shown the results during all classifiers for five classifier models; it's almost all close and encouraging results. Then we run each of the five classifiers, including feature extraction, to define this important feature extraction which increases each show of the classifiers also decreases classified prediction times, as shown in Fig. 4. Two mothed we used, and we got two different results that were tested in the experiment. Tables 1 show the best result classifiers within all feature selection methods. We achieve the Best accuracy of 0.924 in the Linear SVC classifier, and Best F1

Score 0.942% in the same classifier, the Best Precision 0.975 in the AdaBoost classifier, and the Best Recall 0.977 in Multinomial NB. But this does not mean that the other classifiers are a bed, the results for the five classifier model are almost all close and encourage results by using the TF-IDF model.

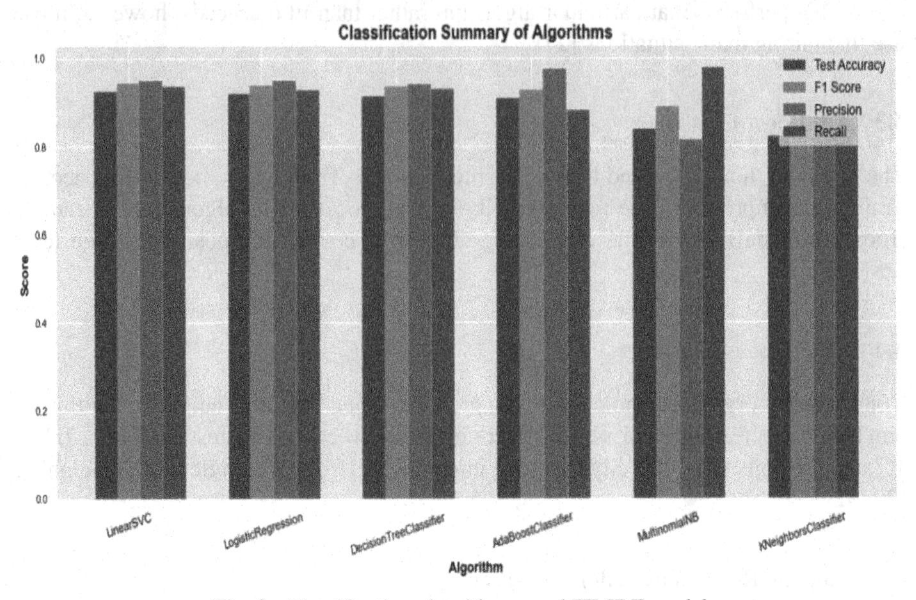

Fig. 3. Classification algorithms used TF-IDF model

Table 1. Best result classification algorithms

Best Accuracy	0.924	Linear SVC
Best F1 Score	0.942	Linear SVC
Best Precision	0.975	AdaBoost classifier
Best Recall	0.977	Multinomial NB

5.1 Training and Prediction Time Used TF-IDF Model

In this section, we will talk about Training and prediction Time by using the TF-IDF model; we have six algorithms such as Decision Tree Classifier, Multinomial NB, Linear SVC, AdaBoost Classifier, Kneighbors Classifier, and Logistic Regression as shown in Fig. 6 and Table 4, we observe some similar, high and worst performance Training and Prediction Time. Here in the Kneighbors classifier, we have seen Best Training Time (0.005) and worst prediction Time (0.005), and Decision Tree classifier (3.252) worst Training Time. Moreover, we have seen the Best prediction Time (0.732) Logistic Regression for prediction time, and we observe the worst prediction Time (24.005) Kneighbors Classifier compared to other algorithms. More details are illustrated about other algorithms in Fig. 6 down (Table 2).

Table 2. Training and prediction time

Best Training Time	0.005	Kneighbors classifier
Worst Training Time	3.252	Decision Tree classifier
Best prediction Time	0.732	Logistic Regression
Worst prediction Time	24.005	Kneighbors classifier

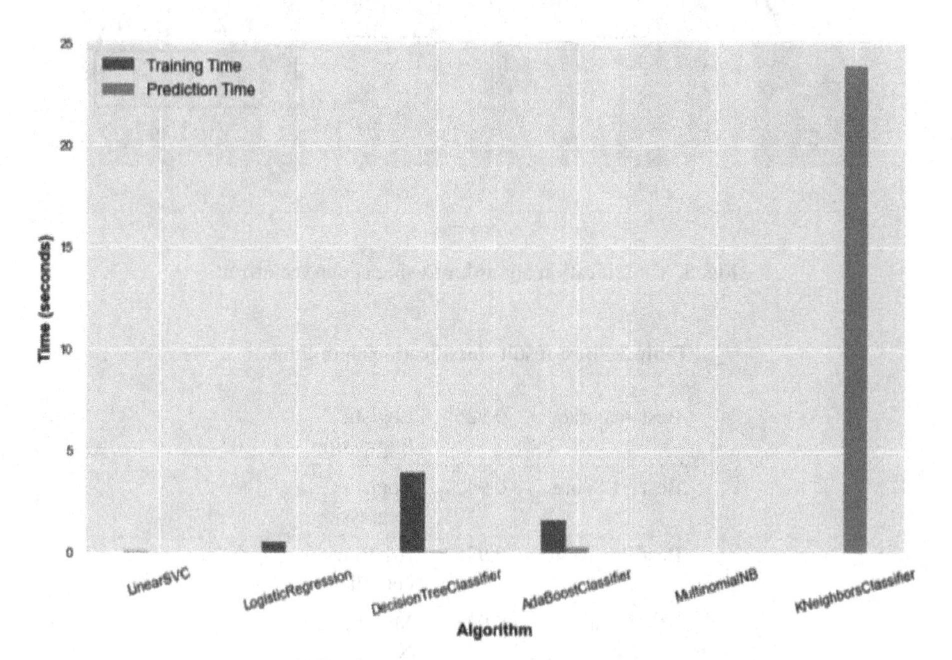

Fig. 4. Training and prediction time

5.2 Count-Vectorizer

The model we used for our scheme work and the dataset were obtained from websites that contain a high rate of bullying content. Here, we used that Machine learning algorithm; as shown in the figure down, we applied six Classification Algorithms using Count-Vectorizer models. Then we make a comparative study of the models considering different values Count-Vectorizer normalization approaches.

Next, tuning each model providing these most reliable results, and we achieved the Best accuracy 0.925 in the Logistic Regression classifier, and Best F1 Score 0.942% in the same classifier, best Precision 0.976 in AdaBoost classifier, and Best Recall 0.941 in Multinomial NB as shown in Table 3 and Count-Vectorizer upon evaluating it on test data (Fig. 5).

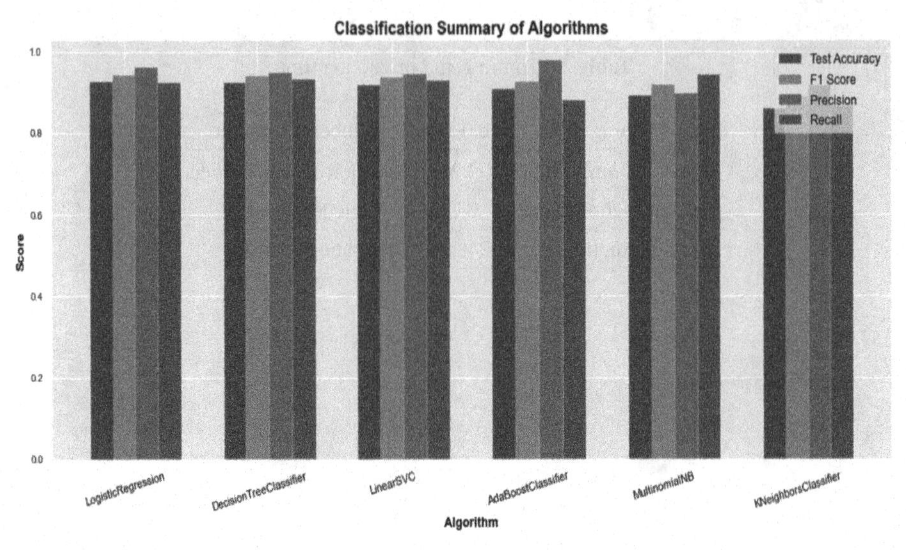

Fig. 5. Classification algorithms using count-vectorizer

Table 3. Best result classification algorithms

Best Accuracy	0.925	Logistic Regression
Best F1 Score	0.942	Logistic Regression
Best Precision	0.976	AdaBoost classifier
Best Recall	0.941	Multinomial NB

5.3 Training and Prediction Time Used Count-Vectorizer Model

In this section, we will talk about Training and prediction Time by used the Count-Vectorizer model; we have six algorithms such as Decision Tree Classifier, Multinomial NB, Linear SVC, AdaBoost Classifier, Kneighbors Classifier, and Logistic Regression as shown in Fig. 6 and Table 4, we observe some similar, high and worst performance Training and Prediction Time. Here in the K-neighbors classifier, we have seen Best Training Time (0.005) and worst prediction Time (0.005), and the Decision Tree classifier (3.252) worst Training Time.

Moreover, we have seen the Best prediction Time (0.732) Logistic Regression for prediction time, and we observe the worst prediction Time (24.005) Kneighbors Classifier compared to other algorithms. More details are illustrated about other algorithms in Fig. 6 down.

Table 4. Training and prediction time

Best Training Time	0.005	Kneighbors classifier
Worst Training Time	3.252	Decision Tree classifier
Best prediction Time	0.732	Logistic Regression
Worst prediction Time	24.005	Kneighbors classifier

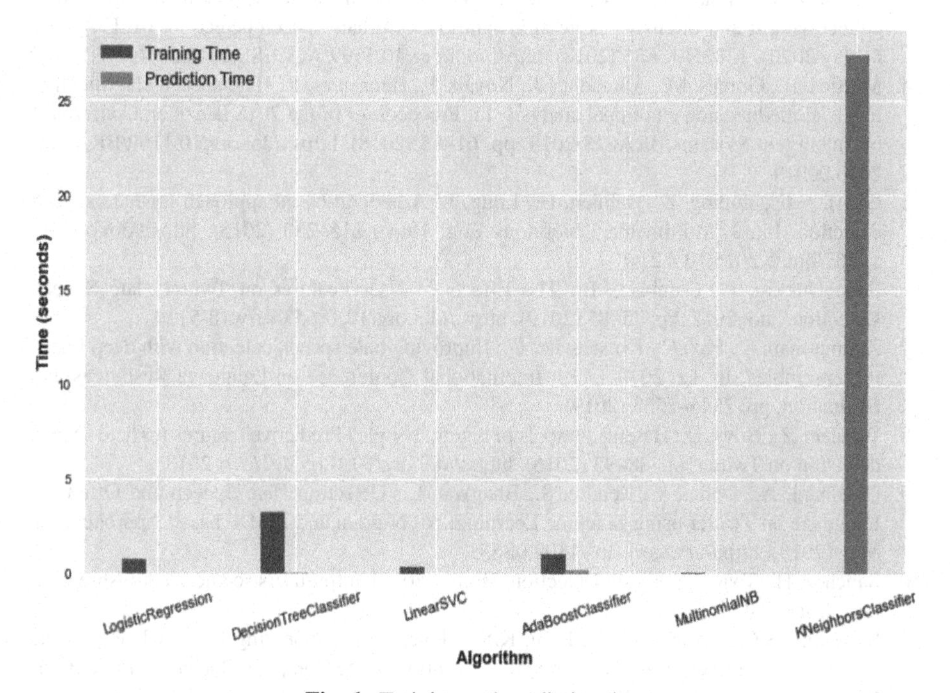

Fig. 6. Training and prediction time

6 Conclusion

In this paper, we introduced two models for feature extraction, which can effectively detect cyberbullying. Here built our model that predicted tweets as Offensive and non-offensive. The result is the possibility of a comment being offensive to users. We achieve in (TFIDF) We achieve the Best accuracy, 0.924 in the Linear SVC classifier, the Best F1

Score 0.942% in the same classifier, and the best Precision 0.975 AdaBoost classifier, and Best Recall 0.977 in Multinomial NB. In the other model(Count-Vectorizer), we achieve the Best accuracy, 0.925 in the Logistic Regression classifier, the Best (F1 Score) 0.942% in same classifier, the best (Precision) 0.976 in AdaBoost classifier, and Best Recall 0.941 in Multinomial NB. Also, we get high-performance Training and prediction time, and all results were satisfactory.

References

1. Abreu, D.A.C., Souza, A.D.E., De Souza, A.: Automatic offensive language detection from Twitter data using machine learning and feature selection of metadata (2020)
2. Konduri, V., Padathula, S., Pamu, A., Sigadam, S.: Hate Speech Classification of social media posts using Text Analysis and Machine Learning, pp. 1–7 (2020)
3. Watanabe, H., Bouazizi, M., Ohtsuki, T.: Hate speech on Twitter: a pragmatic approach to collect hateful and offensive expressions and perform hate speech detection. IEEE Access **6**(May 2020), 13825–13835 (2018). https://doi.org/10.1109/ACCESS.2018.2806394
4. Martins, R., Gomes, M., Almeida, J.J., Novais, P., Henriques, P.: Hate speech classification in social media using emotional analysis. In: Proceedings of the 2018 Brazilian Conference on Intelligent Systems BRACIS 2018, pp. 61–66 (2018). https://doi.org/10.1109/BRACIS.2018.00019
5. Gitari, N.D., Zuping, Z., Damien, H., Long, J.: A lexicon-based approach for hate speech detection. Int. J. Multimedia Ubiquitous Eng. **10**(4), 215–230 (2015). https://doi.org/10.14257/ijmue.2015.10.4.21
6. Fehn Unsvåg, E., Gambäck, B.: The Effects of User Features on Twitter Hate Speech Detection," no. 2012, pp. 75–85 (2019). https://doi.org/10.18653/v1/w18-5110
7. Zimmerman, S., Fox, C., Kruschwitz, U.: Improving hate speech detection with deep learning ensembles. In: Lr. 2018 - 11th International Conference on Language Resources and Evaluation, pp. 2546–2553 (2019)
8. Waseem, Z., Hovy, D.: Hateful symbols or hateful people? Predictive features for hate speech detection on Twitter, pp. 88–93 (2016). https://doi.org/10.18653/v1/n16-2013
9. Gaydhani, A., Doma, V., Kendre, S., Bhagwat, L.: Detecting Hate Speech and Offensive Language on Twitter using Machine Learning: An N-gram and TFIDF based Approach, no. May (2018). http://arxiv.org/abs/1809.08651
10. Sanchez, H.: Twitter Bullying Detection, Homo (2011). http://users.soe.ucsc.edu/~shreyask/ism245-rpt.pdf
11. Satapathy, S.C., Govardhan, A., Raju, K.S., Mandal, J.K.: Emerging ICT for Bridging the Future - Proceedings of the 49th Annual Convention of the Computer Society of India (CSI) Volume 2. Advances in Intelligent Systems and Computing, vol. 338, pp. I–IV (2015). https://doi.org/10.1007/978-3-319-13731-5
12. Chatzakou, D., et al.: Detecting cyberbullying and cyberaggression in social media. ACM Trans. Web **13**(3) (2019). https://doi.org/10.1145/3343484
13. Reynolds, K., Kontostathis, A., Edwards, L.: Using machine learning to detect cyberbullying. In: Proceedings of the 10th International Conference on Machine Learning and Applications and Workshops. ICMLA 2011, vol. 2, no. October 2019, pp. 241–244 (2011). https://doi.org/10.1109/ICMLA.2011.152
14. Sap, M., Card, D., Gabriel, S., Choi, Y., Smith, N.A.: The risk of racial bias in hate speech detection. In: Proceedings of the Conference on ACL 2019 - 57th Annual Meeting of the Association Computational Linguistics, pp. 1668–1678 (2020)

15. Biere, S., Master Business Analytics: Hate Speech Detection Using Natural Language Processing Techniques," Vrije University, Amsterdam, p. 30 (2018)
16. Del Vigna, F., Cimino, A., Dell'Orletta, F., Petrocchi, M., Tesconi, M.: Hate me, hate me not: Hate speech detection on Facebook. In: CEUR Workshop Proceedings, vol. 1816, pp. 86–95 (2017)
17. Watanabe, H., Bouazizi, M., Ohtsuki, T.: Hate speech on twitter: a pragmatic approach to collect hateful and offensive expressions and perform hate speech detection. IEEE Access **6**(c), 13825–13835 (2018). https://doi.org/10.1109/ACCESS.2018.2806394
18. Arango, A., Pérez, J., Poblete, B.: Hate speech detection is not as easy as you may think: a closer look at model validation (extended version). Inf. Syst. (2020). https://doi.org/10.1016/j.is.2020.101584
19. Pereira-Kohatsu, J.C., Quijano-Sánchez, L., Liberatore, F., Camacho-Collados, M.: Detecting and monitoring hate speech in Twitter. Sens. (Switz.) **19**(21), 1–37 (2019). https://doi.org/10.3390/s19214654
20. Zhang, Z., Luo, L.: Hate speech detection: a solved problem? The challenging case of long tail on Twitter. Semant. Web **10**(5), 925–945 (2019). https://doi.org/10.3233/SW-180338
21. Robinson, D., Zhang, Z., Tepper, J.: Hate speech detection on Twitter: feature engineering v.s. feature selection. In: Gangemi, A., et al. (eds.) ESWC 2018. LNCS, vol. 11155, pp. 46–49. Springer, Cham (2018). https://doi.org/10.1007/978-3-319-98192-5_9
22. Pitsilis, G.K., Ramampiaro, H., Langseth, H.: Effective hate-speech detection in Twitter data using recurrent neural networks. Appl. Intell. **48**(12), 4730–4742 (2018). https://doi.org/10.1007/s10489-018-1242-y
23. Faret, J., Reitan, J.: Twitter Sentiment Analysis-Exploring the Effects of Linguistic Negation, no. June (2015). http://brage.bibsys.no/xmlui/handle/11250/2353488
24. Gaydhani, A., Doma, V., Kendre, S., Bhagwat, L.: Detecting Hate Speech and Offensive Language on Twitter using Machine Learning: An N-gram and TFIDF based Approach (2018). http://arxiv.org/abs/1809.08651
25. Gröndahl, T., Pajola, L., Juuti, M., Conti, M., Asokan, N.: All you need is 'love': evading hate speech detection. In: Proceedings of the ACM Conference on Computer and Communications Security, pp. 2–12 (2018). https://doi.org/10.1145/3270101.3270103
26. de Sousa, J.G.R.: Feature extraction and selection for automatic hate speech detection on Twitter, pp. 1–77 (2019). https://repositorio-aberto.up.pt/bitstream/10216/119511/2/326963.pdf
27. Huang, Q., Zhang, L., Cheng, Y., Li, P., Li, W.: Enantioselective construction of vicinal sulfur-containing tetrasubstituted stereocenters via organocatalyzed mannich-type addition of rhodanines to isatin imines. Adv. Synth. Catal. **360**(17), 3266–3270 (2018). https://doi.org/10.1002/adsc.201800642
28. Do, H.T.-T., Huynh, H.D., Van Nguyen, K., Nguyen, N.L.-T., Nguyen, A.G.-T.: Hate speech detection on Vietnamese social media text using the bidirectional-LSTM model, pp. 4–7 (2019). http://arxiv.org/abs/1911.03648
29. Malmasi, S., Zampieri, M.: Detecting hate speech in social media. In: International Conference Recent Advances in Natural Language Processing. RANLP, vol. 2017-Septe, no. March, pp. 467–472 (2017). https://doi.org/10.26615/978-954-452-049-6-062
30. Gröndahl, T., Pajola, L., Juuti, M., Conti, M., Asokan, N.: All You Need is, pp. 2–12 (2018). https://doi.org/10.1145/3270101.3270103
31. Ravi B., N., Kamthe, C.R., Shravan, A.M., Mahesh, K.A., RDTC SCSCOE: Cyber bullying Detection & Prevention for Social Media Using Data Mining, vol. 3, no. 4, pp. 281–283 (2018)
32. Al-Hassan, A., Al-Dossari, H.: Detection of Hate Speech in Social Networks: a Survey on Multilingual Corpus, pp. 83–100 (2019). https://doi.org/10.5121/csit.2019.90208

33. Siddiqui, T., Khan, N.A., Khan, M.A.: PMKBEA : A Process Model Using Knowledge Base Software Engineering Approach, pp. 5–7 (2011)
34. Yahya, A., Amer, A., Siddiqui, T.: Detection of Covid-19 fake news text data using random forest and decision tree classifiers abstract. Int. J. Comput. Sci. Inf. Secur. (IJCSIS) **18**(12), 88–100 (2020)
35. Siddiqui, T., Amer, A.Y.A., Khan, N.A.: Criminal activity detection in social network by text mining: comprehensive analysis. In: 2019 4th International Conference on Information Systems and Computer Networks, ISCON 2019, pp. 224–229 (2019). https://doi.org/10.1109/ISCON47742.2019.9036157

Predicting the Stock Market Trend: An Ensemble Approach Using Impactful Exploratory Data Analysis

Nusrat Rouf(✉), Majid Bashir Malik, and Tasleem Arif

Baba Ghulam Shah Badshah University, Rajouri, Jammu and Kashmir 185236, India
nusratrouf@bgsbu.ac.in

Abstract. The Stock Market Prediction (SMP) has been a fascinating and challenging problem. The involvement of noisy, non-linear, and sparse features and poor feature selection degrades the prediction accuracies. The improved feature quality with an enhanced feature selection mechanism can increase the accuracy of prediction. This research article focuses on the Exploratory Data Analysis (EDA) of the Nifty50 index data of the National Stock Exchange (NSE). It proposes an enhanced hybrid feature engineering mechanism to extract the most relevant features that significantly impact SMP accuracy. This work employs Ensemble Regression Models viz. Random Forest Regression (RFR) and Extreme Gradient Boost Regression (XGBR) to predict the market trend. From the results we conclude that the proposed models achieve an improved R- squared values of 0.97 using RFR and 0.98 using XGBR, respectively.

Keywords: Stock market prediction · Exploratory data analysis · Ensemble regression models

1 Introduction

Technological advancements and innovations are proliferating and day by day are becoming more eminent, subsequently, has impacted almost all the areas of study. The integration of technological innovations with the financial markets has profoundly increased capital gains. The financial world has adapted to the advancements that have opened doors for innovative interdisciplinary research. SMP is a challenging problem due to involvement chaotic features and volatility [1]. Every day, a considerable amount of money is invested in the market, and every investor hopes to get benefitted. People search for tools and techniques that would increase profit and reduce risk [2]. Investing in stock is nothing but committing your money in an investment company with the expectation of getting an optimal profit.

Randomly chosen stocks for investment may lead to a significant disaster. A wise investment is not possible without prior knowledge of stocks. A proper analysis is an essential step before investing in any stock. SMP is complex and challenging since Stock Markets (SM) are dynamic, non-linear, stochastic, and chaotic [3]. It is considered an

M. Bhattacharya et al. (Eds.): ICICCT 2021, CCIS 1417, pp. 223–234, 2021.
https://doi.org/10.1007/978-3-030-88378-2_18

example of time series forecasting that promptly examines previous data and estimates future data values. The traditional approach was to apply simple statistical analysis to the market data and get valuable insights, but due to the increase in the number of investment companies, data volume has increased significantly. It has become imperative to use machine learning algorithms to huge amounts of market data. Machine learning in SMP is usually used to discover valuable data patterns [4]. The prediction of stock market trend has been of concern for the researchers due to chaotic features and volatility [1]. The market data analysis using machine learning tools and techniques helps in intelligent decision making. Data scientists use these self-learning algorithms efficiently and generate valuable insights from available market data. This research article focuses on EDA of the Nifty50 index data and proposes a hybrid feature engineering mechanism for the extraction of most relevant attributes. It employs RFR and XGBR, and evaluates the performance using the R-squared metric.

2 Related Work

This section briefly overviews the SMP studies based on regression approaches. Regression is a predictive approach that models the correlation between a dependent variable and independent variables [5]. Below are some related works based on various regression approaches used.

In [6], linear regression and SVR approach is employed to Coca-Cola stock price to predict the trend. Linear regression and SVR models achieve an accuracy of 73% and 93%, respectively. In [7], the Tata Consultancy Service (TCS) dataset's stock price is predicted used linear regression, polynomial regression, and SVR with a radial basis kernel function. Linear regression resulted in the highest accuracy of 97%, and other regression models achieve lesser accuracy scores.

In [8], authors employ a multiple regression approach to three variables viz. open, high and close to predict the stock market price. Various pre-processing techniques were applied to data, and an accuracy of 89% was achieved.

In [9], the decision tree regression model was used to predict the stock index. The features such as Price, Volume, Moving Average and Position were considered. Results indicated that the Moving Average indicator has a significant effect on the stock index prices. Decision tree model predicted with an accuracy of 70%.

In [10], the authors employed logistic regression to classify a company as "good" or "poor" based on the return rate. The dataset consisted of 30 companies with eight financial ratios. The model classified the companies with an accuracy of 74.6%.

In [11], the authors used the WEKA tool for SM trend prediction. The stock data from the Bursa Malaysia stock exchange was used. The data was generated from corporate annual reports. The authors compared various regression techniques, and Sequential Minimal Optimization (SMO) regression outperformed all other approaches. They have also changed the data type from real-valued to ordinal that increased model performance.

In [12], the authors predicted a company stock price known as the Dhaka Stock Exchange (DSE) ACI group. Different windowing operators were combined with SVR. The results indicate that SVR with rectangle window and flatten window operators produce acceptable results for one day, two day ahead and twenty-two day ahead predictions.

3 Methodology

The methodology implements ensemble regression techniques using revised and pre-processed features to predict the closing price trend of the Nifty50 index. The proposed steps for the method is shown in Fig. 1, and each step is explained in the following sections.

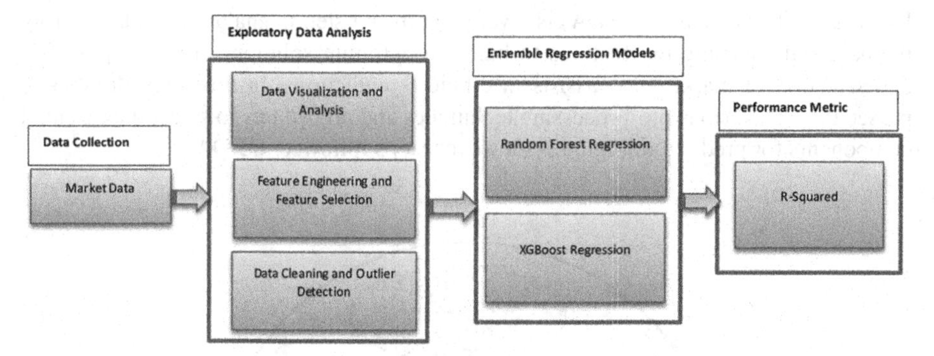

Fig. 1. Methodology overview

3.1 Data Collection

Since the SM data is freely available, the historical market data of the National Stock Exchange (NSE) index Nifty50 for 13 years (from 2007 to 2020) was downloaded from www.nseindia.com. Nifty50 measures the weighted mean performance of the largest 50 companies in NSE. The features of the dataset are temporal, and each attribute is a time series in itself. The extracted features are explained as follows. The 'Open' attribute

	A	B	C	D	E	F
1	Date	Open	High	Low	Close	Volume
2	12/31/2007	6095	6167.75	6095	6138.6	0
3	1/1/2008	6136.75	6165.35	6109.85	6144.35	0
4	1/2/2008	6144.7	6197	6060.85	6179.4	0
5	1/3/2008	6184.25	6230.15	6126.4	6178.55	0
6	1/4/2008	6179.1	6300.05	6179.1	6274.3	0
7	1/7/2008	6271	6289.8	6193.35	6279.1	0
8	1/8/2008	6282.45	6357.1	6221.6	6287.85	0
9	1/9/2008	6287.55	6338.3	6231.25	6272	0
10	1/10/2008	6278.1	6347	6142.9	6156.95	0
11	1/11/2008	6166.65	6224.2	6112.55	6200.1	0
12	1/14/2008	6208.8	6244.15	6172	6206.8	0
13	1/15/2008	6226.35	6260.45	6053.3	6074.25	0
14	1/16/2008	6065	6065	5825.75	5935.75	0

Fig. 2. Snapshot of extracted raw data

indicates the opening price of a stock on a particular date. The 'Low' and 'High' indicate the lowest and highest stock price for a specific date. 'Close' indicates the final cost of the index for a particular day. The 'Volume' attribute indicates the quantity of traded stocks on a specificied date. The snapshot of raw data is shown in Fig. 2.

3.2 Exploratory Data Analysis

Exploratory Data Analysis (EDA) is a very significant step in analyzing and exploring the data and choosing the features to build an optimum solution. Proper exploration and selection of market data assists in efficient predictions. In this step, the data is passed through various pre-processing techniques and procedures to select the essential components for prediction. Figure 3 shows the steps followed in EDA.

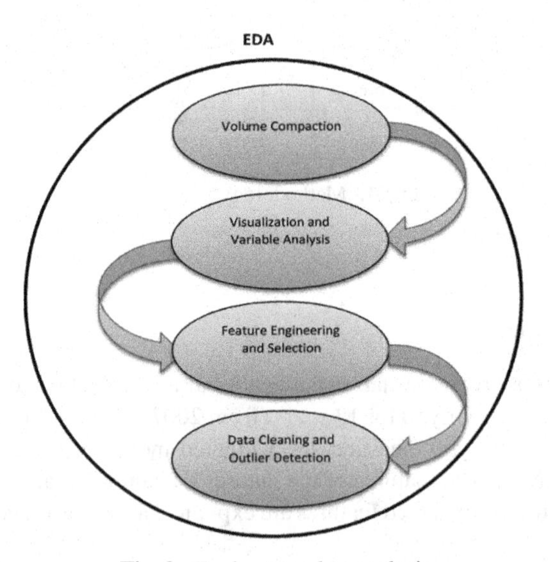

Fig. 3. Exploratory data analysis

Volume Compaction

As indicated by [13], volume reduction has a considerable effect on prediction models. The raw dataset is a daily based time series of sparse and low quality parameters. In order to reduce sparsity and inconsistencies, the monthly average of each parameter is taken. The generalized equation to calculate the average is:

$$Attribute_{avg} = \sum Values_i \Big/ Total\_Value_i \tag{1}$$

Here, $\text{Attribute}_{\text{avg}}$ represents the average value of a feature for a month, $\sum Values_i$ represents the sum total of the values of i instance of an attribute, and Total_Value_i is the number of instances of the ith attribute in a month for which stock transactions occur.

Data Visualization and Variable Analysis

The data is visualized by plotting lines, bars, histograms, or box plots. We look for the hidden patterns in each feature by analyzing the distribution of data. The correlation between variables is calculated to extract the most relevant features. A correlation heat map investigates how strongly the variables are related to each other. This work uses the Pearson correlation to measure the relationship between two variables given as:

$$R_{x}y = \frac{\sum_{i=1}^{n}(x_i - \overline{x})(y_i - \overline{y})}{\sqrt{\sum_{i=1}^{n}(x_i - \overline{x})^2}\sqrt{\sum_{i=1}^{n}(y_i - \overline{y})^2}} \quad (2)$$

Where R_{xy} is the correlation coefficient,x_i represents the value of an ith instance of x variable, $\overline{x}$ represents value of i instances of x variable. y_i is the value of an ith instance of y variable, $\overline{y}$ is the mean of i instances of y variable. The correlation matrix in Fig. 4 shows that 'Volume' and 'Close' are positively correlated. The line plot in Fig. 5 indicates the month wise distribution of the closing price of the Nifty50 index for 13 years. The graph is divided into two subplots for easy readability, as it would have been difficult to visualize the distribution of closing price for 156 months (2007–2020) in a single graph.

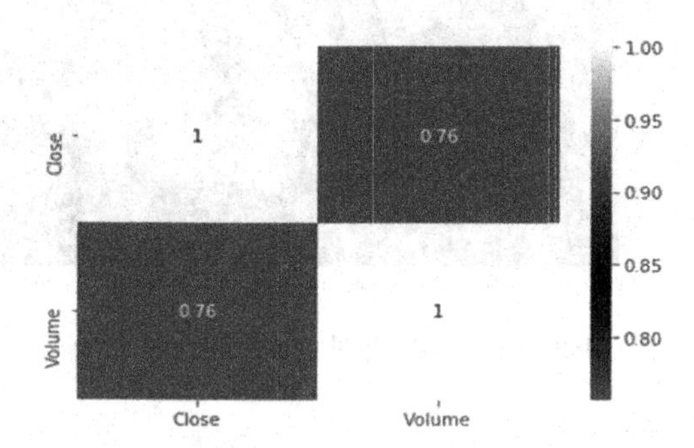

Fig. 4. Correlation heat map for Volume vs Close features

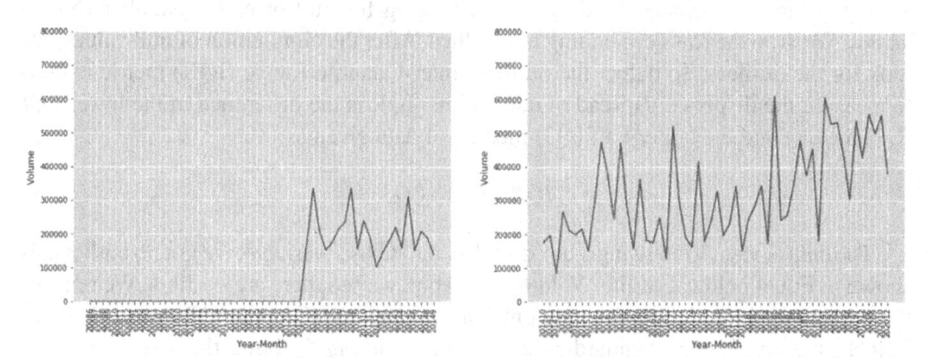

Fig. 5. Monthly volume distribution of Nifty50 index

Feature Engineering and Selection

Sometimes training the model using the default features of a dataset result in poor predictions. To increase the quality of input data, we derive more relevant attributes from existing features and extend feature space by including external data. It is the most critical step as it directly impacts the prediction accuracy. Historical lags or stock price across the last 'n' days is an important stock indicator that affects stock price. It could be an essential feature for the prediction models. Lag is used to measure the strength of a trend. The methodology has used six lag-based features to analyze the effect of closing prices of the last six months on the current closing price. It can be inferred from Fig. 6 that 'Volume' and lag-based features have a reasonable correlation with the 'Close' attribute.

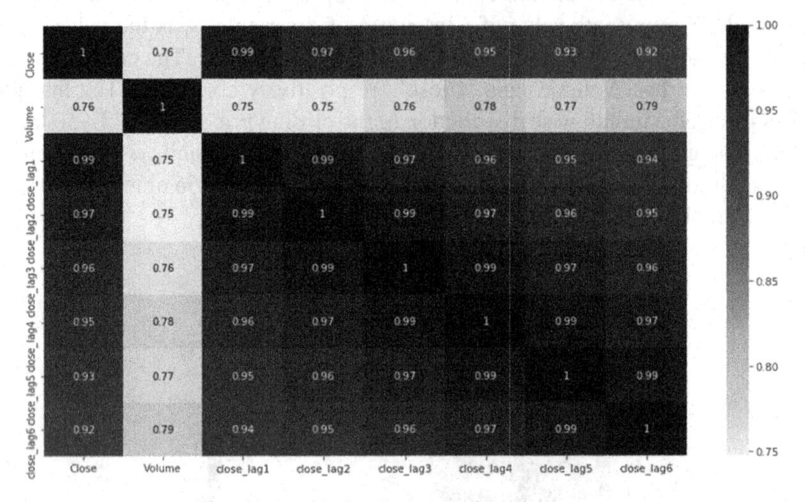

Fig. 6. Correlation heat map for all features

Data Cleaning and Outlier Detection

This step further explores the dataset by checking for null or Not a Number (NAN) values. We drop the rows containing null values. After the elimination of null values, we look for the outliers. To detect the outliers, Inter-Quantile Range (IQR) metric is used to measure the dispersion/spread of data points. IQR is the difference in the lower half (Q1) median and the upper half (Q3) median of data given as:

$$IQR = Q1 - Q3 \tag{3}$$

The data is visualized using a box plot. For the 'Close' variable, we got no outliers, as shown in Fig. 4 below. For the 'Volume' variable, we see there are outliers. We remove the outliers by considering the data only in the range of (Q1–1.5 IQR) and Q3 + 1.5 IQR. Lastly, the feature standardization is done. The Fig. 7 shows the visualization of the 'Volume' feature with outliers, and Fig. 8 shows the 'Close' and cleaned 'Volume' feature. Figure 9 is the snapshot of the prepared dataset.

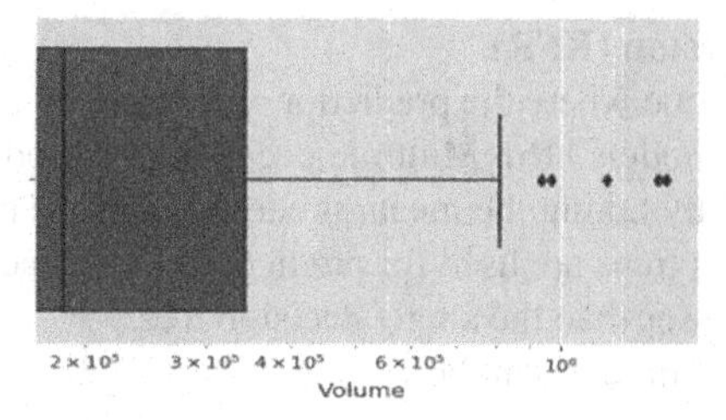

Fig. 7. Box plot for volume feature containing outliers.

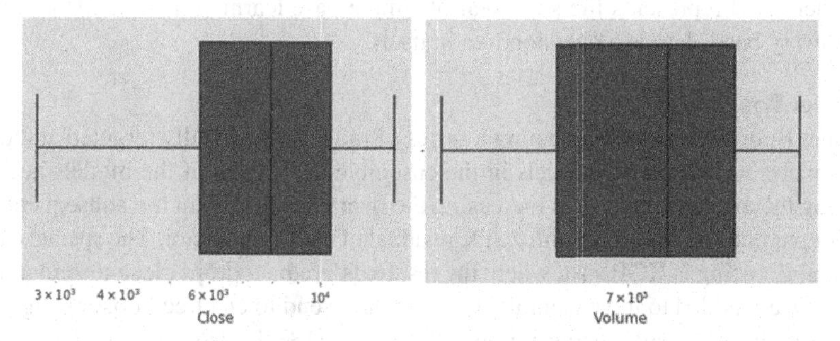

Fig. 8. Box plots for close and volume without outliers.

	A	B	C	D	E	F	G
1	Volume	close_lag1	close_lag2	close_lag3	close_lag4	close_lag5	close_lag6
2	-0.95817	-1.16671	-1.06014	-1.22585	-1.03147	-1.06298	-0.65866
3	-0.95817	-1.48454	-1.17603	-1.05483	-1.22777	-1.02802	-1.07213
4	-0.95817	-1.37251	-1.50105	-1.1721	-1.05459	-1.22672	-1.0366
5	-0.95817	-1.36215	-1.38649	-1.50096	-1.17333	-1.05142	-1.23855
6	-0.95817	-1.53027	-1.37589	-1.38504	-1.50635	-1.17162	-1.06038
7	-0.95817	-1.92703	-1.54781	-1.37432	-1.38897	-1.50871	-1.18255
8	-0.95817	-1.97703	-1.95356	-1.54827	-1.37811	-1.38989	-1.52516
9	-0.95817	-1.89886	-2.00469	-1.95882	-1.55426	-1.3789	-1.4044
10	-0.95817	-1.93117	-1.92474	-2.01056	-1.96999	-1.55721	-1.39323
11	-0.95817	-1.97376	-1.95779	-1.92966	-2.02238	-1.97802	-1.57445
12	-0.95817	-1.2274	-1.09224	-1.25455	-1.26713	-1.4069	-1.35649
13	-0.95817	-0.99722	-1.09132	-1.01039	-1.02119	-1.15041	-1.17762
14	-0.95817	-0.97611	-1.0027	-1.08639	-1.00959	-1.01761	-1.16099

Fig. 9. Prepared dataset

3.3 Ensemble Regression Models

Ensemble models work by combining multiple model's outputs such that each model complements the other [14]. The ensemble models significantly increase prediction accuracies by reducing variance and bias [15]. The proposed model is based on two ensemble regression models viz. RFR and XGBR. The features fed to the models are 'Volume', 'Close_lag1', 'Close_lag2', 'Close_lag3', 'Close_lag4', 'Close_lag5', and 'Close_lag6'.

Random Forest Regression (RFR)
It is the ensemble technique where the predictions are made by combining the decisions made by base learning models [16]. Multiple decision tree models are trained on data, and output is generated by taking the mean/ mode from all the predictors. RFR give us diversity since numerous trees are used for prediction. Since each tree is independently grown, its variance is lesser than the single decision tree.

Formally the class of models can be written as:

$$g(x) = f_0(x) + f_1(x) + f_2(x) + \dots \quad (4)$$

Where g(x) represents the sum total of simple base learning models $f_i(x)$. Furthermore, every base class is a decision tree in itself.

Gradient Boosting
Gradient Boosting works by creating a series of models specifically targeted at the data points incorrectly by other models in the ensemble [17]. Each of the models keeps on reducing the average error, thus increasing the overall accuracy. In the subsequent step, another predictor is built to minimize the residual of a prior predictor. The special case of Gradient Boosting is XGBoost, where the residuals/gradient drops close to zero as more predictors are added to the ensemble. It is a scalable end to end tree boosting algorithm that gives optimal results using minimal resources [18]. It also uses regularization to avoid overfitting [19]. Some parameters of XGBR are:

- Learning rate/shrinkage: the learning rate shows how aggressively the gradients are reduced in the subsequent predictors. It ranges from 0 to 1.
- N_Estimators: represents the number of trees required to generate accurate results.
- Sub-sampling: it is the fraction of data on which the estimators will be built.
- Gamma: it controls the regularization parameter, which helps in the construction of shallow trees.
- Max_depth: improves computational performance by pruning the tree backwards.

Our model learning rate is 0.1, N_estimator is 100, the max_depth is 3 by default, and the Gamma is 0.

4 Results and Evaluation

This section discusses the experimental outcomes of the work carried out as well as the benefits of using Ensemble Regression Models for SMP.

4.1 Performance Analysis.

The performance measure used for the evaluation of the proposed models is the R-squared metric. The basic requirement is to minimize the difference between the actual value Y_{act} and the predicted value Y_{pred} given in the regression problem. The generalized formula for R-squared is:

$$R^2 = 1 - {SS_{res}}/{SS_{tot}} \quad (5)$$

Where SS_{res} is the residual sum of squares of the difference between actual and predicted values given as:

$$SS_{res} = sum(Y_{iact} - YY_{ipred})^2 \tag{6}$$

And SS_{tot} is the sum of squares of the difference between the actual and predicted values given as the R-squared values for both the models is presented in Table 1:

$$SS_{tot} = sum(Y_{iact} - Y_{iavg})^2 \tag{7}$$

Table 1. R-Squared values of RFR and XGBR for train and test sets

Models	Data	R-Squared
RFR	Train	0.99
	Test	0.97
XGBR	Train	0.99
	Test	0.98

The key achievements are:

1. Both RFR and XGBR obtained the R-squared value of .97 and .98, respectively, which shows that models are performing well.
2. EDA has been a critical step in achieving significant results.
3. The models are not overfitted, as can be determined by considering the R- squared value for both the train and test datasets.
4. Visualization of important feature for both the models have revealed that Close_lag1 is an essential feature and Volume is the least important feature. The inclusion of lag based features has dramatically affected the results. Figure 10 and 11 shows the importance of individual features in both the RFR and XGBR models respectively.

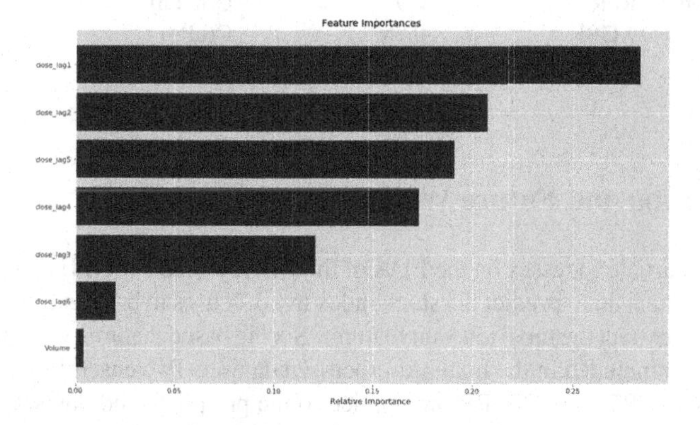

Fig. 10. Feature importance RFR

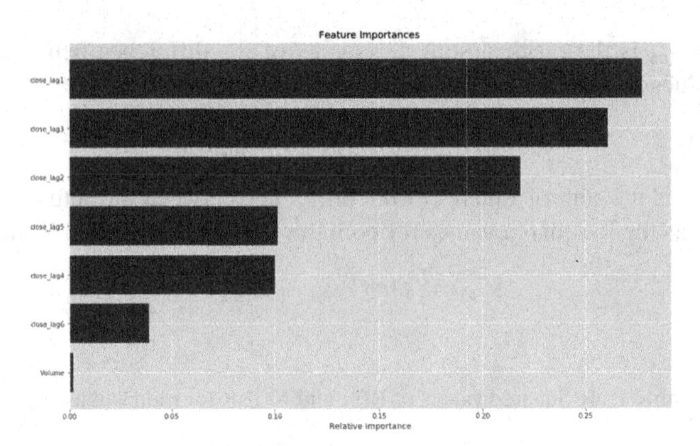

Fig. 11. Feature importance XGBR

4.2 Comparison with the Previous Works

Table 2 presents the comparative analysis with the previous works. These works have employed different types of regression methods. The results reveal that using the ensemble approaches with EDA more accurate predictions can be done. As presented in the table our models have achieved better R-Squared values as compared to previous works. As complexity plays a vital role in any analysis, we evaluated complexity based on the number of features involved.

Table 2. Comparison with the previous works

Models	Regression models	R-squared values	Complexity/number of features(P)
[7]	Linear Regression	0.97	$O(P^2n + P^3)$
[6]	Linear Regression SVR	0.73 0.93	$O(P^2n + P^3)$ $O(n^2P + n^3)$
[20]	Linear Regression	0.93	$O(P^2n + P^3)$
Proposed model	RFR XGBR	0.97 0.98	$O(n^2Pn)$ $O(nPn)$

5 Conclusion and Future Work

This research article focusses on the EDA of the Nifty50 index of NSE and proposes an enhanced mechanism to predict the stock index trend. It uses hybrid feature engineering mechanism to extract the most relevant features. Six lag based features viz. Close_lag1 to Close_lag6 are included in the final pre-processed dataset. Two ensemble-based regression models viz. RFR and XGBR are applied to the pre-processed dataset. The results indicate that both models perform very well. The models achieve an improved R- squared values of 0.97 (RFR) and 0.98 (XGBR), respectively. The proposed method gives comparatively better results. However, various limitations do exist, such as follows: The proposed methodology only uses the historical data with six lag-based indicators. However, the stock market is driven by other factors such as public sentiments [21], weather [22], political events [23], etc. Also, the proposed method predicts monthly stock index trend only.

This work can be further extended by including other market indicators such as Relative Strength Indicator, Moving Average, Volatility, etc. The model can be extended to predict the stock trend on a daily and weekly basis. Also, factors such as public sentiments, weather, political events, etc., can be incorporated for more efficient predictions.

References

1. Binkowski, M., Marti, G., Donnat, P.: Autoregressive convolutional neural networks for asynchronous time series. In: 35th International Conference on Machine Learning ICML 2018, vol. 2, pp. 933–945 (2018)
2. Upadhyay, A., Bandyopadhyay, G., Dutta, A.: Forecasting Stock Performance in Indian Market using Multinomial Logistic Regression (2019)
3. Avenue, N.: Biological Brain-Inspired Genetic Ccomplementry Learning for Stock Market and Bank Failure Prediction, vol. 23 (2007)
4. Asyraf, A.S., Abdul-Rahman, S., Mutalib, S.: Mining textual terms for stock market prediction analysis using financial news. Commun. Comput. Inf. Sci. **788**, 293–305 (2007)
5. Zhang, L., Zhang, L.L., Teng, W., Chen, Y: Based on information fusion technique with data mining in the application of finance Early-Warning. Procedia Comput. Sci. **17**, 695–703 (2013)
6. Gururaj, V., Shriya, V.R., Ashwini, K.: Stock market prediction using linear regression and support vector machines. Int. J. Appl. Eng. Res. **14**, 1931–1934 (2019)
7. Bhuriya, D., Kaushal, G., Sharma, A., Singh, U.: Stock market predication using a linear regression. In: Proceedings of the International Conference on Electronics, Communication and Aerospace Technology ICECA 2017. 2017-January, pp. 510–513 (2017)
8. Kamley, S., Jaloree, S., Thakur, R.S.: Multiple regression: a data mining approach for predicting the stock market trends based on open, close and high price of the month. Int. J. Comput. Sci. Eng. Inf. Technol. Res. **3**, 173–180 (2013)
9. Yuan, J., Luo, Y.: Test on the Validity of futures Market's high frequency volume and price on forecast. In: International Conference on Management of e-Commerce and e-Government, pp. 28–32 (2014)
10. Dutta, A.: Prediction of stock performance in the indian stock market using logistic regression. Int. J. Bus. Inf. **7**, 105–136 (2012)
11. Siew, H.L., Nordin, M.J.: Regression techniques for the prediction of stock price trend. ICSSBE 2012 - Proceedings, International Conference on Statistics in Science, Business and Engineering: "Empowering Decision Making with Statistical Sciences", pp. 99–103 (2012). https://doi.org/10.1109/ICSSBE.2012.6396535
12. Meesad, P., Rasel, R.I.: Predicting stock market price using support vector regression. In: 2013 International Conference Informatics, Electronics and Vision, ICIEV 2013. (2013). https://doi.org/10.1109/ICIEV.2013.6572570
13. Asghar, M.Z., Rahman, F., Kundi, F.M., Ahmad, S.: Development of stock market trend prediction system using multiple regression. Comput. Math. Organ. Theory **25**(3), 271–301 (2019). https://doi.org/10.1007/s10588-019-09292-7

14. Seni, G., Elder, J.F.: Ensemble Methods in Data Mining: Improving Accuracy Through Combining Predictions (2010)
15. Nti, I.K., Adekoya, A.F., Weyori, B.A.: A comprehensive evaluation of ensemble learning for stock-market prediction. J. Big Data **7**(1), 1–40 (2020). https://doi.org/10.1186/s40537-020-00299-5
16. Pavlov, Y.L.: Random forests. Random For, pp. 1–122 (2019) https://doi.org/10.1201/9780367816377-11
17. Khanday, A.M.U.D., Rabani, S.T., Khan, Q.R, Rouf, N., Mohi Ud Din, M.: Machine learning based approaches for detecting COVID-19 using clinical text data. Int. J. Inf. Technol. **12**, 1–9 (2020)
18. Chen, T., Guestrin, C.: XGBoost: A scalable tree boosting system. In: Proceedings of ACM SIGKDD International Conference on Knowledge Discovery and Data Mining, 13–17-August, pp. 785–794 (2016). https://doi.org/10.1145/2939672.2939785
19. XGBoost Algorithm: Long May She Reign! | by Vishal Morde | Towards Data Science. https://towardsdatascience.com/https-medium-com-vishalmorde-xgboost-algorithm-long-she-may-rein-edd9f99be63d. Accessed on 07 Feb 2021
20. Ivanovski, Z., Ivanovski, Z., Narasanov, Z.: The regression analysis of stock returns at MSE. J. Mod. Account. Audit. **12**(4) (2016). https://doi.org/10.17265/1548-6583/2016.04.003
21. Bhardwaj, A., Narayan, Y., Vanraj, Pawan, Dutta, M.: Sentiment analysis for indian stock market prediction using sensex and nifty. Procedia Comput. Sci. **70**, 85–91 (2015). https://doi.org/10.1016/j.procs.2015.10.043
22. Khanthavit, A.: Investors' Moods and the Stock Market (2018)
23. Kaleem, A.: Impact of political events on stock market returns :empirical evidence from Pakistan.(2014). https://doi.org/10.1108/JEAS-03-2013-0011

Can Machine Learning Predict an Employee's Mental Health?

Gaganmeet Kaur Awal(✉) and Kunal Rao

USME, Delhi Technological University, Delhi, India
gaganmeetkaur.awal@dtu.ac.in, krducic@gmail.com

Abstract. Mental health has always been a critical parameter for determining the well-being of a human being, especially working individuals. The people in the IT sector face huge work pressure and are under tremendous stress which affects their personal life and eventually their mental state. A lot of people suffer from mental trauma, depression, and anxiety which sometimes lead to extreme steps such as suicides. There have been various Mental Disorders by which the employees get affected and lack of awareness leads to serious problems in later stages. In our work, the aim is to study and analyze the 2014 and 2019 Open Source Mental Illness Survey Dataset. This research also discovered the hidden pattern that can determine what factors affect mental well-being and result in mental illness. We have performed studies and descriptive data analysis along with machine learning classification prediction to extensively study how mental disorders affect the career growth prospects of an employee. The proposed model also provides the solution to different challenges like class imbalance which is quite prevalent in these datasets. The experimental results clearly demonstrate that this study would aid mental health professionals to effectively deal with mental illness patients to treat them efficiently before they are majorly affected.

Keywords: Mental health · Work pressure · Mental illness · Disorder · Machine learning · Imbalance learning

1 Introduction

Mental Health is the least discussed illness publicly discussed. The people disclose their physical health issues or other things with other people in their known circle but feel reluctant to share their mental illness with their dearest ones. Many people in the past working in corporate setup have faced mental issues for a long time. Especially the people working in the IT sector and employed have faced stress for working for long hours or overtime which has affected their mental health and also personal family issues. Further, this COVID-19 era has resulted in a high increase in the number of people having mental health issues, many people have lost their jobs or faced salary cuts. In times when people are facing difficulty in finding jobs and also to the unemployment worldwide is increasing drastically, the people have more stress-related issues. Allan et al. [1] proposed a model to examine correlation between the workers meaningful work and mental health and

M. Bhattacharya et al. (Eds.): ICICCT 2021, CCIS 1417, pp. 235–247, 2021.
https://doi.org/10.1007/978-3-030-88378-2_19

observed that depression and stress are negatively correlated with job satisfaction and meaningful work.

The American and European nations have been addressing this issue not on a global scale but yet working to deal with it by organizing events, seminars to spread awareness about this issue. Andrews et al. [2] analysed the cross-sectional dataset for 1235 US nurse workforce, to evaluate the causal relationships between job strain, workplace environment for voluntary turnover and intervention strategies. They concluded that strategies pertaining to mental health support and reinforces the copying role for retention efforts for the nurses. The largest section of the world population resides in the South Asian regions and has the greatest share of mentally ill people yet they have not been vocal about this issue. But the past few years have seen that more prominent celebrities have put their foot forward to discuss mental health issues. To effectively deal with this issue using the Artificial Intelligence toolkit is the biggest motivation behind this research. We have developed the models to better judge the people suffering from mental health issues.

In the past, many researchers have proposed different solutions using analytics and artificial intelligence tools to aid the diagnosis of the mental health illness of the patients. Many researchers have been done on one of the most popular mental illness datasets ie., the OSMI dataset, and other researchers have deployed their models on other mental disorders datasets. Islam et al. [10] presented their work on the Kaggle dataset whereas Elmunsyah [7] applied ML algorithms to study the relations between the parameters for the OSMI dataset. Similarly, other researchers [21] have proposed their models to better analyze the stress pattern for the workable force in the IT sector.

In this research, we have presented an analysis for the most common OSMI dataset for the 2014 survey and one of the recent datasets for the 2019 survey which was collected just before the COVID-19 onset. To the best of our knowledge recent work does not provide us with the detailed results for the 2019 survey dataset and we aim to provide results useful for the mental health professionals for the better diagnosis for the patients currently dealing with mental disorders or who are susceptible to mental illness in future. Further 2019 dataset provides us with complex problems like handling imbalanced datasets and dealing with many features and choosing the most important and meaningful one from them. We have explored many imbalanced learning techniques Our work tries to present a holistic approach to address the problems in the OSMI dataset and also the comparison of the various ML techniques and detailed analysis for the two datasets.

The rest of the paper is organized as follows; Sect. 2 describes the literature work done before in this area of study; Sect. 3 describes our research framework which explains various steps that we followed while conducting this study; Sect. 4 introduces the experimental setup including modeling using ML techniques and evaluation metrics used in this research; Sect. 5 includes all the experiments we performed and their respective results, and finally Sect. 6 presents conclusions and future work.

2 Literature Survey

Machine learning has played an important role in the development of efficient heart attack prediction models. Researchers have applied various ML techniques in this domain and they come up with various models for this problem. Mohr et al. [13] proposed a predictive model using the data collected from the patient's daily life including their behavior, feelings, traits, and others using the sensors in smartphones, wearable devices, and others. They provide detailed work on mobile health interventions which provide a good base for further research in this domain. Shatte et al. [19] synthesize the literature work of the researches done in the mental health domain. They analyzed that the most popular research domains are detection and diagnosis, treatment and support, and research and clinical administration.

Srividya et al. [20] proposed a machine learning model to study the relationship between the factors like stress levels, disorders, and work productivity. Classification algorithms like support vector machines, decision trees, naïve bayes classifier, K-nearest neighbor classifier, and logistic regression are deployed on the dataset collected from the questionnaire collected from high school students, college students, and working professionals. Islam et al. [10] have represented the basic requirements of developing some IS solutions for some common mental health problems. They have used the dataset of IT professionals from Kaggle for their study. They have analyzed in detail various common mental health disorders and have done a detailed analysis on the relation between work parameters and these disorders and have developed useful Insights.

Elmunsyah et al. [7] have used the survey conducted by the Open Sourcing Mental Illness(OSMI) and applied various ML algorithms in order to determine the relationship between work and mental illness and also to decide whether the employee needs immediate treatment. Reddy et al. [17] deployed machine learning algorithms to analyze stress levels in employees working at a firm. The ensemble learning methods like boosting yielded better performance as compared to other algorithms used. At 75% accuracy, it outperformed other models to better predict employee's stress and mental health illness. Laijawala et al. [11] have used the online available OSMI data which consists of details about working individuals (questionnaire). They have applied ML algorithms to develop a model that gives knowledge to employees about their mental health and the necessity of immediate treatment.

3 Proposed Model

The research work was carried out in the below-discussed framework consisting of the following steps (see Fig. 1).

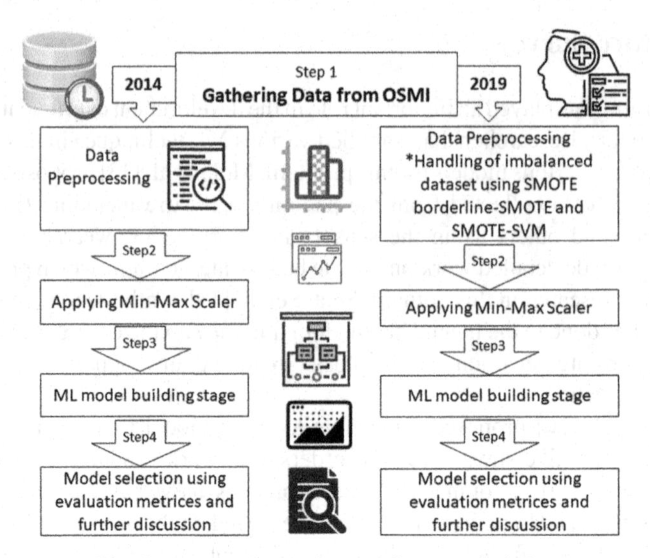

Fig. 1. Proposed experimental framework design

3.1 Dataset Preparation

The first and foremost task was gathering a suitable dataset for the proposed research work. The Open Sourcing Mental Illness (OSMI) dataset [15] is one of the popular datasets and the 2014 survey gave researchers a dataset with wider objects which can be achieved from the dataset. The OSMI provides us with a dataset from 2014 onwards and the most recent being the 2019 and 2020 datasets.

The **2014** survey dataset was retrieved from the OSMI repository consisting of 1200 records and 27 attributes. The detailed set of parameters explanation is provided in Sect. 4.1. The 2014 dataset was preprocessed and the detailed steps are explained in Sect. 4.2.

The **2019** survey dataset was retrieved from the OSMI repository consisting of 350 records and 82 attributes. The detailed set of parameters explanation is provided in Sect. 4.1. The 2019 dataset was preprocessed and the important feature vector of disorders was modeled from the dataset which gave the description of the disorder from which the patient was suffering. The detailed steps are explained in Sect. 4.2.

Handling of Imbalanced dataset using SMOTE, Borderline SMOTE, and SMOTE-SVM:
The 2019 dataset was pre-processed and it was observed that the target attribute of people suffering from mental illness was having imbalanced instances for the multi-labels. 3 different approaches were deployed for Handling imbalance target attributes.

SMOTE: Chawla et al. [4] proposed an oversampling approach by creating synthetic examples. SMOTE or Synthetic Minority Oversampling Technique is applied to address the problem of overfitting, creating synthetic records from minority classes based on records already existing. SMOTE uses the k-nearest neighbor approach to find each minority instance, among these instances it randomly selects one. Next, it computes the difference between the chosen sample and the selected nearest neighbor, then multiplies the computed difference with a random number between 0 and 1 to interpolate the instances selected to produce a new minority record in close vicinity. Similarly, it creates newer minority records to match the count of majority records.

Borderline SMOTE: A Borderline SMOTE learning model [9] generates the synthetic instances for the minority classes similar to the ones nearer to the borderline from the instances in minority classes using the classification models such as a k-nearest neighbor.

Borderline SMOTE SVM: A modified Borderline SMOTE SVM learning model [14] generates the synthetic instances for the minority classes by using a decision boundary to locate the instances for minority classes close to support vectors.

Applying Min-Max Scaler: The 2014 and 2019 dataset was normalized for better model building using Min-Max Scaler. We have applied the Min-Max scaler to normalize the features into the 0–1 range.

$$Xscaled = \frac{X - min(X)}{max(X) - min(X)} \tag{1}$$

where x represents the single feature vector.

3.2 ML Model Building and Selection

ML techniques used in this study are logistic regression (LR), K-nearest neighbor (KNN), Bernoulli Naive Bayes (BNB), Decision tree (DT), Gaussian Process Classifier (GPC), AdaBoost Classifier (ADA), Extreme Gradient Boost Classifier (XGB). Cover et al. [6] used KNN to classify the datasets. Quinlan [16] proposed the idea of building a Decision Tree (DT) from the examples in a robust way. Chen and Guestrin [5] gave the idea for XGB to swift the executing speed of the model training and compared to other classification techniques it holds a significant advantage in performance. Freund and Schapire [8] introduced the ADA. Lawrence et al. [12] have proposed a complete framework using a kernel method for sparse Gaussian Process methods that use forward selection. Rish [18] proposed the BNB algorithm using Monte Carlo simulations to understand the data characteristics. Berkson [3] introduced the Logit function for the classification problem. We applied all the above techniques for the prediction of Employee's Mental Health in this study. We employed the standard 10-fold cross-validation scheme to fit the ML models on the datasets to evaluate their performance.

The following evaluation metrics were evaluated on the fitted dataset: Accuracy, F1 score, recall, Precision, ROC-AUC. Accuracy signifies the percentage of the number of correctly classified data instances upon in a dataset. F1-score is based on the harmonic mean of the recall and precision values. Precision is the ratio of the predicted True Positive data points and the total actual Positive data points. The recall measures the model's accuracy in terms of predicting True Positive data points correctly. ROC is nothing but a probability curve. AUC also tells the extent to which the parameters are separable.

For evaluation of the Imbalanced ML model the metrics evaluated are Accuracy, Balanced Score, F1 weighted, and Geometric means score. Accuracy signifies the percentage of the number of correctly classified data instances upon in a dataset. A balanced Score is used to evaluate the accuracy score for the imbalanced learned ML model. F1 weighted is the balance of precision and recall in harmonic mean calculation controlled by the weighted coefficient. Geometric means is the square root for the multiplicative value of Sensitivity and Specificity.

Using the values obtained for the evaluative metrics in the balanced 2014 dataset and 2019 dataset, the model with better values across models is finally selected.

4 Experimental Setup/Research Methodology

4.1 Dataset Description

The Dataset for this Study is obtained from Open Sourcing Mental Illness (OSMI). Open Sourcing Mental Illness (OSMI) is a non-profit organization with a primary objective of promoting awareness towards mental illness, disorders in the workspace, and the main agenda is to eradicate the stigma surrounding distracted employees. One important function of OSMI is helping workplaces in the identification of the best resources to help their employees to overcome mental health issues. The OSMI mental health dataset [15] is publicly available. We have retrieved the 2014 dataset and the 2019 dataset.

The OSMI Mental Health in Tech 2014 Survey contained 1200 responses from employees working in the tech industry and non-tech industry. Some of the respondents were self-employed while others were employed in a company. Most of the respondents were residing in the USA and the rest of them were from other parts of the world. The responses were recorded for in total 27 attributes related to demographics, previous working experiences, and support regarding mental illness from supervisors and co-workers.

The important parameters in the 2014 dataset are *Age of the employee, Work Country, Gender, Sector in which employment, how comfortable the employee is in sharing his/her mental health state with coworkers and supervisors, sought mental health treatment before or not*, and, among other attributes.

Similarly, the OSMI Mental Health in Tech 2019 Survey contained 350 responses from respondents in similar sectors. In this dataset, some respondents were from tech and others from non-technical backgrounds. Respondents were required to fill in 82 questions spanning personal and professional life.

The important parameters in the 2019 dataset are *Age of the employee, Work Country, Gender, Sector in which employment, how comfortable the employee is in sharing his/her mental health state with coworkers and supervisors, sought mental health treatment before or not*, details related to the previous and present employer, whether he/she would be comfortable in discussing the mental situation in future, the disorder for which employee has sought treatment in future, views related to mental health, whether he would consult OSMI help for mental illness and, among other attributes.

There are two datasets used in the study which are completely different from the year in which they were collected, the respondents filling the responses, and finally the data recorded in each survey.

4.2 Preliminary Descriptive Analysis

The dataset for this study is obtained from Open Sourcing Mental Illness (OSMI). There are two datasets used in the study which are completely different. The following Sect. 4.3 provides a detailed description of the pre-processing steps involved in this research paper.

Preliminary Descriptive Analysis

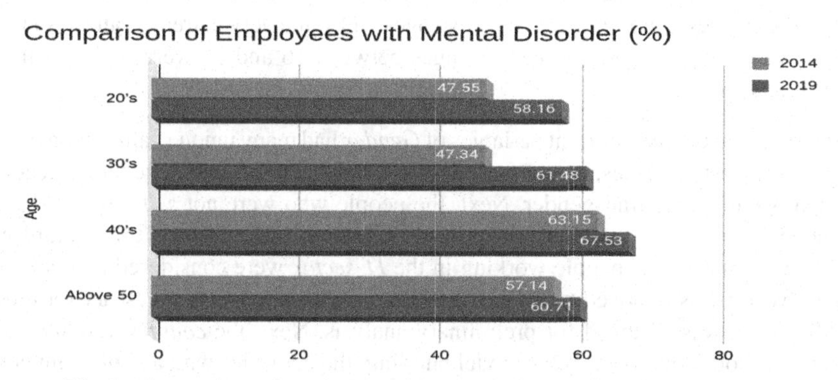

Fig. 2. Comparison of employees with mental disorder from 2014 and 2019

The motivation behind the conduct of this research can be explained through Fig. 2, which clearly shows that the percentage of employees suffering from mental illness and other disorders has shown a significant increase from the 2014 OSMI survey dataset to the one of the most recent 2019 OSMI survey dataset. Every age group suggests that people are more prone to mental related illness in their early 20's and also to within their late 50's. Thus this becomes an interesting problem to study the reasons which should aid the diagnosis of such patients so that people get better treatment from mental health professionals.

4.3 Data Preprocessing

2014 analysis: The first variable of *Age* had outlier values in it with some negative ages and some very extreme values, so these values were removed and people with ages between 16 and 75 were included in the further analysis.

Further one of the important variables of *Gender* had many unique values with people filling very absurd values, so they needed to be categorized into 3 gender categories of men, women, and transgender. Next, the people who were not *self-employed* were included in the further analysis and after considering these records this feature was dropped. Similarly, the *people working in the IT sector* were considered and next this feature vector was removed too. The *Timestamp* variable was not necessary for model building so it was removed for preliminary analysis. Next, the *country* variable has a count of various countries, so for model building, the countries with a significant count of more or equal to 10 values were included, for a total of 8 different countries were finally chosen.

Next for the features where values in character or String format the label encoding process was done to transform it into a numeric value and then the numeric feature vectors were combined with transformed variables to create the final pre-processed data. The target attribute for the 2014 dataset is *whether the employee has sought treatment from a mental health professional.*

2019 analysis: 2019 dataset had 352 records and 82 features. The first variable of *Age* had outlier values in it with some negative ages and some very extreme values, so these values were removed and people with ages between 16 and 75 were included in the further analysis.

Further one of the important variables of *Gender* had many unique values with people filling very absurd values, so they needed to be categorized into 3 gender categories of men, women, and transgender. Next, the people who were not *self-employed* were included in the further analysis and after considering these records this feature was dropped. Similarly, the people working in the *IT sector* were considered and next this feature vector was removed too. The *Timestamp* attribute was not necessary for model building so it was removed for preliminary analysis. Next, the *country* variable has a count of various countries, so for model building, the countries with a significant count of more or equal to 10 values were included, for a total of 3 different countries were finally chosen.

Next to many feature vectors had a lot of null values and they were imputed with altogether different values of "didn't reveal" or combined with existing sub-categories or fed with the mean values. Also, the features of comments or open-ended discussion were removed for the model building process. Another important variable of the *type of mental disorder* was processed during the model building process. Now we had 13 attributes with different disorders from which employees sought treatment. So pre-processing removed around 30 feature vectors that didn't add much value to the model building process. The target attribute for the 2019 dataset is *whether the employee thinks mental illness affects the career growth prospects.* Finally, we were left with 149 records with 63 features for the ML model-building task. The imbalance ratio in the dataset is 14.0.

5 Results

In this section, we present the result of our study for the 2 datasets. In order to develop the models, we applied ML techniques discussed in Sect. 4 on both datasets. For fitting the model the 10-fold cross-validation method is applied.

5.1 2014 Analysis

Table 1 describes the comparative analysis of the ML models developed for the 2014 pre-processed dataset. It is to be noted that in Tables 1, A*refers to as Accuracy score, P* refers to as Precision score, R* refers to as Recall score, F1* refers to the F1 score and AUC refers to AUC-ROC.

Table 1. Comparative Analysis of Various ML Algorithms on 2014 Dataset

ML	A*	P*	R*	F1*	AUC
LR	0.83	0.83	0.84	0.83	0.90
KNN	0.83	0.82	0.85	0.84	0.89
BNB	**0.85**	0.83	**0.88**	**0.85**	0.90
DT	0.77	0.77	0.77	0.77	0.76
GPC	0.84	**0.84**	0.86	0.84	**0.91**
ADA	0.82	0.84	0.82	0.82	0.89
XGB	0.83	**0.84**	0.83	0.83	**0.91**

For the OSMI 2014 survey dataset, we found that the BNB algorithm is the best model as it gives the highest accuracy of 85%. Other metrics are also comparatively higher for Bernoulli NB Model like Recall-88%, F1 score-85%, Recall-88%. GPC and XGB model gave higher results for Precision-84% and AUC-ROC of 85%. So BNB and GPC are the two most promising models.

5.2 2019 Analysis

Tables 2, 3, 4 and 5 describes the comparative analysis of the ML models developed for the 2019 pre-processed dataset.

It is to be noted that in Tables 2, 3, 4, and 5, O*refers to as Original Dataset without normalization, O_S* refers to as SMOTE learned Original dataset, MM* refers to as Dataset after Min-Max Scaler, MM_S1* refers to the Scaled SMOTE learned Data, MM_S2* refers to the Scaled Border-line SMOTE learned Data and MM_S3* refers to the Scaled Borderline SMOTE SVM learned Data.

Table 2. Accuracy values for ML Models on 2019 Dataset

ML	O	O_S	MM	MM_S1	MM_S2	MM_S3
LR	0.846	0.85	0.91	0.91	0.992	0.992
KNN	0.845	0.5	0.92	0.80	0.94	0.94
BNB	0.88	0.91	0.88	0.91	0.92	0.92
DT	**0.97**	0.93	**0.97**	**0.97**	0.996	0.996
GPC	0.84	0.578	0.845	0.91	0.995	-
ADA	**0.97**	**0.944**	**0.97**	**0.97**	**0.997**	**0.998**
XGB	0.95	0.932	0.95	**0.97**	**0.997**	0.997

Table 2 presents the comparative analysis for various ML algorithms in terms of Accuracy values for various Models mentioned above in Sect. 5.2. We found that the Bernoulli Naive Bayes algorithm is the best performing model as it outperforms almost all of their other models. Decision Trees and XGB models are the next best models.

Table 3. Balanced Score values for ML Models on 2019 Dataset

ML	O	O_S	MM	MM_S1	MM_S2	MM_S3
LR	0.54	0.52	0.625	0.61	0.992	0.992
KNN	0.35	0.41	0.71	0.71	0.94	0.94
BNB	0.64	0.67	0.67	0.67	0.921	0.921
DT	**0.91**	0.714	**0.94**	**0.83**	0.93	0.995
GPC	0.46	0.478	0.35	0.61	0.995	-
ADA	0.85	**0.78**	0.86	**0.84**	0.997	0.998
XGB	0.81	0.697	0.81	0.82	**0.99**	**0.999**

Table 3 presents the comparative analysis for various ML algorithms in terms of Balanced Score values for various Models mentioned above in Sect. 5.2. We found that the Decision Trees algorithm is the best performing model as it outperforms almost all of their other models. AdaBoost model and XGBoost are the next best models.

Table 4. F1 weighted values for ML Models on 2019 Dataset

ML	O	O_S	MM	MM_S1	MM_S2	MM_S3
LR	0.83	0.83	0.90	0.89	0.992	0.992
KNN	0.78	0.59	0.90	0.83	0.94	0.94
BNB	0.886	0.90	0.89	0.89	0.92	0.92
DT	**0.96**	0.92	**0.975**	**0.97**	0.995	0.995
GPC	0.80	0.62	0.775	0.89	0.995	-
ADA	**0.96**	**0.933**	0.96	0.96	0.997	0.998
XGB	0.95	0.924	0.95	0.95	**0.999**	**0.999**

Table 4 presents the comparative analysis for various ML algorithms in terms of F1 weighted values for various Models mentioned above in Sect. 5.2. We found that the Decision Trees is the best performing model as it outperforms almost all of their other models. AdaBoost and XGB models are the next best models.

Table 5. Geometric Mean values for ML Models on 2019 Dataset

ML	O	O_S	MM	MM_S1	MM_S2	MM_S3
LR	0.87	0.87	0.97	0.98	0.98	0.98
KNN	0.85	0.53	0.925	0.81	0.885	0.884
BNB	0.93	0.95	0.92	0.94	0.85	0.85
DT	**0.998**	0.993	**0.999**	**0.996**	0.992	**0.99**
GPC	0.82	**0.64**	0.84	0.98	0.99	-
ADA	0.996	0.995	0.996	**0.996**	0.994	**0.996**
XGB	0.994	**0.994**	0.994	0.994	**0.998**	**0.998**

Tables 5 presents the comparative analysis for various ML algorithms in terms of Geometric Mean values for various Models mentioned above in Sect. 5.2. We found that the Decision Trees is the best performing model as it outperforms almost all of their other models. AdaBoost and XGB models are the next best models.

Thus, we observe that all the different models give significantly better results progressively as we apply the SMOTE, Borderline SMOTE, and SMOTE SVM for the imbalanced dataset. Across the various OSMI 2019 survey dataset, the Decision Trees gives better results and outperforms other models in the majority of models.

6 Conclusions and Future Work

In this study we have done a detailed analysis on the 2014 and 2019 OSMI Dataset and also have applied various Machine Learning Algorithms on both the datasets and also predicted their respective accuracies. In the preliminary analysis, we have identified the countries in which the highest number of employees are having Mental illness or prone to mental illness. We have also done the age-wise analysis in order to know the significant age group in which employees are more prone to have a mental illness, which clearly suggests that the count and percentage of people suffering from mental disorders are rising as compared observed from 2014 to 2019 survey results and impacting the performance and health of the employees. Tables 1, 2, 3, 4 and 5 showcase the results obtained for data-modeled for the 2014 and 2019 OSMI survey. For the OSMI 2014 survey dataset, the Bernoulli Naive Bayes is the best performing model followed by Gaussian Process Classifier and XG Boost Classifier. For the OSMI 2019 survey dataset, many data models were taken and ML algorithms were deployed on the same and evaluated using 4 evaluation metrics and the results show that Decision Trees is the best performing models across the various data models and followed by AdaBoost Classifier and XG Boost Classifier.

Since the detailed analysis and interpretation of the models have been presented on the 2014 and 2019 survey dataset. Further researches can be based on the 2020 dataset. The neural network models can also be applied to mental health data. Other imbalanced learning mechanisms can also be tried on the OSMI dataset to deal with class imbalance problems. Newer work can be based on the trend analysis of the factors that impact mental patients around the globe which will broaden the scope of the mental health prediction domain.

References

1. Allan, B.A., Dexter, C., Kinsey, R., Parker, S.: Meaningful work and mental health: job satisfaction as a moderator. J. Mental Health **27**(1), 38–44 (2018)
2. Andrews, D.R., Wan, T.T.: The importance of mental health to the experience of job strain: an evidence-guided approach to improve retention. J. Nursing Manage. **17**(3), 340–351 (2009)
3. Berkson, J.: Why I prefer logits to probits. Biometrics **7**(4), 327–339 (1951)
4. Chawla, N.V., Bowyer, K.W., Hall, L.O., Kegelmeyer, W.P.: SMOTE: synthetic minority over-sampling technique. J. Artif. Intell. Res. **16**, 321–357 (2002)
5. Chen, T.. Guestrin, C.: Xgboost: a scalable tree boosting system. In: Proceedings of the 22nd Acm Sigkdd International Conference on Knowledge Discovery and Data Mining, pp. 785–794 (2016)
6. Cover, T., Hart, P.: Nearest neighbor pattern classification. IEEE Trans. Inf. Theor **13**(1), 21–27 (1967)
7. Elmunsyah, H., Mu'awanah, R., Widiyaningtyas, T., Zaeni, I., Dwiyanto, F.A.: Classification of employee mental health disorder treatment with k-nearest neighbor algorithm. In: 2019 International Conference on Electrical, Electronics and Information Engineering (ICEEIE), vol. 6, pp. 211–215. IEEE (2019)
8. Freund, Y., Schapire, R., Abe, N.: A short introduction to boosting. J.-Japan. Soc. Artif. Intell. **14**, 771–780 (1999)
9. Han, H., Wang, W.-Y., Mao, B.-H.: Borderline-SMOTE: a new over-sampling method in imbalanced data sets learning. In: Huang, D.-S., Zhang, X.-P., Huang, G.-B. (eds.) ICIC 2005. LNCS, vol. 3644, pp. 878–887. Springer, Heidelberg (2005). https://doi.org/10.1007/11538059_91
10. Islam, M.R., Miah, S.J., Kamal, A.R.M., Burmeister, O.: A design construct of developing approaches to measure mental health conditions. Australasian J. Inf. Syst. **23** (2019)
11. Laijawala, V., Aachaliya, A., Jatta, H., Pinjarkar, V.: Classification algorithms based mental health prediction using data mining. In: 2020 5th International Conference on Communication and Electronics Systems (ICCES), pp. 1174–1178. IEEE (2020)
12. Lawrence, N., Seeger, M., Herbrich, R.: Fast sparse Gaussian process methods: the informative vector machine. In: Proceedings of the 16th Annual Conference on Neural Information Processing Systems, no. CONF, pp. 609–616 (2003)
13. Mohr, D.C., Zhang, M., Schueller, S.M.: Personal sensing: understanding mental health using ubiquitous sensors and machine learning. Ann. Rev. Clin. Psychol. **13**, 23–47 (2017)
14. Nguyen, H.M., Cooper, E.W., Kamei, K.: Borderline over-sampling for imbalanced data classification. Int. J. Knowl. Eng. Soft Data Paradigms **3**(1), 4–21(2011)
15. OSMI Mental Health in Tech Survey Dataset. https://osmihelp.org/research. Accessed on 02 Feb 2021
16. Quinlan, J.R.: Induction of decision trees. Mach. Learn. **1**(1), 81–106 (1986)
17. Reddy, U.S., Thota, A.V., Dharun, A.: Machine learning techniques for stress prediction in working employees. In: 2018 IEEE International Conference on Computational Intelligence and Computing Research (ICCIC), pp. 1–4. IEEE (2018)
18. Rish, I.: An empirical study of the naive bayes classifier. In: IJCAI 2001 Workshop on Empirical Methods in Artificial Intelligence, vol. 3, no. 22, pp. 41–46 (2001)

19. Shatte, A.B., Hutchinson, D.M., Teague, S.J.: Machine learning in mental health: a scoping review of methods and applications.: Psychol. Med. **49**(9), 1426–1448 (2019)
20. Srividya, M., Mohanavalli, S., Bhalaji, N.: Behavioral modeling for mental health using machine learning algorithms. J. Med. Syst. **42**(5), 1–12 (2018). https://doi.org/10.1007/s10916-018-0934-5
21. Vorbeck, J.: Algorithms and Anxiety: An Investigation of Mental Health in Tech (2020)

Predicting the Decomposition Level of Forest Trees Through Ensembling Methods

S. Jeyabharathy(✉) and Padmapriya Arumugam

Department of Computer Science, Alagappa University, Karaikudi, India
padmapriyaa@alagappauniversity.ac.in

Abstract. Forest provides the biodiversity necessary for human livelihood and thousands of life forms. To preserve the species in the forest and environment it is necessary to measure the decomposition level of the trees and the increase in carbon emission caused due to decaying. The increase in carbon emission causes environmental pollution and climate change. This paper elaborates on the usage of data mining techniques on big data to predict the level of the decomposition of trees and their carbon content. To get an accurate prediction, the data should be classified correctly. In this paper, we apply classification techniques like Support Vector Machine, Naive Bayes, and Random forest and ensemble methods. The Classification algorithms are applied to the training dataset and classified the data into two decay classes. Then the accuracy of the classification algorithms is measured. This paper extends with a comparative evaluation of these algorithms to find the suitable algorithm for tree decay classification. From these results, the amount of carbon emitted by the decomposed tree is predicted using a regression algorithm. Finally, the research work concludes with the prediction of the decomposition level of the tree and its carbon content by applying the selected classification and prediction algorithm on the test data.

Keywords: Data mining · Big data · Pre-processing · Classification · Prediction · Machine learning · SVM · Naïve Bayes · Random forest · Ensemble · Regression

1 Introduction

Forest provides us air to breathe, livelihood, watershed protection, prevent soil erosion, and softens the climatic change. It also serves as a buffer in case of any natural disaster. The forests are mainly affected by tree damage and forest fire. For preserving the forest it is necessary to regenerate new trees, fill the gap by removing the old and dead trees. Coarse Woody Debris (CWD) refers to both standing and fallen dead trees [17]. These dead trees may cause natural disasters such as forest fires, fallen off without any use and it also causes Environment pollution. Forest has a major role to play in the protection of the global carbon cycle. Forest acts as a carbon sink, when the trees get decomposed it starts to emit carbon back to the atmosphere causes air pollution and forest fire. As prevention, the decayed trees are to be identified and removed to avoid the carbon emission by the tree back to the atmosphere. Reforestation and afforestation could contribute to

M. Bhattacharya et al. (Eds.): ICICCT 2021, CCIS 1417, pp. 248–262, 2021.
https://doi.org/10.1007/978-3-030-88378-2_20

reducing the atmospheric carbon concentrations [16]. This paper offers a solution to the issue of carbon emission by the dead trees, the dead trees classification, and predicting its carbon emission. This can be done using machine learning algorithms and classification algorithms. To resolve this issue the research work is carried out in three phases namely pre-processing, classification, and prediction.

2 Related Work

In the research work [1], the authors presented estimation on carbon emission by the dead tree exceeds than carbon emission by vehicles. The LTER-2 proposal [2] states that reforestation can help to reduce carbon emission by decayed trees to the atmosphere. Harmon et al. [3] represented that reduction in carbon emission avoids deforestation and forest degradation can be eliminated. This paper focuses on preventing carbon emission by identifying the decayed trees and suggests removing before it changes as deforestation causes. For doing so, classification [4] states that ensemble classification methods provide greater predictive accuracy than individual-based classifiers. Since the trees are not evenly decayed after a certain number of years the dataset is imbalanced. For handling the imbalanced dataset, the research work [5] stated that ensemble methods enhance the performance of the classifier models for imbalanced datasets. Researchers have recognized that the diameter of CWD [4] is an important factor related to ecological functions. The decay process of the wood is identified by measuring its diameter. Using the value of height, volume, density mass of the wood is calculated [1] (i.e. mass/volume). The wood density is derived from the mass & volume of the wood. Thus the variations in the wood density help us to predict its decay class. Decay class 0 refers to the tree that is not decayed and Decay class 1 refers to the decayed trees. The decomposition rate is calculated for the species [3]. The wood can be decomposed by leaching, fragmentation, transport, respiration, biological transformation [4]. Prediction is the best solution to identify the decaying tree in the forest [10, 13]. For identifying this, the classification algorithms SVM, Naive Bayes, Random forest, Ensemble methods are chosen. All these algorithms can be used for both classification and regression is most suitable for prediction [15].

3 Architecture Diagram

The architecture diagram is shown in Fig. 1 which consists of three phases. In the first phase, the forest dataset is pre-processed using feature selection techniques to choose the relevant features which improve wood density prediction accuracy. Wood density of the tree is predicted using machine learning algorithm multiple linear regression. In the second phase, trees are classified into binary class Decay Class 1 (DC1) and Decay class 0 (DC0) using various classification algorithms. Decay Class1 refers to the decayed trees. Decay class 0 refers to the trees not yet decayed. Here the wood density of the tree and the species type plays a vital role to classify the trees into Decay class 1 & Decay class 0. In the third phase, the carbon emitted by the tree under decay class 1 is predicted using a regression algorithm. Prediction is based on the Wood density of the decomposed tree and the type of species. Information about individual trees is collected and their wood density is predicted. The dataset has been taken from the USDA data repository. We have chosen 4 species namely PSME, TSHE, THPL, ABAM. The decomposition level of the tree can be classified into two classes, namely Class 0 refers to the tress is not decayed; Class 1 refers to a decayed tree.

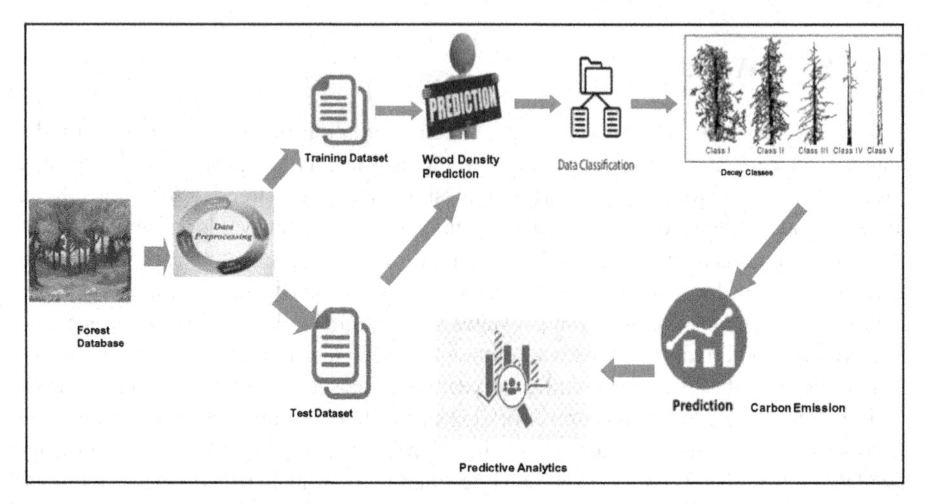

Fig. 1. Architecture diagram for predicting the decay class

The classification models are trained to classify the available trees into two different ordinal decay classes. The decay class refers to that the tree is not alive, it is ready to cut off because leaving the decayed tree will destroy the environment and it will no more absorb Carbon dioxide (CO_2) from the atmosphere. The result of the classification model helps us to identify the decomposition level of the trees. With these results, environmental pollution and forest fire can be avoided [7, 11, 12]. The trained model is used to predict the decay class of the tree species in the test dataset.

4 Materials and Methods

4.1 Dataset

The dataset is taken from the USDA repository. Two datasets are used as training datasets and test dataset. Training Dataset has the threshold value of the Wood density. The training dataset consists of the decay class and wood density of the respective species. Test data set consists of 21 attributes and 54,000 instances. Test data contains the details such as diameter, year, volume, dry weight, moisture, etc. for 4 species (Table 1).

4.2 Wood Density Prediction

In this section, the wood density of the forest trees is predicted using the multiple linear regression method. The density of the tree's wood must first be determined to predict the

Table 1. Dataset description

S. No	Attributes	Description
1.	Log num	It is a number given to every log
2.	Species	Kind of the tree, Here 4 species
3.	Time	Number of years for the tree
4.	Year	Year Label
5.	Subtype	Type of the trees such as Hard, Soft
6.	Radpos	The position where measurement taken
7.	D1	Diameter of the tree
8.	D2	Diameter of the tree at various pos
9.	D3	Diameter of the tree at various pos
10.	D4	Diameter of the tree at various pos
11.	VOL1	The volume of the Tree
12.	VOL2	The volume of the Tree
13.	WetWt	Weight of the wetness in the tree
14.	DRYWT	The dry weight of the tree
15.	MOIST	Moisture content in the wood
16.	Decay	Decomposition level of the tree
17.	WDENSITY	Wood density of the tree concerning vol1
18.	Den2	Wood density of the tree concerning Vol2
19.	Knot Vol	The volume of the wood at a knot
20.	Sample Date	The date on which the data is collected
21.	Comments	About any special features of the tree

amount of carbon released by the decaying tree [8]. The wood density is predicted using a multiple linear regression because it is dependent on several explanatory variables and a target variable. The wood density of the dataset is preprocessed using Feature selection techniques and among 21 attributes 6 attributes are chosen. The six attributes chosen are Species, Diameter, Volume, Wet weight, Dry Weight, Decay class. These attributes give better accuracy in the prediction of tree wood density.

4.2.1 Multiple Linear Regression

Multiple linear regression (MLR) is a statistical technique that uses several explanatory variables to predict the result of a response variable. MLR is designed to model the linear relation between the explanatory (independent) variables and the (dependent) response variable.

- Here is Multiple Linear regression the explanatory variables are

  Ø Species
  Ø Diameter
  Ø Volume
  Ø Wet Weight
  Ø Dry Weight
  Ø Decay

- Target Variable is

  Ø Wood density of the tree

Prediction Equation

$$W_i = \beta_0 + \beta_1 Species + \beta_2 \text{Diameter} + \beta_3 \text{Volume} + \beta_4 \text{Wetwt} + \beta_5 \text{Drywt} + \beta_6 \text{Decay} + \varepsilon$$

Where for n observations:
yi = dependent variable
Species, Diameter, Volume, Wetwt, DryWt, Decay = explanatory variables
β0 = y-intercept (constant term)
βj = slope coefficients for each explanatory variable (j indicates attribute index)
ϵ = the model's error term (also known as the residuals)

4.3 Classification Algorithm

In this section, we described the two groups of classification algorithms, Individual classifiers, and Ensemble classifiers. These classification algorithms have been implemented for classifying the trees into Decay Classes.

4.3.1 Random Forest

Random Forest builds a decision tree for the given input. Each tree predicts a class and it is a supervised algorithm that combines the decision of each tree and gives the output based on the majority. Creating multiple decision trees gives accurate results. It is high in accuracy and performance. It is suitable for both classification and regression using categorical values. It creates 10 decision trees for this work [9] (Fig. 2).

4.3.2 Support Vector Machine

Support Vector Machine (SVM) draws a perpendicular bisector between the data without overlapping, that perpendicular bisector is known as the hyperplane. SVM builds a hyperplane in the multidimensional space to segregate the dataset into different classes. It can be used for binary and multiclass classification. SVM method can be implemented for classification and prediction and the dataset used in this method can handle continuous and categorical values. The maximum space between two hyperplanes is the best split [12] (Fig. 3).

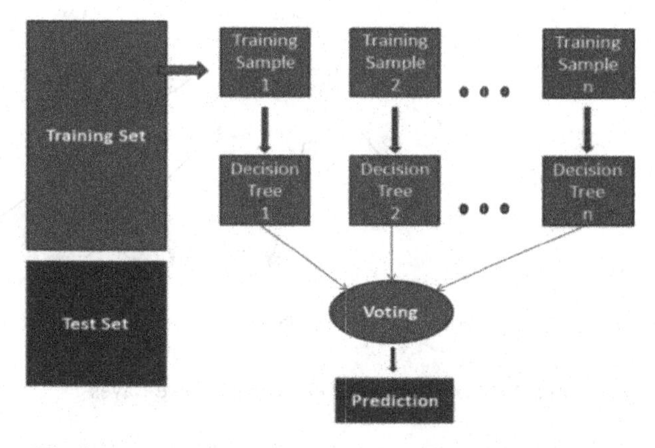

Fig. 2. Random Forest for classifying decay class of the tree

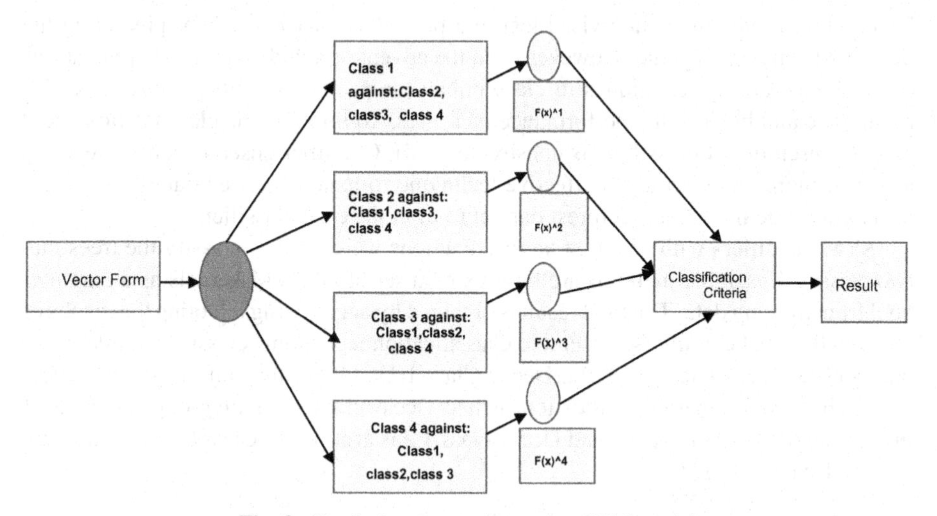

Fig. 3. Predicting decay class using SVM model

4.3.3 Naive Bayes

Naive Bayes classifies the data by giving the value of the feature, the prediction is done. It works on the principles of conditional probability. It is well suited for multiclass prediction. It finds the probability of the event occurring by giving the probability of an event already occurred; it expects the presence of a particular feature in a class. In this paper, the decay class of the tree is predicted by giving the feature, Wood density [11] (Fig. 4).

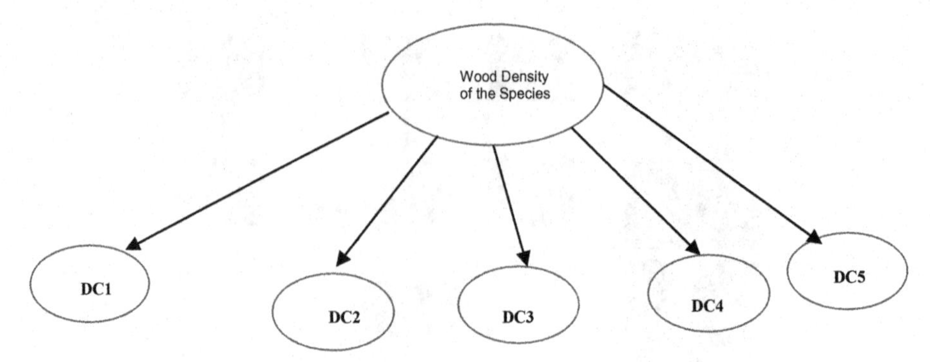

Fig. 4. Predicting decay class using Naïve Bayes

4.3.4 Ensembling Method Classification

Ensembling Methodis a supervised learning procedure that is used for predicting the outcome from existing data. However, with the advent of a wide variety of applications of machine learning techniques to class imbalance problems further focus is needed to improve and balance the performance measures. To improve the classification accuracy an ensemble of classifiers is constructed [14]. Classifier ensembles have recently achieved more attention as an effective technique to handle skewed data [18]. In this paper ensemble model uses a forest dataset to train the base classifiers.

SVM classifiers with different kernel functions were used to classify the trees into two classes. Ensemble methods are built using a set of all SVM kernels as a classifier pool from training data. The final result is obtained by aggregating or voting the results of all the individual classifiers. Ensemble Classification algorithms classify the tree under binary class Decay Class 0(DC0)&Decay Class 1(DC1). For this, multiclass classification is changed to binary classification such as Decay class 3,4,5 are grouped to Class 1 means that the tree is decayed and Decay class 1,2 is grouped to Class 0 means the trees are not decayed (Fig. 5).

- Decay Class 0 – Trees not yet decayed
- Decay Class 1 – Decayed Tress

The framework of the ensemble model can be constructed as either dependent or independent. We constructed the ensemble model using independent base classifiers SVM kernel and the results are combined using the averaging method.

In this work, ensemble SVM with 8 different kernels are used and listed below (Fig. 6).

- SVM-Linear
- SVM-RBF
- SVM-Sigmoid
- SVM-Polynomial
- SVM-Bayesian
- SVM-Multiquadratic

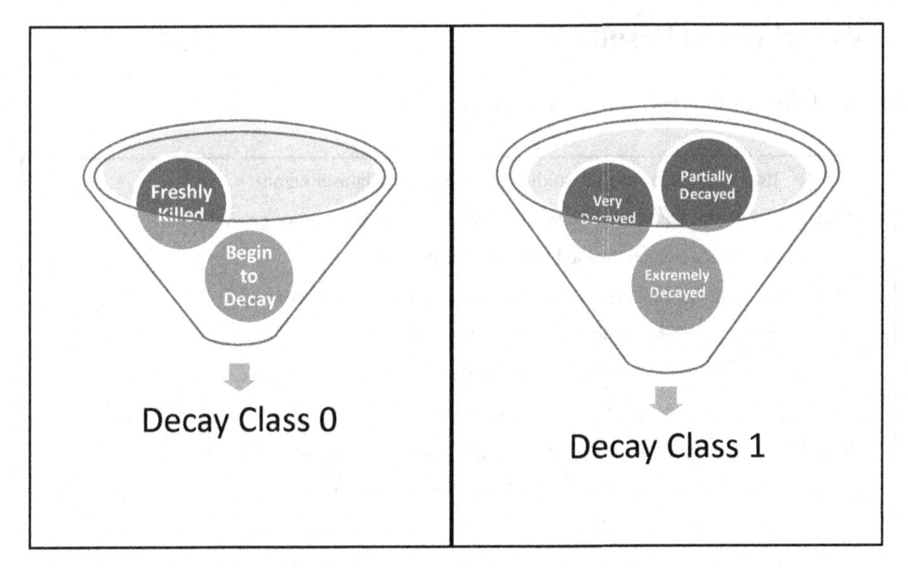

Fig. 5. MultiClass into binary class fusion

- SVM-Exponential
- SVM-Log

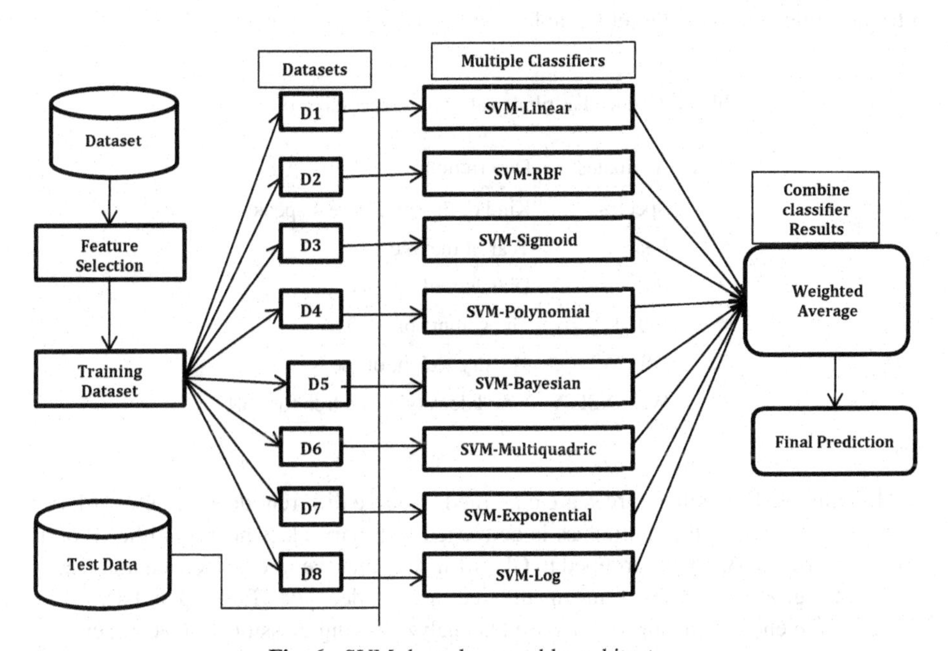

Fig. 6. SVM- kernels ensemble architecture

5 Experimental Results

The workflow of the research work is given below.

Step1: Import the dataset to R studio using read.csv (dataset name)
Step 2: Select the features (independent variables) which help for prediction.
Attributes Selected based on the correlation matrix.
Step 3: Change the data type of the attribute decay as a Factor variable.
Step 4: Train the SVM model with the training dataset.
Step 5: Test the model with the testing dataset.
Step 6: Predict the output with the random dataset.
Step 7: Calculate the model Accuracy, Here Accuracy is calculated for every 5 years.
Step 8: Repeat steps 5 to 7 for the random forest algorithm.

Dataset is pre-processed before implementing the algorithm. The dataset is preprocessed using the feature selection technique back elimination to choose the optimal subset feature which contributes maximum to predict the wood density of the tree. Among the 21 attributes, we have taken 6 attributes that are necessary for prediction. For the experiment, we have trained the model using the full dataset. For the testbed, we have taken a different percentage of dataset such as 60 * 40, 80 * 20, 70 * 30. Test data has 6 attributes relevant to the target variable wood density is given in Table 2.

Table 2. Reduced attributes after pre-processing the dataset

S.no	Attributes	Description
1.	Species	Kind of the tree, Here 4 species
2.	Year	Year of the tree
3.	D1	Diameter of the tree
4.	VOL1	The volume of the Tree
5.	DRYWT	The dry weight of the tree
6.	WDENSITY	Wood density of the tree for vol1

In individual classifiers, trees are classified into five different decay classes [6]. For ensemble classifiers, the multiclass is converted to binary classification either 0 or 1. Here Decay class 3,4,5 are grouped to Class 1 means that the tree is decayed and Decay class 1,2 is grouped to Class 0 means the trees are not decayed (Tables 3 and 4).

The efficiency of the algorithms used is analyzed using classification accuracy.

$$\textbf{Accuracy} = (\textbf{TP} + \textbf{TN})/(\textbf{TP} + \textbf{TN} + \textbf{FP} + \textbf{FN})$$

Table 3. Details of the decay class

Decay class	Description
1	Freshly Killed
2	Beginning to Decay
3	Partially Decayed
4	Very Decayed
5	Extremely Decayed

Table 4. Details of the species

Species name	Label
PSME	1
TSHE	2
THPL	3
ABAM	4

True Negative (TN): It refers to the number of predictions where the classifier correctly predicts the negative class as negative.
True Positive (TP): It refers to the number of predictions where the classifier correctly predicts the positive class as positive.
False Positive (FP): It refers to the number of predictions where the classifier incorrectly predicts the negative class as positive.
False Negative (FN): It refers to the number of predictions where the classifier incorrectly predicts the positive class as negative (Fig. 7 and Table 5)

Table 5. Accuracy results for the classification models in percentage

Year	No of records	SVM	Random	Naïve Bayes	Ensemble
1985	2920	92.53	88.77	59.28	**95.9**
1990	2796	91.38	87.05	52.72	**94.7**
1995	2159	94.35	89.86	65.4	**96.1**
2001	2817	87.54	84.38	65.53	**94.4**
2005	3417	93.77	93.91	71.14	**95**
2015	1103	95.29	92.11	83.41	**96.5**

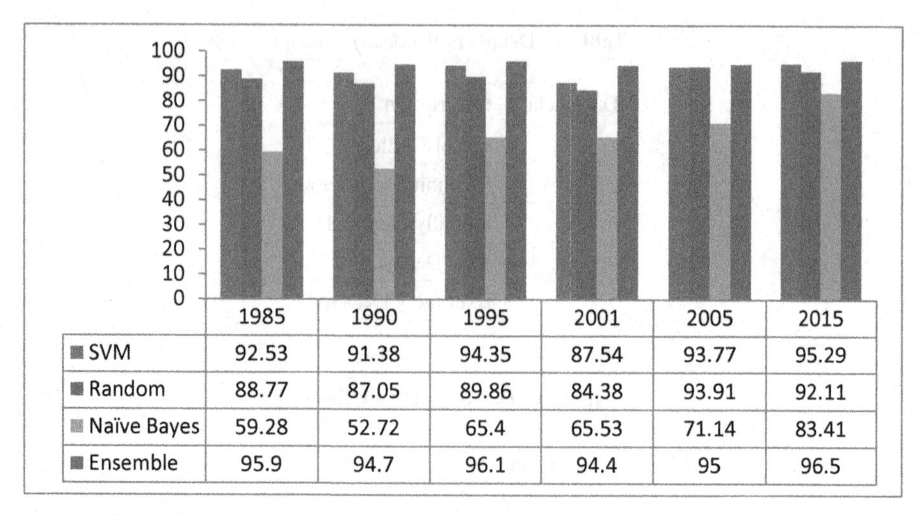

	1985	1990	1995	2001	2005	2015
SVM	92.53	91.38	94.35	87.54	93.77	95.29
Random	88.77	87.05	89.86	84.38	93.91	92.11
Naïve Bayes	59.28	52.72	65.4	65.53	71.14	83.41
Ensemble	95.9	94.7	96.1	94.4	95	96.5

Fig. 7. Performance Comparison of individual classifiers against Ensemble methods.

5.1 Carbon Emission Prediction

Carbon emitted by the decayed trees is predicted using a regression algorithm. 50% of dry weight is the carbon content of the tree. For prediction, the attributes dry weight, species, Decay and year are the independent variable contribute more to the target variable carbon. The trees classified under decay class 1 are used for this prediction. The results show that the amount of carbon content that can be emitted back to the atmosphere after decomposition. Linear regression [15] can be applied to the classified dataset to get an accurate prediction on carbon stock.

5.1.1 Linear Regression

Linear regression is a technique that shows the relationship between two variables. In this paper linear regression predicts the carbon emission for the given year. The instances under decay class 0 are filtered and predicted the amount of carbon emitted by these trees. In [19], the authors explain how the carbon content present in the trees is calculated. Using this, the carbon present in the trees is calculated and the results are compared shows that the carbon content in species 4 is high compared to the other three species.

$$Y = B0 + B1.X$$

Where, Y – Dependent variable, X – Independent variable, and B0, B1 – Regression parameter (Fig. 8 and Table 6).

Table 6. Estimation of carbon content in each species.

Species	Carbon content in the tree
1	2481
2	3770
3	1472
4	7543

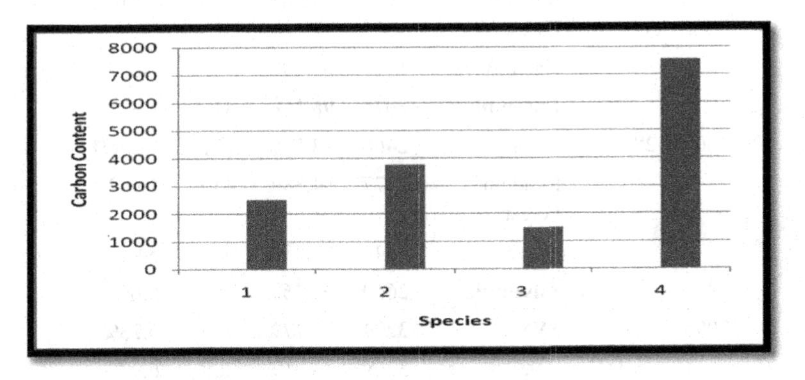

Fig. 8. Carbon content available in each species

6 Evaluating the Techniques

In this section, the performance of the ensemble classifier is compared against the individual base classifiers. The observations are listed in Table 7.

Table 7. Evaluation results among 3 techniques

Year	No of records	Classification techniques	Correctly classified instances		Incorrectly classified instances	
1985	2920	SVM	2702	92.53%	218	7.47%
		Random Forest	2592	88.77%	328	11.23%
		Naïve Bayes	1731	59.28%	1189	40.72%
		Ensemble	2803	**95.9%**	117	4%
1990	2796	SVM	2555	91.38%	241	8.62%
		Random Forest	2434	87.05%	362	12.95%

(*continued*)

Table 7. (*continued*)

Year	No of records	Classification techniques	Correctly classified instances		Incorrectly classified instances	
		Naïve Bayes	1474	52.72%	1322	47.28%
		Ensemble	2648	**94.7%**	147	10.5%
1995	2159	SVM	2037	94.35%	122	5.65%
		Random Forest	1940	89.86%	219	10.14%
		Naïve Bayes	1412	65.40%	747	34.60%
		Ensemble	2075	**96.1%**	84	3.9%
2001	2817	SVM	2466	87.54%	351	12.46%
		Random Forest	2377	84.38%	440	15.62%
		Naïve	1846	65.53%	971	34.47%
		Ensemble	2660	**94.4%**	157	5.6%
2005	3417	SVM	3204	93.77%	213	6.23%
		Random Forest	3209	93.91%	208	6.09%
		Naïve Bayes	2431	71.14%	986	28.86%
		Ensemble	3249	**95%**	168	5%
2015	1103	SVM	1051	95.29%	52	4.71%
		Random Forest	1016	92.11%	87	7.89%
		Naïve Bayes	920	83.41%	183	16.59%
		Ensemble	1065	**96.5**	38	3.4%

The above table depicts the accuracy results of the four classification techniques. It is evident from the table that SVM-Ensemble has the highest classification accuracy (96.50%) where 1065 instances have been correctly classified and 38 instances are classified incorrectly. The second-highest classification accuracy for the SVM algorithm is 95.29% in which 1051 instances have been correctly classified and 52 instances are classified incorrectly.

The Naïve Bayes algorithm results in the lowest classification accuracy which is 52.74% among the three algorithms. The Naïve Bayes was not able to classify 1322 instances correctly.

7 Conclusion

This research work focuses on the classification of forest trees under the decay classes using four classification algorithms. All the classification techniques Naive Bayes, SVM, and Random forest, Ensemble were applied to the dataset to predict the decay class of a tree. The results on the test data show that the Ensemble method outperforms the others with high Accuracy. In the Ensemble method, the trees are more appropriately classified than the rest three algorithms. All 4 techniques make a relationship between the dependent variable (decay) with other independent variables (Species, Wood Density, and Year). With the classification results, the carbon emission is predicted using regression model and estimated that after 8 years the carbon content in the 4th species is higher than the other 3 species. This paper concludes with an estimation of carbon emission by the decayed trees if we kept them uncleared in the forest. This paper suggests for reforestation to plant a new tree after removing the decayed tree and prevent environmental pollution and forest fire.

Acknowledgement. The authors thankfully acknowledge the Department of Science and Technology for the financial assistance rendered through the PURSE Programme [Phase 2: SR/PURSE Phase 2/38 (G)].

References

1. Harmon, M.E., Fasth, B., Woodall, C.W., Sexton, J.: Carbon concentration of standing and downed woody detritus: effects of tree taxa, decay class, position, and tissue type. For. Ecol. Manage. **291**, 259–267 (2013)
2. In LTER – 5 proposal – details of the tree dataset are mentioned
3. Harmon, M.E., et al.: Production, respiration, and overall carbon balance in an old-growth Pseudotsuga-Tsuga forest ecosystem. Ecosystems **7**(5), 498–512 (2004)
4. Harmon, M.E., Krankina, O.N., Sexton, J.: Decomposition vectors: a new approach to estimating woody detritus decomposition dynamics. Can. J. For. Res. **30**(1), 76–84 (2000)
5. Salunkhe, U.R., Mali, S.N.: Classifier ensemble design for imbalanced data classification: a hybrid approach. Procedia Comput. Sci. **85**, 725–732 (2016)
6. Yan, E., Wang, X., Huang, J.: Concept and classification of coarse woody debris in forest ecosystems. Front. Biol. China **1**(1), 76–84 (2006)
7. Acker, S.A., Kertis, J., Bruner, H., O'Connell, K., Sexton, J.: Dynamics of coarse woody debris following wildfire in a mountain hemlock (Tsugamertensiana) forest. For. Ecol. Manage. **302**, 231–239 (2013)
8. Campbell, J.L., et al.: Estimating uncertainty in the volume and carbon storage of downed coarse woody debris. Ecol. Appl. **29**(2) (2019). https://doi.org/10.1002/eap.1844
9. Everingham, Y., Sexton, J., Skocaj, D., Inman-Bamber, G.: Accurate prediction of sugarcane yield using a random forest algorithm. Agron. Sustain. Dev. **36**(2), 1–9 (2016). https://doi.org/10.1007/s13593-016-0364-z
10. Lakshmi, B.N., Indumathi, T.S., Ravi, N.: A comparative study of classification algorithms for predicting gestational risks in pregnant women. In: 2015 International Conference on Computers, Communications, and Systems (ICCCS), pp. 42–46. IEEE (2015)
11. Campero-Jurado, I., Robles-Camarillo, D., Simancas-Acevedo, E.: Problems in pregnancy, modeling fetal mortality through the Naïve Bayes classifier. Int. J. Combinat. Opt. Prob. Inf. **11**(3) (2020)

12. Madge, S., Bhatt, S.: Predicting stock price direction using support vector machines. Independent work report spring, vol. 45 (2015)
13. Sharma, A.K., Sahni, S.: A comparative study of classification algorithms for spam email data analysis. Int. J. Comput. Sci. Eng. **3**(5), 1890–1895 (2011)
14. Bahl, N., Bansal, A.: Balancing performance measures in classification using ensemble learning methods. In: International Conference on Business Information Systems, pp. 311–324. Springer Cham (2019). https://doi.org/10.1007/978-3-030-20482-2_25
15. Asri, H., Mousannif, H., Al Moatassime, H., Noel, T.: Using machine learning algorithms for breast cancer risk prediction and diagnosis. Procedia Comput. Sci. **83**, 1064–1069 (2016)
16. Pawar, K.V., Rothkar, R.V.: Forest conservation & environmental awareness. Procedia Earth Planetary Sci. **11**, 212–215 (2015)
17. Herrmann, S., Kahl, T., Bauhus, J.: Decomposition dynamics of coarse woody debris of three important central European tree species. For. Ecosyst. **2**(1), 1–14 (2015). https://doi.org/10.1186/s40663-015-0052-5
18. Zhang, Y., Fu, P., Liu, W., Chen, G.: Imbalanced data classification based on scaling kernel-based support vector machine. Neural Comput. Appl. **25**(3–4), 927–935 (2014). https://doi.org/10.1007/s00521-014-1584-2
19. New Mexico's Flagship University | The University of New Mexico (unm.edu)/Calculating tree carbon.pdf

Author Index

Printed in the United States
by Baker & Taylor Publisher Services